Power Electronics and Control Techniques for Maximum Energy Harvesting in Photovoltaic Systems

INDUSTRIAL ELECTRONICS SERIES

Series Editors:
Bogdan M. Wilamowski & J. David Irwin

PUBLISHED TITLES

Power Electronics and Control Techniques for Maximum Energy Harvesting in Photovoltaic Systems, *Nicola Femia, Giovanni Petrone, Giovanni Spagnuolo, and Massimo Vitelli*

Extreme Environment Electronics, *John D. Cressler and H. Alan Mantooth*

Renewable Energy Systems: Advanced Conversion Technologies and Applications, *Fang Lin Luo and Hong Ye*

Multiobjective Optimization Methodology: A Jumping Gene Approach, *K.S. Tang, T.M. Chan, R.J. Yin, and K.F. Man*

The Industrial Information Technology Handbook, *Richard Zurawski*

The Power Electronics Handbook, *Timothy L. Skvarenina*

Supervised and Unsupervised Pattern Recognition: Feature Extraction and Computational Intelligence, *Evangelia Micheli-Tzanakou*

Switched Reluctance Motor Drives: Modeling, Simulation, Analysis, Design, and Applications, *R. Krishnan*

FORTHCOMING TITLES

Industrial Wireless Sensor Networks: Applications, Protocols, Standards, and Products, *Vehbi Cagri Gungor and Gerhard P. Hancke*

Smart Grid Technologies: Applications, Architectures, Protocols, and Standards, *Vehbi Cagri Gungor, Carlo Cecati, Gerhard P. Hancke, Concettina Buccella, and Pierluigi Siano*

Multilevel Converters for Industrial Applications, *Sergio Alberto Gonzalez, Santiago Andres Verne, and Maria Ines Valla*

Data Mining: Theory and Practice, *Milos Manic*

Granular Computing: Analysis and Design of Intelligent Systems, *Witold Pedrycz*

Electric Multiphase Motor Drives: Modeling and Control, *Emil Levi, Martin Jones, and Drazen Dujic*

Sensorless Control Systems for AC Machines: A Multiscalar Model-Based Approach, *Zbigniew Krzeminski*

Next-Generation Optical Networks: QoS for Industry, *Janusz Korniak and Pawel Rozycki*

Signal Integrity in Digital Systems: Principles and Practice, *Jianjian Song and Edward Wheeler*

FPGAs: Fundamentals, Advanced Features, and Applications in Industrial Electronics, *Juan Jose Rodriguez Andina and Eduardo de la Torre*

Dynamics of Electrical Machines: Practical Examples in Energy and Transportation Systems, *M. Kemal Saioglu, Bulent Bilir, Metin Gokasan, and Seta Bogosyan*

Power Electronics and Control Techniques for Maximum Energy Harvesting in Photovoltaic Systems

Nicola Femia • Giovanni Petrone
Giovanni Spagnuolo • Massimo Vitelli

CRC Press
Taylor & Francis Group
Boca Raton London New York

CRC Press is an imprint of the
Taylor & Francis Group, an **informa** business

CRC Press
Taylor & Francis Group
6000 Broken Sound Parkway NW, Suite 300
Boca Raton, FL 33487-2742

Printed in the United States of America on acid-free paper
Version Date: 20121112

International Standard Book Number: 978-1-4665-0690-9 (Hardback)

Visit the Taylor & Francis Web site at
http://www.taylorandfrancis.com

and the CRC Press Web site at
http://www.crcpress.com

To Giuseppe,

Nicola, Maria, Patrizia, and Nicola Jr.

Angelo, Pina, Simonetta, and Valeria

Felice, Annamaria, and Alessia

Contents

Preface

Photovoltaic (PV) systems produce a significant amount of the electrical energy used around the world. PV technology will be capable of offering a great deal of support in the future to the rate of growth of advanced economies as well as developing countries. The incentives provided at a first stage by the European governments have resulted in the rapid growth of the PV market and an increase in the number and quality of products offered by industry. PV modules by many producers are now commercially available, and a number of power electronic systems have been put on the market for processing the electric power produced by PV systems, especially for grid-connected applications.

The scientific literature concerning PV applications has been characterized by a strong quantitative and qualitative growth in the past decade. A huge number of papers have been written and continue to be published in many journals, and there are many high-impact scientific journals specifically devoted to PV systems. A significant number of scientific papers are dedicated to control of the PV source. A simple search on the Reuters Thomson website reveals that at the end of May 2012, about 600 papers include *maximum power point tracking* (MPPT) among their keywords. Many authors have contributed to the scientific field of circuits and systems, ensuring the best operation of the photovoltaic generator, but a reference in this field is still lacking.

Some books that assess the most significant improvements concerning the connection of PV systems to the grid have been published recently. The most recent advances in this field and the various solutions offered to researchers as a starting point for their activities and to the industries for developing new products are overviewed.

The aim of this book is to fill the gap in the field of control circuits, systems, and techniques dedicated to the maximization of the electrical power produced by a PV source. In the first part of the book, an overview of the methods allowing a PV array working in uniform and mismatched conditions to be modeled is given. Next, the ways in which the best MPPT performances can be achieved are discussed. The design of the parameters affecting the algorithm performances is treated. The maximization of the energy harvested in mismatched conditions is then discussed, in terms of both power architecture and control algorithms. Last, the design of the DC/DC converter, which usually performs the MPPT function, is discussed, with special emphasis on its energy efficiency.

This first edition of the book is the result of ten years of activity in the field of PV systems. In this period, some of the points that have been investigated in detail, with many relevant scientific and applicative results, were raised from discussions with a number of persons. Researchers and friends from academia, who are acknowledged for these fruitful discussions, and industry design experts, from whom we received encouragement to continue our studies, have had a significant role in helping us to write this book.

We also acknowledge out mentor, Professor Emeritus Luigi Egiziano, who trusted in us and supported our research initiatives and academic growth.

Nicola Femia, Giovanni Petrone, Giovanni Spagnuolo, and Massimo Vitelli

About the Authors

Nicola Femia was born in 1963 in Salerno. He received a laurea degree in industrial technologies engineering with honors from the University of Salerno in 1988. He was assistant professor from 1990 to 1998 and associate professor from 1998 to 2001. Since 2001 he has been a full professor of electrotechnics at the University of Salerno, where he teaches circuit theory and power electronics and leads the Power Electronics and Renewable Sources Laboratory. His research interests cover circuit theory, analysis, simulation and design methods of high-energy-efficiency switching power supplies, control techniques and design optimization, and power electronics for photovoltaic systems. He has been responsible for research and education projects in collaboration with several industries, including Magnetek, National Semiconductor, STMicroelectronics, PowerOne, and Texas Instruments. He is co-author of more than 140 scientific papers published in international journals and in the proceedings of international conferences. He is co-author of five patents regarding control techniques and power converters for photovoltaic applications. He taught courses in power electronics design and optimization organized by National Semiconductor in Europe and the United States in 2006–2007. He was the associate editor of *IEEE Transactions on Power Electronics* from 1995 to 2003. He has been a reviewer for *IEEE Transactions on Circuits and Systems, IEEE Transactions on Industrial Electronics, IEEE Transactions in Power Electronics, IEEE Transactions on Energy Conversion, International Journal of Circuits Systems and Computers, International Journal of Applied Electromagnetics and Mechanics, International Journal of Simulation Modelling Practice and Theory,* and *IET Power Electronics*. He was a member of the administration council of the University of Salerno from 2002 to 2005. He was also a member, and chairman from 2009 to 2010, of the patents council of the University of Salerno.

Giovanni Petrone was born in Salerno in 1975. In 2001 he received the laurea degree (with honours) in electronic engineering from the University of Salerno, Italy, and his PhD in electrical engineering from the University of Napoli "Federico II," Italy, in 2004. Since March 2001 he has collaborated with the Department of Electronic Engineering and Computer Science of the University of Salerno. In the 2005 he joined the same department as assistant professor, where he teaches electrotechnics and power electronic circuits for renewable energy sources. He is a member of the College of the Professors of the Faculty of Engineering of the University of Salerno for the PhD degree in information engineering. His main research interests are in the analysis and design of switching converters for low-power applications, tolerance analysis of electronic circuits, and nonlinear control techniques. His activities are mainly focused in the field of renewable energy sources, and in particular his studies have been oriented to the development of DC/DC switching converters for photovoltaic, fuel cell, and wind applications, stability analysis of MPPT control techniques, and inverters for photovoltaic AC modules. He is involved in several research projects with international companies and institutions and has also assumed responsibility for some Italian research projects supported by public funds. He is co-author of five patents and several scientific papers published in international journals and in the proceedings of international symposia. He is member of the IEEE Industrial Electronics Society Technical Committee on Renewable Energy Systems. He has been track co-chair of several special sessions in international conferences and guest editor for two special issues in *IEEE Transactions on Industrial Electronics*. He is reviewer, within his area of expertise, for the following international journals: *IEEE Transactions on Power Electronics, IEEE Transactions on Industrial Electronics, IEEE Transactions on Industry Application, IEEE Transactions on Control Systems Technology*, and *International Journal on Progress in Photovoltaics: Research and Applications.*

Giovanni Spagnuolo was born in Salerno, Italy, on September 12, 1967. He received the MSc degree in electronic engineering from the University of Salerno (Italy) in 1993 and the PhD degree in electrical engineering from the University "Federico II" of Naples in 1998. In 1998 and 1999 he received a postdoctoral scholarship from the University of Salerno. From November 1999 to December 2003 he was assistant professor of electrical engineering at the University of Salerno where, since January 1, 2004, he has been associate professor. Since 1999 he has been with the Department of Electronic and Computer Engineering (DIEII) of the University of Salerno. He is a senior member of the IEEE and the "PV system control" editor of the *IEEE Journal of Photovoltaics*. He is associate editor of *IEEE Transactions on Industrial Electronics*, and for this journal he has been guest editor of four special issues focused on renewable energy systems. He is associate editor of the *International Journal of Industrial Electronics and Drives*, Inderscience Publishers Ltd. For the DIEII, he is the project leader of European project Leonardo Da Vinci, European project FP7, and a PRIN 2008 project financed by the Italian Ministry of the University and of Scientific Research. Moreover, he is the coordinator of research contracts financed by National Semiconductors Corporation (Santa Clara, California), Matrix S.r.l. in Conversano (BA-Italy), and Bitron Industrie S.p.A. (Torino, Italy). All these projects are focused on power electronics and control of renewable energy systems. He is co-author of six patents, two of them filed by Power One and Bitron, respectively. He has been a member of the scientific committee of the Electrimacs 2011 conference, Paris, June 2011. He was publication chair of the 2010 IEEE International Symposium on Industrial Electronics, Bari, July 2010. He is track chair of "Power Systems, PHEV, and Renewable Energy" for the IEEE International Symposium on Industrial Electronics, Hangzhou, May 2012, and track chair of "Renewable Energy Systems" for the IEEE International Conference on Industrial Technology (ICIT), Cape Town, March 2013. He was invited to give plenary lectures at the Energy Day of IEEE ETFA 2009 and at Electrimacs 2011. He organized several special sessions on renewable energy systems in the frame of IEEE ISIE and IEEE ICIT. He is reviewer of research projects for the United States–Israel Binational Science Foundation and the Scientific Research Support Fund, Jordan. He serves as reviewer for several international journals, transactions, and conferences, especially from IEEE. He is a member of three technical committees of three IEEE societies. He is author/co-author of about 40 papers published in international journals and about 100 in proceedings of international conferences.

Massimo Vitelli was born in Caserta, Italy, in 1967. He received the laurea degree with honors in electrical engineering from the University of Naples "Federico II," Italy, in 1992. In 1994 he joined the Department of Information Engineering of the Second University of Naples as a researcher. In 2001 he was appointed associate professor and in 2006 full professor in the Faculty of Engineering of the Second University of Naples, where he teaches electrotechnics and power electronics. He has been engaged, in cooperation with researchers of the University of Salerno and University of Naples Federico II, in a number of scientific national projects financed by the Italian Ministry of University and of Scientific and Technological Research (MURST) and by the National Science Foundation (CNR). His main research interests concern maximum power point tracking techniques in photovoltaic applications; power electronics circuits for renewable energy sources; methods for analysis and design of switching converters; calculation of steady-state solutions; tolerance analysis/design by means of interval arithmetic, affine arithmetic, and genetic algorithms; optimization of switching converters; tolerance analysis/design in electromagnetic field problems, electromagnetic compatibility, and electromagnetic characterization of new insulating and semiconducting materials. He has served as a reviewer for the international journals *IEEE Transactions on Electromagnetic Compatibility*, *IEEE Transactions on Circuits and Systems 1: Fundamental Theory and Applications*, *IEEE Transactions on Power Electronics*, and *IEEE Transactions on Industrial Electronics*, and for the following international IEEE conferences: ISCAS, ISIE, ICIT, and EPE-PEMC. He was referee for CIVR (Committee for Evaluation of Research for the Italian Ministry of Education, University and Research) of research products submitted for the evaluation of scientific research in Italy during the three-year period 2001–2003. He has also been engaged in a number of research contracts with industries (C.R.I.S. Ansaldo, ELCON s.r.l., MagneTek s.p.a., Power One, Astrid Energy Enterprises S.p.A., Matrix s.r.l, National Semiconductor Corporation, Elettronica Santerno S.p.A, and Bitron S.p.A). Since 2003 Professor Vitelli has been an associate editor of the *IEEE Transactions on Power Electronics*. He has been guest editor for *IEEE Transactions on Industrial Electronics*—special section "Industrial

Electronics for Photovoltaic Systems" (2008). He is a component of the Technical Committee on Renewable Energy Systems (RE) of the Industrial Electronics Society. He is a component of the Power Electronics Technical Committee IEEE-ISCAS. He has been guest editor for *IEEE Transactions on Industrial Electronics*—special sections "Efficient and Reliable Photovoltaic Systems" (2008) and "Fuel Cells Power Processing and Control" (2009). He is co-author of six national and international patents.

1

PV Modeling

1.1 From the Photovoltaic Cell to the Field

Photovoltaic (PV) generators are made of a number of PV cells connected in series and in parallel. The type of connection depends on the voltage and current levels at which it is desired that the power processing system dedicated to the PV generator works. The right choice of the values of such electrical variables is of fundamental importance in determining the efficiency of the switching converters that condition the power produced by the PV generator in order to feed the AC mains or recharge a battery. Unfortunately, in usual applications the voltage and current levels of the PV generator cannot be referred to a single PV cell. In fact, cells are arranged into PV panels, which contain some tens of them connected in series. The choice of connecting the cells in series comes from the fact that their operating voltage is few hundreds of millivolts, while the current they generate at high irradiation levels is of some amperes. As a consequence, the cells series connection leads to PV panels working at few tens of volts and some amperes.

In PV power plants, panels are connected in series to form strings in order to reach a voltage level that meets the input requirements of the power processing system. The desired power level of the PV plant is reached by connecting a number of strings, each one composed of the same number of panels, in parallel, thus leading to an increase of the current level of the PV field.

The nominal current vs. voltage (I-V) characteristic of the whole PV field is obtained by stretching that of the single cell, namely by multiplying the voltage values by the number of cells in series into each panel and then by the number of panels of each string, and by multiplying the cell current values by the number of strings of panels connected in parallel.

Such a voltage and current scaling up holds only under the hypotheses that all the cells are exactly equal and that they work in exactly the same operating conditions, especially in terms of irradiance and temperature. Unfortunately such conditions do not occur in practice because of manufacturing tolerances and aging, so that physical parameters of the cells are different, and because of shadowing, of the different orientation of the cells with respect to the sun rays, due to leaves, bird droppings, and so on.

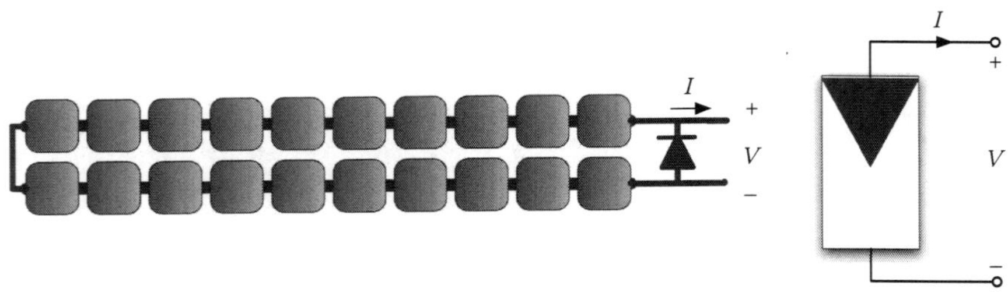

FIGURE 1.1
PV module made of some cells and the bypass diode.

Nonuniform operating conditions are usually referred to as *mismatching*, and they can lead to significant drops in the energy produced by the PV field. This mechanism will be explained in much more detail in this chapter, but a simple example can be instructive now. If, due to an accident, one cell is broken or simply disconnected from the contiguous ones, then the current of all the cells in series in the string is interrupted and the power contribution of the whole string is missed.

To reduce the effect of such events, PV panel manufacturers connect a bypass diode in parallel to short strings of series connected cells, usually two or three depending on the power rating forming the PV panels. In this way, a PV panel is made of two or three PV modules, structured as shown in Figure 1.1. In the sequel, if the modules the PV panel is made of join the same operating conditions and are perfectly equal, the terms *module* and *panel* will be assumed as synonyms.

The diodes are physically mounted into a junction box on the rear side of the panel and are normally inactive. Each of them fires as soon as the group of cells it is dedicated to generates a current lower than the one of the other subpanels in the string, so that it bypasses the exceeding current. Such a mechanism will be described in this chapter, showing the effects the bypass diode may have on the electrical characteristics of the whole string and field, as well as on the cells, owing to the bypassed group. The electrical structure of the PV panel just described changes the level of granularity at which the PV generator is considered in this book. Indeed, the PV generator becomes a parallel connection of strings, each one made of a series connection of PV modules, as shown in Figure 1.2.

In order to understand the way in which current and voltages are shared among the PV modules, especially in mismatched conditions, a model of the generator is required. The model is as very useful, as its level of complexity allows embedding it into circuits and systems simulation and computing programs of general use, like PSIM®, SPICE®, and MATLAB®/Simulink®.*

* MATLAB® is a registered trademark of The MathWorks, Inc.

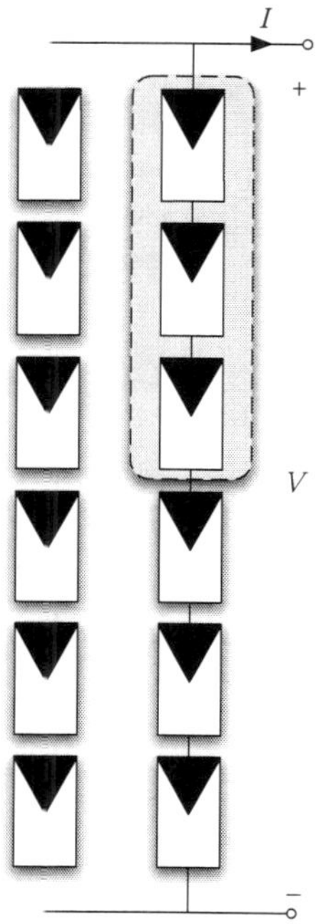

FIGURE 1.2
From module to field.

This chapter is devoted to the description of the PV generator operation, in both uniform and mismatched conditions, by means of nonlinear models. The models and the conclusions they lead to will be useful for the full comprehension of the concepts described in the following chapters of the book.

1.2 The Electrical Characteristic of a PV Module

The electrical characteristics of a PV panel working in uniform conditions are plotted in Figures 1.3 to 1.6 in normalized units. T_{max} and G_{max} can be assumed as the values used for testing the panel in standard conditions. Such curves are a qualitative example because, as will be much more clear in the sequel, their characteristics depend on the type of cells, materials, and technical solutions adopted for manufacturing the panel.

The plots put into evidence the presence of a maximum power point (MPP). Figures 1.5 and 1.6 show that the curves exhibit three particular points:

- The short-circuit (SC) condition, characterized by a zero voltage at the PV module terminals and by a short-circuit current I_{SC}

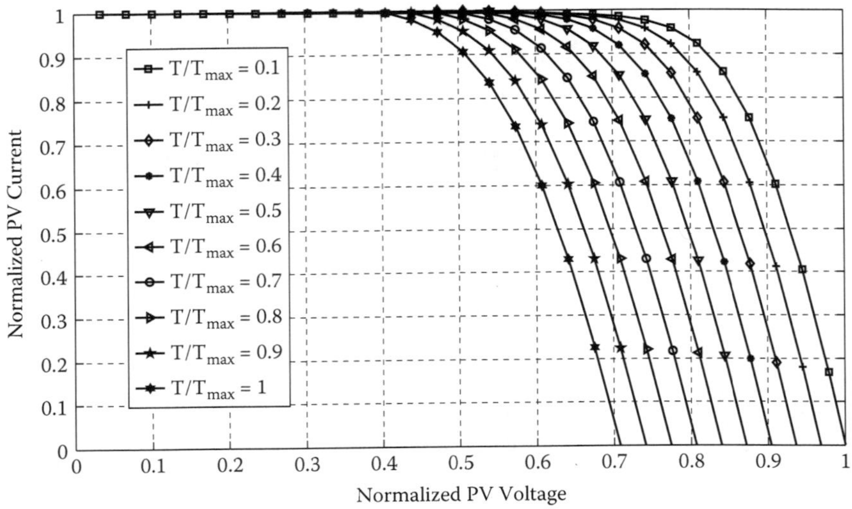

FIGURE 1.3
Current vs. voltage characteristic of a PV panel: effect of temperature *T*.

- The open-circuit (OC) condition, characterized by a zero current in the PV panel terminals and by an open-circuit voltage V_{OC}
- The MPP, at which the current value is I_{MPP}, the voltage value is V_{MPP}, and the power $P_{MPP} = V_{MPP} \cdot I_{MPP}$ is the maximum the PV panel is able to deliver in the temporary operating conditions

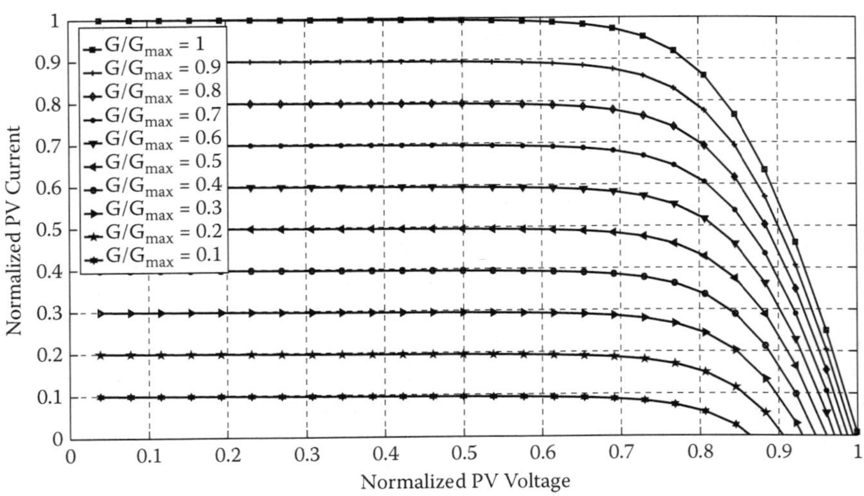

FIGURE 1.4
Current vs. voltage characteristic of a PV panel: effect of irradiation *G*.

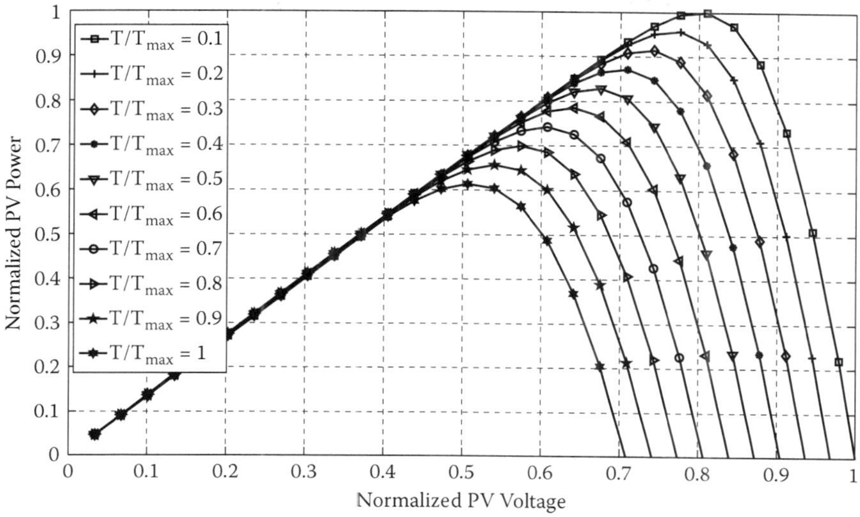

FIGURE 1.5
Power vs. voltage characteristic of a PV panel: effect of temperature *T*.

Figures 1.3 to 1.6 put into evidence the strong dependence of the panel performances on the temperature and the irradiance level. The temperature has a significant effect on the open-circuit voltage value, as shown in Figures 1.3 and 1.5. On the contrary, the temperature has a negligible effect on the short-circuit current value. The cell temperature, here indicated as *T*, is calculated

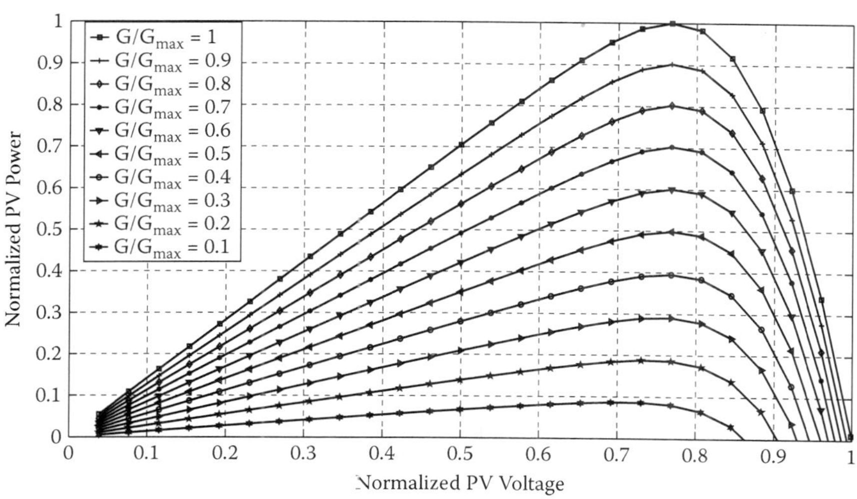

FIGURE 1.6
Power vs. voltage characteristic of a PV panel: effect of irradiation *G*.

from the ambient temperature T_a by using the *nominal operating cell temperature* (NOCT)*:

$$T = T_a + \frac{NOCT - 20}{800} \cdot G \tag{1.1}$$

It is worth noting that the temperature usually changes quite slowly, so that the temperature value is often considered a constant with respect to the variation the irradiation level can be subjected to during the day. This simplifying assumption will be also adopted in this book.

The irradiance variation has dual effects on the electrical characteristics with respect to the temperature. The PV module open-circuit voltage is almost independent of the irradiation: in literature it is stated that such a dependence is logarithmic. On the contrary, the short-circuit current is linearly dependent on the irradiance.

Electrical characteristics of PV modules are given by manufacturers in specific operating conditions, which are worldwide defined as standard test conditions (STC). Such conditions are defined by the cell temperature T_{STC} = 25°C, irradiation level G_{STC} = 1000 W/m², and the air mass value AM = 1.5. The latter gives a measure of the effect of the air mass between a surface and the sun on the spectral distribution and intensity of sunlight. In fact, the path length of the solar radiation through the atmosphere affects the light deviation and absorption [1, 2].

The irradiance variation is considered the main perturbing factor to be faced, because of its unpredictability: its effects will be widely discussed in this book. The irradiation rate of change is another unpredictable variable that must be taken into account: According to [3], the usual irradiance slope is $\dot{G} = 30\, W/m^2/s$. Such value is referred to classical stationary PV applications; more recent applications of PV systems, e.g. sustainable mobility, require performing the analysis with much more critical values of $\dot{G}$. For this reason, in this book higher values will also be considered.

The best way to analyze the behavior of the PV generator in a simulation environment is to adopt an equivalent circuit model and relevant equations describing it. In the sequel, such models are introduced and a special emphasis is devoted to the identification of the parameters either on the basis of the data obtained from experimental measurements or starting from data sheet information.

* The nominal operating cell temperature (NOCT) is defined as the temperature reached by open-circuit cells in a module under the following conditions: irradiance on cell surface G = 800 W/m², air temperature T_a = 20°C, wind velocity = 1 m/s, mounting = open back side.

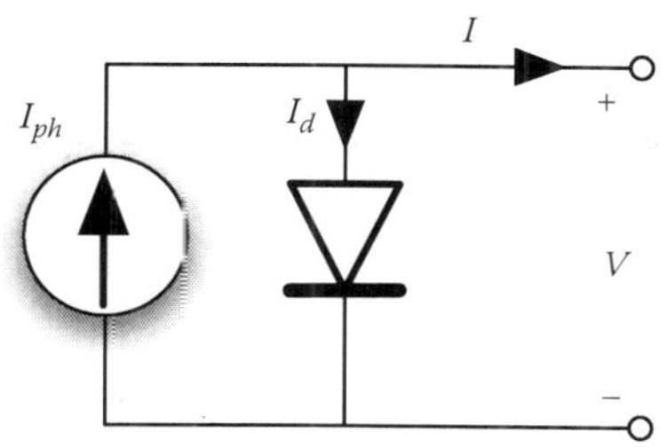

FIGURE 1.7
Ideal equivalent circuit of a PV generator.

1.3 The Double-Diode and Single-Diode Models

Looking at the figures shown in the previous section reveals that the I-V curve of a PV generator is obtained by subtracting a diode current, which is nonlinearly dependent on its voltage, from a constant current value. As a consequence, the ideal equivalent circuit shown in Figure 1.7 is made of a current generator in parallel with a diode.

The diode takes into account the physical effects taking place at the silicon p-n junction of the cell. The current generator represents the photo-induced current, which is dependent on the characteristics of the semiconductor material used for the cell, and especially, it is linearly dependent on the cell area, irradiation level, and temperature. Equation (1.2) gives the dependency on the two latter exogenous variables:

$$I_{ph} = I_{ph,STC} \cdot \frac{G}{G_{STC}} \cdot [1 + \alpha_I \cdot (T - T_{STC})] \tag{1.2}$$

where α_I is the temperature coefficient of the current, defined in STC as follows:

$$\alpha_I = \left.\frac{dI}{dT}\right|_{STC} \tag{1.3}$$

As a consequence, the I-V characteristic takes the form shown in (1.4).

$$I = I_{ph} - I_{sat} \cdot \left(e^{\frac{V}{\eta V_t}} - 1 \right) \tag{1.4}$$

where, as usual, V_t is the thermal voltage (1.5):

$$V_t = \frac{k \cdot T}{q} \tag{1.5}$$

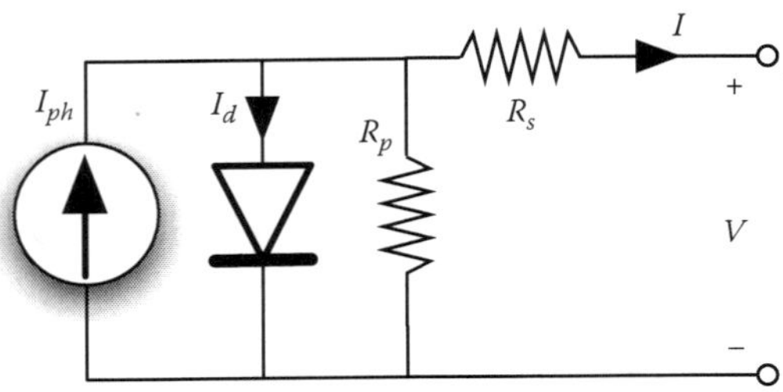

FIGURE 1.8
Single-diode model accounting for ohmic losses.

$k = 1.3806503 \cdot 10^{-23}$ J/K is the Boltzmann constant, $q = 1.60217646 \cdot 10^{-19}$ C is the electron charge, and η is the ideality factor. The saturation current is expressed as follows:

$$I_{sat} = C \cdot T^3 \cdot e^{\left(-\frac{E_{gap}}{kT}\right)} \tag{1.6}$$

where E_{gap} is the band gap of the semiconductor material* and C is the temperature coefficient [2].

Such a model does not take into account either the loss mechanisms taking place in the cell due to the metallic ribbon ensuring the current continuity between each cell and the two cells before and after it in the sequence or the diffusion and recombination characteristics of the charge carriers in the semiconductor. Such losses are introduced in the model by adding a series R_s and a parallel R_p resistance in order to take into account internal cell resistances and contact resistances, as well as the effect of leakage currents, respectively. As a consequence, the model becomes the one shown in Figure 1.8.

The series resistance R_s mainly affects the slope of the I-V curves in Figures 1.3 and 1.4 at high voltage levels, namely, approaching the open-circuit voltage: the worse the cell quality, the lower the curve slope, because of a high voltage drop across the series resistance. As a consequence, the approximated definition (1.7) can be adopted:

$$R_s \approx -\left.\frac{dV}{dI}\right|_{V=V_{OC}} \tag{1.7}$$

On the other hand, R_p affects the curve slope at current levels close to the short-circuit one: the lower the R_p value, the higher the current drawn by the parallel resistance, which is subtracted from the net output

* For crystalline silicon: $E_{gap} = 1.124$ eV $= 1.8 \cdot 10^{-19}$ J.

current, witnessed by a smaller slope of the curve at low voltage values, where it is

$$R_p \approx -\left.\frac{dV}{dI}\right|_{I=I_{SC}} \tag{1.8}$$

The values of these two resistances have an immediate effect on the cell energy productivity expressed through the index called *fill factor*, which is defined as the ratio between the product of the voltage and current values at the MPP and the product of the short-circuit current and open-circuit voltage values:

$$FF = \frac{V_{MPP} \cdot I_{MPP}}{V_{OC} \cdot I_{SC}} \tag{1.9}$$

In order to account for the ohmic losses, Equation (1.4) must be modified. According to Kirchhoff's laws, Equation (1.10) is obtained.

$$I = I_{ph} - I_{sat} \cdot \left(e^{\frac{V + I \cdot R_s}{\eta V_t}} - 1 \right) - \frac{V + I \cdot R_s}{R_p} \tag{1.10}$$

It is worth noting that both (1.4) and (1.10) are nonlinear, but the latter is also in implicit form, because neither the current nor the voltage can be explicitly expressed as functions of the other electrical variable. As will be pointed out in Section 1.6, this represents a strong limitation for many theoretical and numerical analyses.

According to the Shockley theory, the equivalent circuit shown in Figure 1.3 describes diffusion and recombination characteristics of the charge carriers in the semiconductor, but it neglects the recombination in the space-charge zone. If such effect must to be taken into account, a two-diode model must be used.

The consequent I-V characteristic, expressed by Equation (1.11) and describing the equivalent circuit shown in Figure 1.9, is more involved:

$$I = I_{ph} - I_{sat} \cdot \left(e^{\frac{V + I \cdot R_s}{\eta V_t}} - 1 \right) - I_{sat,2} \cdot \left(e^{\frac{V + I \cdot R_s}{\eta_2 V_t}} - 1 \right) - \frac{V + I \cdot R_s}{R_p} \tag{1.11}$$

where the second saturation current shows the following nonlinear dependence on temperature:

$$I_{sat\ 2} = C_2 \cdot T^{\frac{5}{2}} \cdot e^{\left(-\frac{E_{gap}}{2kT} \right)} \tag{1.12}$$

In this book, the single-diode model shown in Figure 1.8 and described by the implicit equation (1.10) is used. Such a model holds for a single PV crystalline cell, and it can be scaled up suitably in order to also be used for a PV module, panel, string, or for the whole PV field.

Under the assumption that all the cells are equal and that they work in the same operating conditions, all the voltages will be multiplied by n_s, if this is the number of the series connected cells, and all the currents will be multiplied by n_p, if this is the number of parallel connected strings. Consequently, the series resistance R_s and the parallel resistance R_p increase by a factor n_s and are divided by a factor n_p. In symbols:

$$I_{ph,field} = n_p \cdot I_{ph,cell} \tag{1.13}$$

$$I_{sat,field} = n_p \cdot I_{sat,cell} \tag{1.14}$$

$$V_{t,field} = n_s \cdot V_{t,cell} \tag{1.15}$$

$$R_{s,field} = \frac{n_s}{n_p} \cdot R_{s,cell} \tag{1.16}$$

$$R_{p,field} = \frac{n_s}{n_p} \cdot R_{p,cell} \tag{1.17}$$

It is worth noting that according to the active sign convention chosen for the current and voltage in Figure 1.8, the first quadrant of the plots shown in Figures 1.3 and 1.4 refers to an operating condition in which the PV system generates electrical power. Nevertheless, as will be demonstrated in Section 1.6, in some conditions the PV generator can be polarized with a reverse voltage, which is negative with respect to the reference used in Figure 1.8. In this case, if the PV system is illuminated, the current keeps its direction; thus it is positive with respect to the reference used in Figure 1.8, and the

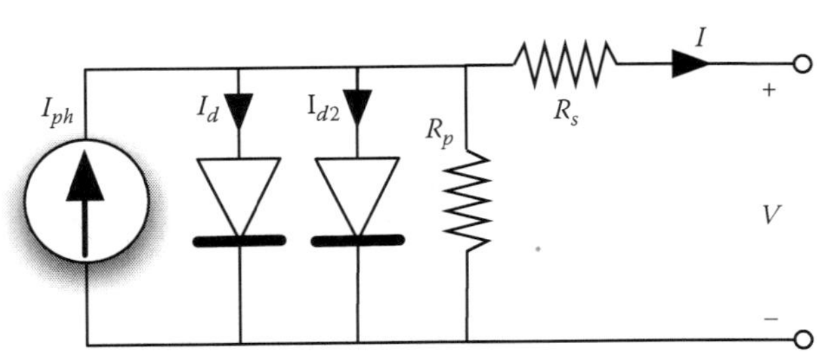

FIGURE 1.9
Two-diode equivalent circuit for a solar cell.

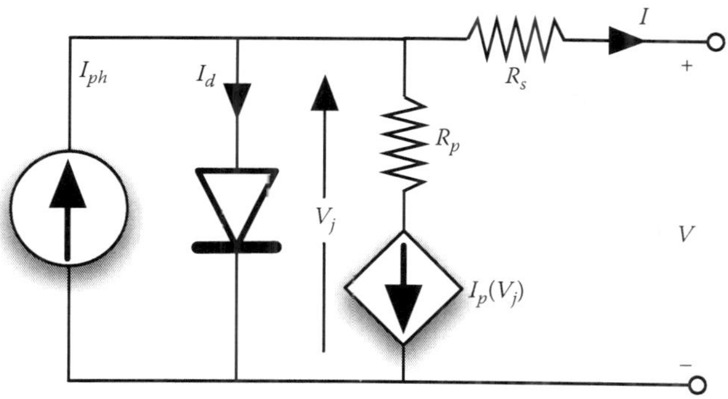

FIGURE 1.10
Bishop equivalent circuit for a counterpolarized solar cell.

system absorbs, rather than generates, electrical power. In this case, the operating point moves in the second quadrant of the Cartesian plot shown in Figures 1.3 and 1.4, and a suitable model of the physical effects taking place in such conditions is needed.

According to [4], the shunt branch in Figure 1.8 must include a voltage-controlled current generator, so that the equivalent circuit becomes the one shown in Figure 1.10.

In this way, the current through the shunting branch is expressed as

$$I_t = \frac{V_j}{R_p} \cdot \left[1 + \alpha \cdot \left(1 - \frac{V + I \cdot R_s}{V_{br}}\right)^{-m}\right] \tag{1.18}$$

where V_{br} is the junction breakdown voltage, α is the fraction of the ohmic current involved in avalanche breakdown, and m is the avalanche breakdown exponent. In some conditions, which will be discussed in Section 1.6, the voltage can assume large negative values, up to some tens of volts: Under a reverse voltage condition the cell can even dissipate a significant amount of electrical power, with a consequent increment in temperature and occurrence of hot spots. This may lead to an irreversible damaging of the cell, with a consequent interruption of the current flux into the whole string it belongs to. It is commonly believed that the bypass diode helps in reducing the probability that the reverse polarization of a cell determines hot spots and power losses, simply because it limits the reverse voltage to its direct polarization voltage, which is usually less than 1 V. This would be true if a bypass diode was used for each PV cell; however, due to the high number of diodes that such a solution would require for a PV panel realization, in practice a bypass diode is dedicated to a string of few tens of cells forming a PV module. In this case, the total reverse voltage of the cells in the PV module is limited by the forward voltage of the bypass diode, so that they are protected from hot spot occurrence, provided that they work in the same operating conditions. A different, and also more involved, study is required whenever mismatched conditions affect the cells of the PV module.

As a result of the discussion above, in Sections 1.4 and 1.6 the model in Equation (1.10) and the equivalent circuit in Figure 1.8 will be assumed to represent the string of PV cells into a PV module or panel, a string or the whole field, provided that the voltage, current, and resistance values are suitably scaled. Until no bypass diode conducts, the PV characteristic of a PV panel is obtained by scaling up that of a single module, so that the words *module* and *panel* have the same meaning.

1.4 From Data Sheet Values to Model Parameters

The circuit shown in Figure 1.8 and the corresponding equation (1.10) give the desired compromise between complexity and accuracy. The next task is the tuning of the model parameters to the PV module of interest. Two approaches are proposed in literature to solve this problem. The first is based on the best fitting of the I-V or P-V curves obtained by means of the model with the experimental measurement data obtained from the PV module testing. The data resulting from the experimental measurements correspond to points acquired at different voltage/current levels and for different irradiation/temperature values. When such data are available, some approaches presented in literature can be applied: For example, the mean square error fitting approach might be an option.

The most puzzling problem is the I-V and P-V curves reconstruction on the unique basis of the data available on the PV module data sheet. The difficulty arises from the fact that only the following information in STC is given therein:

- Operation in open-circuit conditions: $V_{OC,STC}$
- Open-circuit voltage/temperature coefficient: $\alpha_v = \left.\frac{dV_{OC}}{dT}\right|_{STC}$
- Operation in short-circuit conditions: $I_{SC,STC}$
- Short-circuit current/temperature coefficient: $\alpha_i = \left.\frac{dI_{SC}}{dT}\right|_{STC}$
- Operation at the maximum power point: $I_{MPP,STC}$ and $V_{MPP,STC}$

The availability of the set of the parameters listed above allows determining the values of five unknowns. If the single-diode model (1.10) is adopted, the values taken from the data sheet find their counterparts in the unknowns $\{I_{ph}, I_{sat}, \eta, R_s, R_p\}$.

1.4.1 Parameters Identification Assuming $R_p \to \infty$

The problem is afforded in literature by using two approaches: A simplified one considers the value of the parallel resistance R_p so high that it can be neglected. In this case, the equation whose parameters need to be identified is (1.19).

$$I = I_{ph} - I_{sat} \cdot \left[e^{\frac{(V+I\cdot R_s)}{\eta V_t}} - 1 \right] \tag{1.19}$$

In [2], the procedure for parameters identification is divided into steps. The data sheet voltage values are preliminarily divided by the number n_s of series connected cells forming the PV panel. Then the photo-induced current is calculated straightforwardly, because it is significantly higher than the diode current, so that it is assumed to be equal to the short-circuit current in STC:

$$I_{ph,STC} = I_{SC,STC} \tag{1.20}$$

The second step starts by applying (1.19) in the open-circuit condition:

$$0 = I_{ph,STC} - I_{sat,STC} \cdot \left(e^{\frac{V_{OC,STC}/n_s}{\eta V_{t,STC}}} - 1 \right) \approx I_{ph,STC} - I_{sat,STC} \cdot e^{\frac{V_{OC,STC}/n_s}{\eta V_{t,STC}}} \tag{1.21}$$

where the second equivalence has been obtained by neglecting the unitary term with respect to the exponential one. Consequently, it is

$$V_{OC,STC} \approx -\eta \cdot n_s \cdot V_{t,STC} \cdot \ln\left(\frac{I_{sat,STC}}{I_{ph,STC}} \right) \tag{1.22}$$

The open-circuit voltage temperature coefficient can be calculated as follows:

$$\alpha_v = \frac{dV_{OC,STC}}{dT} = \frac{d}{dT}\left[n_s \cdot \eta \cdot V_{t,STC} \cdot \ln I_{ph,STC} - n_s \cdot \eta \cdot V_{t,STC} \cdot \ln I_{sat,STC} \right] \tag{1.23}$$

Accounting for the temperature dependence of the thermal voltage (1.5) yields

$$\alpha_v = \frac{n_s \cdot \eta \cdot V_{t,STC}}{T_{STC}} \cdot \ln\left(\frac{I_{ph,STC}}{I_{sat,STC}} \right) + n_s \cdot \eta \cdot V_{t,STC} \cdot \left[\frac{1}{I_{ph,STC}} \cdot \frac{dI_{ph,STC}}{dT} - \frac{1}{I_{sat,STC}} \cdot \frac{dI_{sat,STC}}{dT} \right] \tag{1.24}$$

The second term in the parentheses can be calculated by means of (1.6)

$$\frac{1}{I_{sat,STC}} \cdot \frac{dI_{sat,STC}}{T_{STC}} = \frac{3}{T_{STC}} + \frac{E_{gap}}{k \cdot T_{STC}^2} \tag{1.25}$$

while the first term is related to the current temperature coefficient α_i, so that

$$\alpha_v = \frac{n_s \cdot \eta \cdot V_{t,STC}}{T_{STC}} \cdot \ln\left(\frac{I_{ph,STC}}{I_{sat,STC}}\right) + n_s \cdot \eta \cdot V_{t,STC} \cdot \left(\frac{\alpha_i}{I_{ph,STC}} - \frac{3}{T_{STC}} - \frac{E_{gap}}{k \cdot T_{STC}^2}\right) \quad (1.26)$$

From (1.22) it is

$$\alpha_v = \frac{V_{OC,STC}}{T_{STC}} + n_s \cdot \eta \cdot V_{t,STC} \cdot \left(\frac{\alpha_i}{I_{ph,STC}} - \frac{3}{T_{STC}} - \frac{E_{gap}}{k \cdot T_{STC}^2}\right) \quad (1.27)$$

thus the diode ideality factor is calculated by means of (1.28):

$$\eta = \frac{\alpha_v - \dfrac{V_{OC,STC}}{T_{STC}}}{n_s \cdot V_{t,STC} \cdot \left(\dfrac{\alpha_i}{I_{ph,STC}} - \dfrac{3}{T_{STC}} - \dfrac{E_{gap}}{k \cdot T_{STC}^2}\right)} \quad (1.28)$$

In order to calculate I_{sat}, the open-circuit conditions can be used as well, so that from (1.21) it is

$$I_{sat,STC} \approx I_{ph,STC} \cdot e^{-\frac{V_{OC,STC}}{\eta \cdot n_s \cdot V_{t,STC}}} \quad (1.29)$$

so that the temperature coefficient C in (1.6) can be evaluated and I_{sat} can be calculated at the desired value of temperature T.

$$C = \frac{I_{sat,STC}}{T_{STC}{}^3 \cdot e^{-\frac{E_{gap}}{k \cdot T_{STC}}}} \quad (1.30)$$

The series resistance can be determined by using the MPP data:

$$\begin{aligned} I_{MPP,STC} &= I_{ph,STC} - I_{sat,STC} \cdot \left[e^{\frac{V_{MPP,STC} + I_{MPP,STC} \cdot R_s}{\eta n_s \cdot V_{t,STC}}} - 1 \right] \\ &\approx I_{ph,STC} - I_{sat,STC} \cdot e^{\frac{V_{MPP,STC} + I_{MPP,STC} \cdot R_s}{\eta n_s \cdot V_{t,STC}}} \end{aligned} \quad (1.31)$$

Substituting (1.29) into (1.31) yields

$$I_{MPP,STC} = I_{ph,STC} - I_{ph,STC} \cdot \left[e^{\frac{(-V_{OC,STC} + V_{MPP,STC} + R_s \cdot I_{MPP,STC})}{\eta n_s \cdot V_{t,STC}}} \right] \quad (1.32)$$

thus the series resistance can be calculated by means of (1.33).

$$R_s = \frac{n_s \cdot \eta \cdot V_{t,STC} \cdot \ln\left(1 - \frac{I_{MPP,STC}}{I_{ph,STC}}\right) + V_{OC,STC} - V_{MPP,STC}}{I_{MPP,STC}} \tag{1.33}$$

Equations (1.20), (1.30), (1.28), and (1.33) allow calculation of the four values of the unknown parameters in (1.19) starting from the data $\{V_{oc}, I_{sc}, \alpha_i, \alpha_v, V_{MPP}, I_{MPP}\}$ in the STC available in the PV panel data sheets provided by the manufacturers.

As shown in Section 1.4.2, if the R_p value is accounted for, the procedure becomes much more involved and not straightforward, as it requires the solution of a nonlinear system of equations. An approximate solution might be obtained if some simplification of small terms with respect to larger ones is done. This result will be shown in Section 1.4.3.

1.4.2 Parameters Identification Including R_p

The approach proposed in the previous section can be extended by accounting for the presence of the parallel resistance. Equation (1.10), applied at the MPP, can be used to express the parallel resistance as a function of the series one:

$$R_p = \frac{V_{MPP,STC} + I_{MPP,STC} \cdot R_s}{I_{ph} - I_{MPP,STC} - I_{sat,STC} \cdot \left(e^{\frac{V_{MPP,STC} + I_{MPP,STC} \cdot R_s}{\eta V_{t,STC}}} - 1 \right)} \tag{1.34}$$

where the module thermal voltage V_t is related to the thermal voltage of the single cell V_T multiplied by the number of cell in series: $V_t = n_s V_T$.

The second equation, which completes the nonlinear system allowing calculation of both R_s and R_p, is obtained by observing that at the MPP the following condition holds:

$$\left.\frac{\partial(V \cdot I)}{\partial V}\right|_{MPP,STC} = 0 \rightarrow I_{MPP,STC} + V_{MPP,STC} \cdot \left.\frac{\partial I}{\partial V}\right|_{MPP,STC} = 0 \tag{1.35}$$

thus using (1.10) yields:

$$\left.\frac{\partial I}{\partial V}\right|_{MPP,STC} = -\frac{\frac{1}{R_p} + \frac{I_{sat,STC}}{\eta \cdot V_{t,STC}} \cdot e^{\frac{V_{MPP,STC} + I_{MPP,STC} \cdot R_s}{\eta V_{t,STC}}}}{1 + \frac{R_s}{R_p} + \frac{Rs \cdot I_{sat,STC}}{\eta \cdot V_{t,STC}} \cdot e^{\frac{V_{MPP,STC} + I_{MPP,STC} \cdot R_s}{\eta V_{t,STC}}}} \tag{1.36}$$

The second equation is obtained by substituting the above result in (1.35), so that

$$I_{MPP,STC}-V_{MPP,STC}\cdot\frac{\dfrac{1}{R_p}+\dfrac{I_{sat,STC}}{\eta\cdot V_{t,STC}}\cdot e^{\frac{V_{MPP,STC}+I_{MPP,STC}\cdot R_s}{\eta V_{t,STC}}}}{1+\dfrac{R_s}{R_p}+\dfrac{Rs\cdot I_{sat,STC}}{\eta\cdot V_{t,STC}}\cdot e^{\frac{V_{MPP,STC}+I_{MPP,STC}\cdot R_s}{\eta V_{t,STC}}}}=0 \tag{1.37}$$

The joint solution of the nonlinear equations (1.34) and (1.37) provides the values or the series and parallel resistances.

1.4.3 Parameters Identification Including R_p: Explicit Solution

Based on the change of variable

$$x=\frac{V_{MPP,STC}+R_s\cdot I_{MPP,STC}}{\eta V_{t,STC}} \tag{1.38}$$

the series and parallel resistances can be written, using (1.38) and (1.34), as follows:

$$R_s=\frac{x\eta V_{t,STC}-V_{MPP,STC}}{I_{MPP,STC}};\qquad R_p=\frac{x\eta V_{t,STC}}{I_{ph}-I_{MPP,STC}-I_{sat,STC}\cdot(e^x-1)} \tag{1.39}$$

After substituting such expressions in (1.37), the term $\eta V_{t,STC}x^2$ appears. A few algebra steps, neglecting the small quantity term R_s^2, lead to

$$\eta V_{t,STC}x^2\approx-\frac{V_{MPP,STC}^2}{\eta V_{t,STC}}+2\cdot V_{MPP,STC}\cdot x \tag{1.40}$$

Consequently, it results that solving (1.37) means solving (1.41) with respect to x:

$$\begin{aligned}&2V_{MPP,STC}\cdot\left(I_{MPP,STC}-I_{ph}-I_{sat,STC}\right)+\left(I_{ph}+I_{sat,STC}\right)\eta V_{t,STC}\cdot x\\&+I_{sat,STC}\cdot e^x\left[-\eta V_{t,STC}x+V_{MPP,STC}\cdot\left(2-\frac{V_{MPP,STC}}{\eta V_{t,STC}}\right)\right]=0\end{aligned} \tag{1.41}$$

The first two terms in (1.41) can be simplified, by accounting for (1.39), as follows:

$$
\begin{aligned}
&2V_{MPP,STC}\cdot(I_{MPP,STC}-I_{ph}-I_{sat,STC})+(I_{ph}+I_{sat,STC})\eta V_{t,STC}x\\
&=V_{MPP,STC}\cdot(2\cdot I_{MPP,STC}+I_{ph}+I_{sat,STC})+(I_{ph}+I_{sat,STC})\cdot(R_s\cdot I_{MPP,STC}-2\cdot V_{MPP,STC})\\
&\approx V_{MPP,STC}\cdot(2I_{MPP,STC}+I_{ph}+I_{sat,STC})-2\cdot(I_{ph}+I_{sat,STC})\cdot V_{MPP,STC}\\
&=V_{MPP,STC}\cdot(2\cdot I_{MPP,STC}-I_{ph}-I_{sat,STC})
\end{aligned}
\tag{1.42}
$$

The approximation adopted in (1.42) is justified by the fact that, for a PV panel, the quantity $R_s\cdot I_{MPP,STC}$ is almost two orders of magnitude smaller than $2\cdot V_{MPP,STC}$. In this way, the nonlinear equation (1.41) can be written as follows:

$$
\begin{aligned}
&V_{MPP,STC}\cdot\left(2\cdot I_{MPP,STC}-I_{ph}-I_{sat,STC}\right)+\\
&I_{sat,STC}e^{x}\left[-\eta V_{t,STC}x+V_{MPP,STC}\cdot\left(2-\frac{V_{MPP,STC}}{\eta V_{t,STC}}\right)\right]=0
\end{aligned}
\tag{1.43}
$$

This equation admits an analytical solution based on the use of the Lambert W function, which is the solution of the equation

$$
f(x)=x\cdot e^{x} \tag{1.44}
$$

Finally, it is

$$
\begin{aligned}
x=lambertW&\left[\frac{V_{MPP,STC}\left(2I_{MPP,STC}-I_{ph}-I_{sat,STC}\right)e^{\frac{V_{MPP,STC}\left(V_{MPP,STC}-2\eta V_{t,STC}\right)}{\eta^2 V_{t,STC}^2}}}{\eta I_{sat,STC}V_{t,STC}}\right]\\
&+2\frac{V_{MPP,STC}}{\eta V_{t,STC}}-\frac{V_{MPP,STC}^2}{\eta^2 V_{t,STC}^2}
\end{aligned}
\tag{1.45}
$$

The value obtained by (1.45) is substituted in (1.39) so that the values of the parallel and series resistances result.

Another approach, also based on the adoption of the Lambert W function but not giving an explicit expression for the two resistances, is presented in [11].

1.4.4 Other Approaches Proposed in Literature

Some other approaches have been proposed in literature to solve the problem of finding the parameters of the equivalent circuit model of PV modules.

The one presented in [5] uses Equation (1.10) applied in the open-circuit, short-circuit, and MPP operating points. The results are

$$0 = I_{ph} - I_{sat,STC} \cdot \left[e^{\frac{V_{OC,STC}}{\eta V_{t,STC}}} - 1 \right] - \frac{V_{OC,STC}}{R_p} \tag{1.46}$$

$$I_{SC,STC} = I_{ph} - I_{sat,STC} \cdot \left[e^{\frac{I_{SC,STC} \cdot R_s}{\eta V_{t,STC}}} - 1 \right] - \frac{I_{SC,STC} \cdot R_s}{R_p} \tag{1.47}$$

$$I_{MPP,STC} = I_{ph} - I_{sat,STC} \cdot \left[e^{\frac{V_{MPP,STC} + I_{MPP,STC} \cdot R_s}{\eta V_{t,STC}}} - 1 \right] - \frac{V_{MPP,STC} + I_{MPP,STC} \cdot R_s}{R_p} \tag{1.48}$$

In order to evaluate the five unknown parameters $\{I_{ph}, I_{sat}, \eta, R_s, R_p\}$, two additional equations are needed. The first one (1.49) expresses the condition occurring at the MPP, where the derivative of the power with respect to the voltage is equal to zero.

$$\begin{aligned} \left.\frac{dP}{dV}\right|_{V=V_{MPP,STC}, I=I_{MPP,STC}} &= \left.\frac{d(V \cdot I)}{dV}\right|_{V=V_{MPP,STC}, I=I_{MPP,STC}} \\ &= I_{MPP,STC} + V_{MPP,STC} \cdot \left.\frac{dI}{dV}\right|_{V=V_{MPP,STC}, I=I_{MPP,STC}} = 0 \end{aligned} \tag{1.49}$$

Unfortunately, the derivative of the current with respect to the voltage in (1.49) cannot be calculated easily, because of the inherent nonexplicit form of Equation (1.10).

The fifth, and last, condition needed for having the same number of unknowns and equations is equivalent to (1.8):

$$\left.\frac{dI}{dV}\right|_{I=I_{SC,STC}} \approx -\frac{1}{R_p} \tag{1.50}$$

By means of simple calculations, (1.46) and (1.47) allow expression of the saturation current and the photo-induced current as functions of the short-circuit current.

$$I_{sat,STC} = \left(I_{SC,STC} - \frac{V_{OC,STC} - R_s \cdot I_{SC,STC}}{R_p} \right) \cdot e^{-\frac{V_{OC,STC}}{\eta V_{t,STC}}} \tag{1.51}$$

$$I_{ph} = I_{SC,STC} \left(1 + \frac{R_s}{R_p} \right) \tag{1.52}$$

Substituting the results of (1.51) and (1.52) in (1.48) provides

$$I_{MPP,STC} = I_{SC,STC} - \frac{V_{MPP,STC} + R_s \cdot (I_{MPP,STC} - I_{SC,STC})}{R_p} + \\ - \left(I_{SC,STC} - \frac{V_{OC,STC} - R_s \cdot I_{SC,STC}}{R_p} \right) \cdot e^{\frac{V_{MPP,STC} + R_s \cdot I_{MPP,STC} - V_{OC,STC}}{\eta V_{t,STC}}} \tag{1.53}$$

Equation (1.49) requires the evaluation of the derivative of the current with respect to the voltage. By writing the right term of (1.53) as $f(I,V)$, it is

$$\frac{dP}{dV} = I_{MPP} + V_{MPP} \frac{\frac{\partial}{\partial V} f(I,V)}{1 - \frac{\partial}{\partial I} f(I,V)} \tag{1.54}$$

Thus it results that

$$I_{MPP,STC} + V_{MPP,STC} \cdot \frac{-\frac{1}{R_p} - \frac{\left(I_{SC,STC} - \frac{V_{OC,STC} - R_s \cdot I_{SC,STC}}{R_p} \right) \cdot e^{\frac{V_{MPP,STC} + R_s \cdot I_{MPP,STC} - V_{OC,STC}}{\eta V_{t,STC}}}}{\eta V_{t,STC}}}{1 + \frac{\left(I_{SC,STC} - \frac{V_{OC,STC} - R_s \cdot I_{SC,STC}}{R_p} \right) \cdot e^{\frac{V_{MPP,STC} + R_s \cdot I_{MPP,STC} - V_{OC,STC}}{\eta V_{t,STC}}}}{\eta V_{t,STC}} + \frac{R_s}{R_p}} = 0 \tag{1.55}$$

Finally, condition (1.50) can be rewritten as follows:

$$-\frac{1}{R_p} = \frac{-\frac{1}{R_p} - \frac{\left(I_{SC,STC} - \frac{V_{OC,STC} - R_s \cdot I_{SC,STC}}{R_p} \right) \cdot e^{\frac{V_{MPP,STC} + R_s \cdot I_{MPP,STC} - V_{OC,STC}}{\eta V_{t,STC}}}}{\eta V_{t,STC}}}{1 + \frac{\left(I_{SC,STC} - \frac{V_{OC,STC} - R_s \cdot I_{SC,STC}}{R_p} \right) \cdot e^{\frac{V_{MPP,STC} + R_s \cdot I_{MPP,STC} - V_{OC,STC}}{\eta V_{t,STC}}}}{\eta V_{t,STC}} + \frac{R_s}{R_p}} \tag{1.56}$$

The system of nonlinear equations made of (1.53), (1.55), and (1.56) must be solved numerically in order to determine R_s, R_h, and η.

Similar approaches have also been proposed by other authors, e.g., [1, 6–9]. As an example, in [1] the optimal tuning of the two resistance values is obtained by means of an iterative procedure that determines the values of R_s and R_p, ensuring that the I-V curve has the desired open-circuit voltage and short-circuit current and that the MPP occurs at the current and voltage values indicated by the PV panel data sheet. The determination of both resistances require in any case the numerical solution of a nonlinear system of equations. This is due to the fact that the single-diode model is described by a nonlinear, but also not explicit, equation, in which neither the voltage nor the current can be expressed as an explicit function of the other variable. Such a limitation can be somehow avoided by using the Lambert W function, which allows us to give an explicit solution to (1.10). This aspect will be discussed in the final part of this chapter.

1.5 Example: PV Module Equivalent Circuit Parameters Calculation

In this section the identification of the PV model parameters for a string of PV cells and a commercial module is proposed. The first one is a string whose nominal characteristics are shown in Table 1.1. The application of the procedure described in Section 1.4.2 allows determination of the values of the five parameters reported in Table 1.2. The thermal voltage assumes the value $V_t = K \cdot T/q = 0.474$ V at T_{STC}.

Table 1.3 shows the main points of the current vs. voltage characteristic referred to operating conditions for the PV module that are different from the STC ones. Figure 1.11 shows the corresponding PV curves in which the voltage and power deratings due to the temperature are evident.

TABLE 1.1

Nominal Operating Conditions for a PV String

PV Array	Values in STC Conditions
Short-circuit current I_{STC}	7.7 A
Open-circuit voltage V_{OC}	10.8 V
MPP current I_{MPP}	7.2 A
MPP voltage V_{MPP}	8.6 V
Temperature coefficient of I_{SC} (α_I)	0.07%/°C
Temperature coefficient of V_{OC} (α_V)	–0.35%/°C
NOCT	46°C
Number of cells in series	18

TABLE 1.2

Parameters of the PV Circuital Model

Parameters	Values
I_{ph}	7.7 A
R_s	0.1166 Ω
R_p	148.9151 Ω
C	373.0739 A/K³
η	1.0255

A second example concerns the Suntech STP245S-20/Wd PV module, whose parameters are listed in Table 1.4.

The model parameters obtained applying the procedure described in Section 1.4.2 are listed in Table 1.5.

With such values, the model discussed in Section 1.3 allows us to reconstruct the current vs. voltage and the power vs. voltage curves with discrepancies limited to few percents with respect to the experimental data provided by the manufacturer in the data sheet and referred to different operating irradiance and temperature values. Figure 1.12 shows the behavior of the PV panel at high, medium, and low irradiance values.

The approach introduced in Section 1.4.3 has been applied to the Suntech module. The results are practically coincident with those obtained by the implicit method described in Section 1.4.2, so that a good reconstruction of the I-V and P-V curves at different irradiation levels has been obtained, as shown in Figures 1.13 and 1.14.

TABLE 1.3

Outdoor Operating Conditions for a Commercial PV Module at the Ambient Temperature $T_a = 35°C$

Parameter Name and Symbol	Values at $G = 1000$ W/m²
Short-circuit current I_{SC}	7.92 A
Open-circuit voltage V_{OC}	9.18 V
MPP current I_{MPP}	7.24 A
MPP voltage V_{MPP}	6.97 V
—	**Values at $G = 100$ W/m²**
Short-circuit current I_{SC}	0.76 A
Open-circuit voltage V_{OC}	9.12 V
MPP current I_{MPP}	0.68 A
MPP voltage V_{MPP}	7.63 V

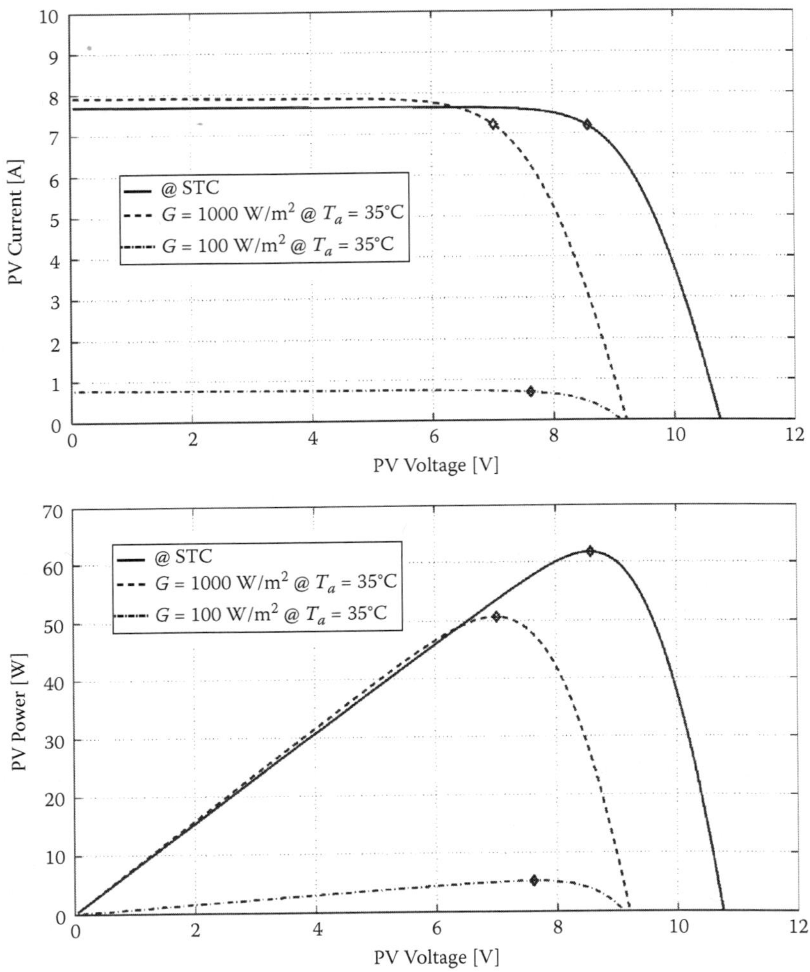

FIGURE 1.11
Current vs. voltage and power vs. voltage curves of a string of PV cells in STC and outdoor environmental conditions.

1.6 The Lambert W Function for Modeling a PV Field

1.6.1 PV Generator Working in Uniform Conditions

The fact that Equation (1.10) does not admit an explicit solution represents a significant limitation, not only in the identification of the model parameters but also in the simulation of PV systems, as the operating point of the PV unit can be obtained only at the price of the solution of the nonlinear equation

TABLE 1.4
Suntech STP245S-20/Wd: Data Sheet Values

PV Array	Values in STC Conditions
Short-circuit current I_{STC}	8.52 A
Open-circuit voltage V_{OC}	37.3 V
MPP current I_{MPP}	8.04 A
MPP voltage V_{MPP}	30.5 V
Temperature coefficient of I_{SC} (α_I)	0.05%/°C
Temperature coefficient of V_{OC} (α_V)	–0.34%/°C
NOCT	45°C
Number of cells in series	60

TABLE 1.5
Parameters of the PV Circuital Model

Parameters	Values
I_{ph}	8.52 A
R_s	0.2584 Ω
R_p	1278.1 Ω
C_0	286.0364 A/K³
η	1.0457

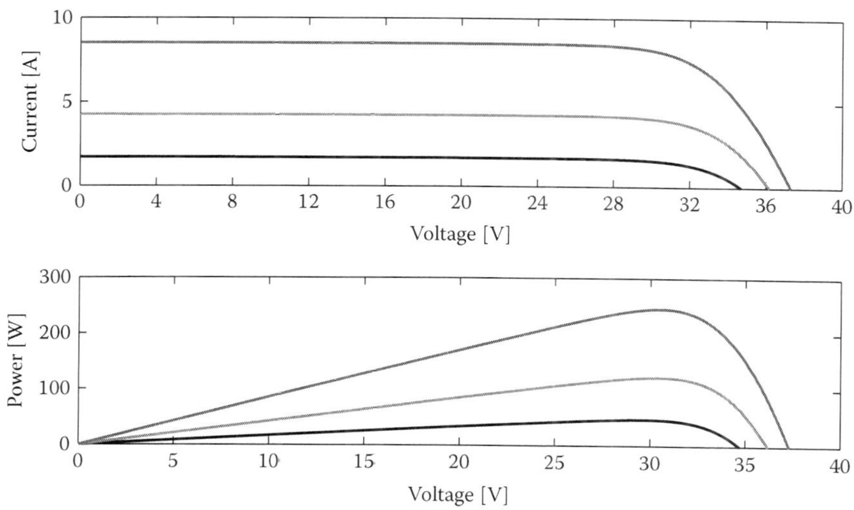

FIGURE 1.12
Current vs. voltage and power vs. voltage curves of the Suntech PV panel: effect of irradiation.

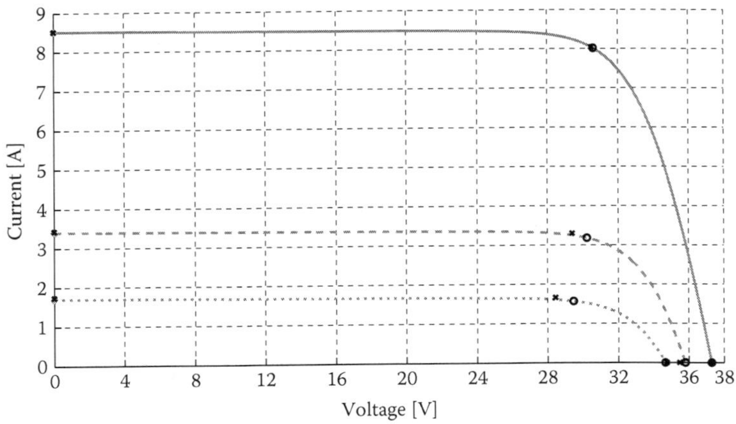

FIGURE 1.13
Current vs. voltage curves of the Suntech PV panel. Continuous line $G = 1000$ W/m^2, dashed line $G = 400$ W/m^2, dash-dotted line $G = 200$ W/m^2. Crosses are experimental points and dots are estimated points.

(1.10). Such limitation can be overcome by using the Lambert W function. In [10] many details and useful references about the Lambert W function can be found. The use of the Lambert W function leads to the following expression:

$$I = \frac{R_p \cdot (I_{ph} + I_{sat}) - V}{R_s + R_p} - \frac{\eta \cdot V_t}{R_s} \cdot LambertW(\theta) \tag{1.57}$$

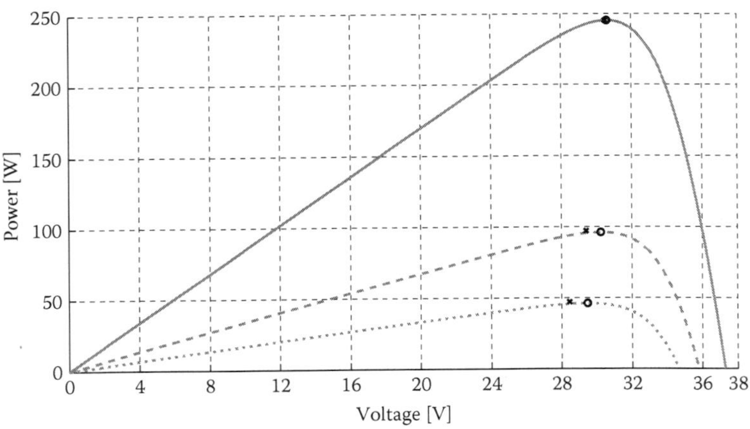

FIGURE 1.14
Power vs. voltage curves of the Suntech PV panel. Continuous line $G = 1000$ W/m^2, dashed line $G = 400$ W/m^2, dash-dotted line $G = 200$ W/m^2. Crosses are experimental points and dots are estimated points.

where

$$\theta = \frac{(R_p // R_s) \cdot I_{sat} \cdot e^{\frac{R_p \cdot R_s \cdot (I_{ph} + I_{sat}) + R_p \cdot V}{\eta \cdot V_t \cdot (R_p + R_s)}}}{\eta \cdot V_t} \tag{1.58}$$

It is easy to see that the expression is now explicit, so that for any value of the PV voltage V the corresponding value of the PV current I can be calculated straightforwardly, the Lambert W function being calculated by means of a series expansion [10]. Similarly, the voltage can be explicitly expressed as a function of the current as well.

Needless to say, such a model can be applied to any level of the PV field because, provided that all the cells in all the modules in all the strings of the PV field operate in exactly the same conditions, the parameters appearing in (1.57) and (1.58) can be scaled accordingly.

Unfortunately, in nonuniform operating conditions, the model and the explicit expression (1.57) are not so simple and scalable, so that a different analysis approach must be adopted.

1.6.2 Modeling a Mismatched PV Generator

As discussed in the preceding sections, mismatching conditions occur when a part of the PV generator works in conditions that are different from those of the remaining part of the generator itself. The occurrence of this situation is due to shadowing or to a significant difference in the physical parameters characterizing the cells. Some approaches have been presented in literature, e.g., in [7, 12, 13], to analyze such situations.

The analysis approach discussed in this section refers to a single PV string made of a number of PV modules connected in series. Each PV module is supposed to be a number of equal cells connected in series and working in the same temperature and irradiance conditions. The cells in the PV module are supposed to be protected by a bypass diode connected in antiparallel, as in Figure 1.1. The approach can be applied to any number of strings connected in parallel.

The group of cells in each module, working in homogeneous conditions, is represented by the equivalent circuit in Figure 1.8 and modeled by the corresponding equation (1.10). Nevertheless, the modules in the string are characterized by different sets of parameters, accounting for nonuniformities in terms of irradiance, temperature, or manufacturing tolerances. In order to have a compact model, it is useful to embed the bypass diode model in a complete model describing the whole PV module, including the cells and the bypass diode, as shown in Figure 1.15.

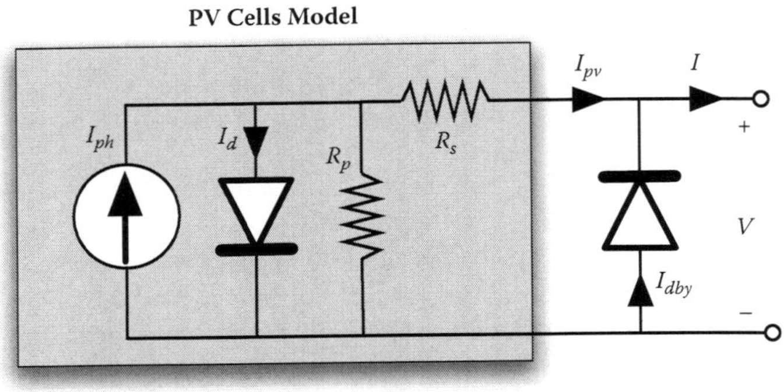

FIGURE 1.15
Equivalent circuit of a PV module with its own bypass diode.

The nonlinear characteristic equation of the bypass diode is given by the Shockley equation:

$$I_{dby} = I_{sat,dby} \cdot \left(e^{-\frac{V}{\eta_{dby} V_{t,dby}}} - 1 \right) \tag{1.59}$$

Merging (1.10) and (1.59) provides the following nonlinear and implicit equation:

$$I = I_{ph} - I_{sat} \cdot \left[e^{\frac{V + I \cdot R_s}{\eta V_t}} - 1 \right] - \frac{V + I \cdot R_s}{R_p} + I_{sat,dby} \cdot \left(e^{-\frac{V}{\eta_{dby} V_{t,dby}}} - 1 \right) \tag{1.60}$$

Once again, the Lambert W function is useful to put (1.60) into explicit form. The whole module, including the group of cells and the bypass diode, is then described by the following explicit equation, which gives the value of the current of the building block shown in Figure 1.15 as a function of the voltage at its terminals:

$$I = \frac{\left[R_p \cdot \left(I_{ph} + I_{sat} \right) - V \right]}{R_s + R_p} + I_{sat,dby} \cdot \left(e^{-\frac{V}{\eta_{dby} V_{t,dby}}} - 1 \right) - \frac{\eta V_t}{R_s} \cdot lambertW(\theta) \tag{1.61}$$

where θ is given in (1.58).

The mismatched string is made of a series of elements depicted as in Figure 1.15, each one working in certain irradiation and temperature conditions and characterized by given physical parameters that are different from those of the other elements of the string. The string is sketched in Figure 1.16, where each block contains one module depicted in Figure 1.15 and described

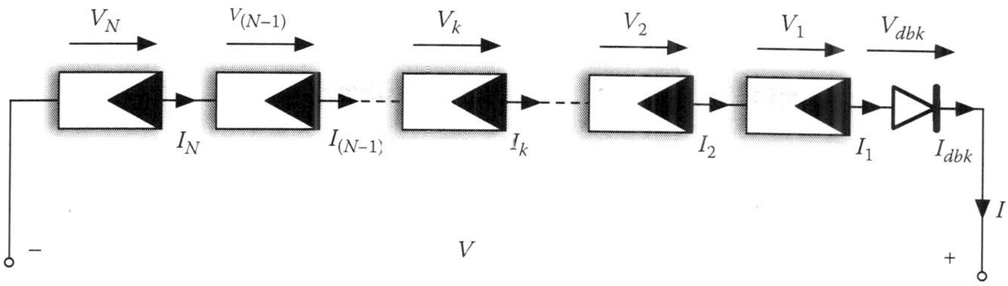

FIGURE 1.16
String of PV modules with blocking diode.

by Equation (1.61). A blocking diode terminates the string, to avoid the current flowing back from the other strings connected in parallel or, especially in stand-alone applications, from the storage element (e.g., battery).

Applying Kirchhoff's laws, the nonlinear system of Equation (1.62) results:

$$\begin{cases} V_1 + V_2 + \cdots + V_k + \cdots + V_{N-1} + V_N + V_{dbk} - V = 0 \\ I_1(V_1) - I_2(V_2) = 0 \\ I_1(V_1) - I_3(V_3) = 0 \\ \cdots \\ I_1(V_1) - I_k(V_k) = 0 \\ \cdots \\ I_1(V_1) - I_{N-1}(V_{N-1}) = 0 \\ I_1(V_1) - I_N(V_N) = 0 \\ I_1(V_1) - I_{dbk}(V_{dbk}) = 0 \end{cases} \tag{1.62}$$

As the model is based on a voltage-driven approach, given any value of the voltage V of the string, the corresponding current has to be calculated through the solution of the system in (1.62). The $(N + 1)$ unknowns $\{V_1, V_2, \ldots, V_k, \ldots, V_{N-1}, V_N, V_{dbk}\}$ are the voltages across the modules and the blocking diode. The N currents appearing in (1.62) are given by (1.61) for the PV modules and by (1.63) for the blocking diode:

$$I_{dbk} = I_{sat,dbk} \cdot \left(e^{\frac{V_{dbk}}{V_{t,dbk}}} - 1 \right) \tag{1.63}$$

All the parameters appearing in the system in (1.62), from the second up to the $(N + 1)$-th equation, are different for each PV panel. The nonlinear system (1.62) can be solved by using the Newton-Raphson method. It is shown below that the way the system has been written brings significant advantages for its numerical solution. In fact, the equations from the second to the $(N + 1)$-th have been all referred to the first module. Such a choice simplifies the expression of the Jacobian matrix of the system of equations (1.62).

Thanks to Lambert W function properties, the first derivative of the panel current with respect to the terminal voltage can be calculated in explicit form starting from Equation (1.61). In particular, Equation (1.64) provides the derivative of the *lambertW*(θ) function with respect to θ,

$$\frac{dlambertW(\theta)}{d\theta} = \frac{1}{[1+lambertW(\theta)]\cdot e^{lambertW(\theta)}} = \frac{lambertW(\theta)}{[1+lambertW(\theta)]\cdot\theta} \tag{1.64}$$

whereas Equation (1.65) provides the expression of the derivative of I with respect to V at the panel terminals.

$$\frac{\partial I}{\partial V} = -\frac{1}{R_s+R_p} - \frac{I_{sat,dby}}{V_{t,dby}}\cdot e^{-\frac{V}{V_{t,dby}}} - \frac{R_p}{R_s\cdot\left(R_s+R_p\right)}\cdot lambertW(\theta) \tag{1.65}$$

In this way, both the PV current and its derivative with respect to the PV voltage have been expressed in closed form as functions of the sole voltage V at the terminals.

Thanks to (1.10) it is possible to obtain each term of the Jacobian matrix J, to be inverted at each iteration of the Newton-Raphson-based algorithm, as a function of the unknowns. The Jacobian matrix structure reported in (1.66) puts in evidence that it is sparse and shows a pattern that is characteristic of doubly bordered and diagonal square matrices, also known as arrow-up matrix. It is

$$J = \left[\begin{array}{c|c} \multicolumn{2}{c}{ones(1,N+1)} \\ \hline c & d \end{array}\right] \tag{1.66}$$

where *ones*(1, N + 1) is a (N + 1) columns row vector with all the elements equal to 1, *c* is a column vector of N rows equal to

$$c = \frac{\partial I_1}{\partial V_1}\cdot ones(N,1) \tag{1.67}$$

ones(N, 1) is a N rows column vector with all the elements equal to 1, and

$$d = diag\left(\frac{\partial I_1}{\partial V_1},\frac{\partial I_2}{\partial V_2},\ldots,\frac{\partial I_k}{\partial V_k},\ldots,\frac{\partial I_{N-1}}{\partial V_{N-1}},\frac{\partial I_N}{\partial V_N},\frac{\partial I_{dbk}}{\partial V_{dbk}}\right) \tag{1.68}$$

is a diagonal matrix of size N.

The first row is composed by (N + 1) constants, while all the other rows require the evaluation of $\partial I_1/\partial V_1$ and the calculation of another derivative. As a whole, the evaluation of the system (1.62) requires N times the use of Equation (1.61) and 1 times Equation (1.63); the calculation of the Jacobian matrix requires N evaluations of Equation (1.65) and one evaluation of Equation (1.69).

$$\frac{\partial I_{dbk}}{\partial V_{dbk}} = \frac{I_{sat,dbk}}{V_{t,dbk}} \cdot e^{-\frac{V_{dbk}}{V_{t,dbk}}} \tag{1.69}$$

Such features are useful in terms of both memory requirements during the simulation and computation time.

Finally, it must be pointed out that Equation (1.65) allows to express in a closed form, and without neglecting any parameter appearing in the model, the differential conductance of the PV generator, which is the slope of the tangent to the I-V curve in the desired operating point. It is a negative value that is of fundamental importance for the best design of the power processing system that takes the PV power at its input and delivers it to the load or the grid. The main role of the differential conductance will be in the determination of the small signal model of the whole system, including the PV generator and the switching converter performing the maximum power point tracking function. This model allows us to design the best dynamic performances for the tracking function. It is also worth noting that the differential conductance (1.65) has a time-varying value, because of its dependence on the irradiance and temperature values.

1.7 Example

In this example the analysis of a PV generator structure is analyzed in presence of nonuniform irradiation conditions. Two identical modules connected in series have been considered, one subjected to a normalized irradiation level equal to $G_1/G_{max} = 1$ and the other to a normalized irradiation level $G_2/G_{max} = 0.1$. If the two modules were subjected to the same $G = 1$, the P-V characteristic would be that shown in Figure 1.17. The plot has been obtained by multiplying by 2 the voltage values of the characteristic curve of one of the two identical modules.

In the presence of a significant difference in the irradiance value, the method described in Section 1.6 has been used. The same analysis has been repeated many times, thus assigning a number of samples to the total array voltage V in (1.62) in the range of normalized voltage between 0 and 1. Figure 1.18 shows how dramatically the P-V curve has changed, with the

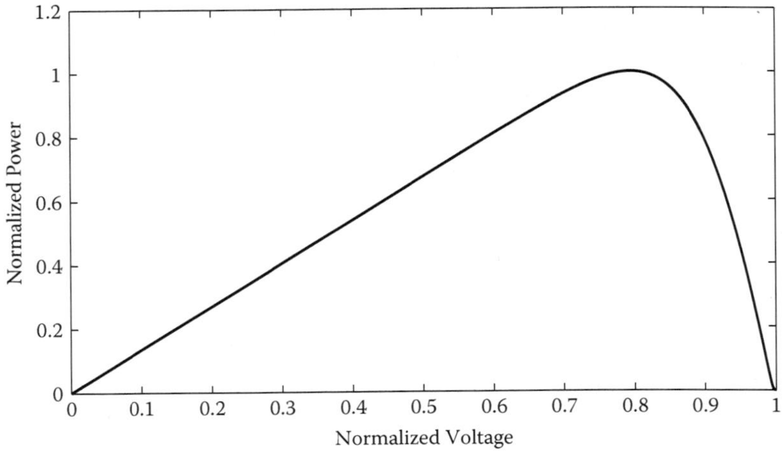

FIGURE 1.17
PV characteristic in uniform conditions.

presence of two maximum power points, the absolute MPP occurring at a low voltage level and the relative MPP being placed at a voltage quite close to that one at which the unique MPP occurred in the uniform case shown in Figure 1.17.

The presence of the two peaks can be explained in a simple way by looking at the I-V characteristic reported in Figure 1.19. At a low voltage level, the current I_1 produced by the string of PV cells receiving the highest irradiation is larger than the I_2 the other string is able to produce, so that the

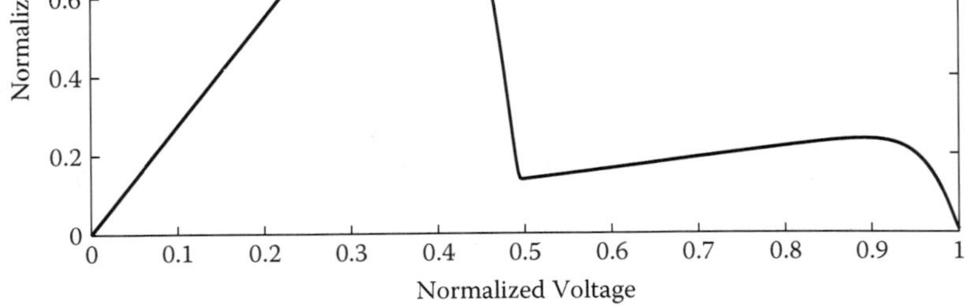

FIGURE 1.18
P-V characteristic in mismatched conditions.

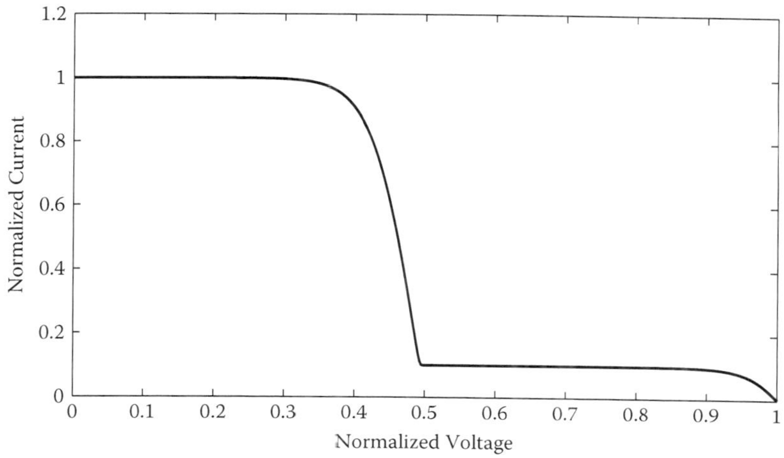

FIGURE 1.19
I-V characteristic in mismatched conditions.

bypass diode of the second module starts to conduct the difference current $I_1 - I_2$. The diode, during its conduction, inverts the voltage at the terminals of the second PV string, which generates a nonzero current. As a consequence, the second string absorbs an electrical power that is the product of the current it generates by the conduction voltage of the bypass diode. Due to the few hundreds of millivolts of the latter, the dissipated power is limited to a low value. In uniform, yet reduced, irradiation conditions over the cells of the module, the less irradiated string is not subjected to hot spot phenomena. In such conditions, the whole system produces an electrical power that is the power produced by the first string from which the power absorbed by the second string and the power dissipated by the second bypass diode must be subtracted.

Truly speaking, the bypass diode is helpful if the *whole* second string is subjected to G_2, but unfortunately, if some cells of this string receive a very low irradiation level, due to a shadowing body lying on its surface, disruptive phenomena can occur. Indeed, as pointed out in literature (e.g., in [2]), if few cells in a module are affected by a deep shadowing effect with respect to many others fully illuminated, hot spot phenomena can take place. The shadowed cells are forced to work at a very negative voltage, but they conduct a significant forward current, imposed by the other cells that receive a high irradiation. The bypass diode does not go into conduction because the large forward voltage of the illuminated cells is partially compensated by the deep negative voltage of the shadowed ones. In such conditions the risk of a permanent damage of the shadowed cells is very high.

As shown in Figures 1.19 and 1.20, at a value of the normalized voltage approximately equal to 0.5, in the example considered herein, the current I_1

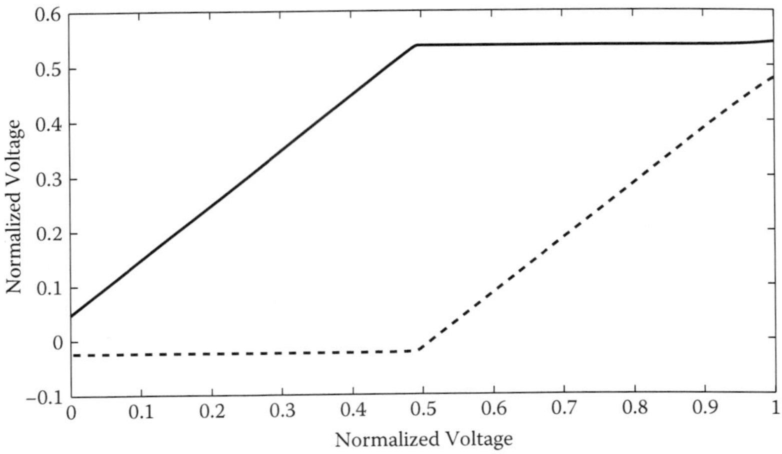

FIGURE 1.20
Voltage across the two modules vs. total voltage. Horizontal axis: voltage on the series connection of the two modules. Vertical axis: voltage on each module. Continuous line: fully irradiated module. Dashed line: shadowed module.

drops and reaches the value I_2, so that the current in the second bypass diode becomes zero and in the two PV strings the same current flows. This condition holds up to the total voltage V being equal to the sum of the open-circuit voltages of the two PV strings.

References

1. M.G. Villalva, J.R. Gazoli, and E.R. Filho. Comprehensive approach to modeling and simulation of photovoltaic arrays. *IEEE Transactions on Power Electronics*, 24(5):1198–1208, 2009.
2. U. Eicker. *Solar technologies for buildings*. Wiley, West Sussex, UK, 2003.
3. B. Bletterie, R. Bruendlinger, and S. Spielauer. Quantifying dynamic MPPT performance under realistic conditions first test results—The way forward. In *Proceedings of 21st European Photovoltaic Solar Energy Conference and Exhibition*, 2006, pp. 2347–2351.
4. J.W. Bishop. Computer simulation of the effects of electrical mismatches in photovoltaic cell interconnection circuits. *Solar Cells*, 25:73–89, 1988.
5. D. Sera, R. Teodorescu, and P. Rodriguez. PV panel model based on datasheet values. In *Proceedings of the IEEE International Symposium on Electronics*, June 2007, pp. 2392–2396.
6. A. Chatterjee, A. Keyhani, and D. Kapoor. Identification of photovoltaic source models. *IEEE Transactions on Energy Conversion*, 26(3):883–889, 2011.

7. Y.-J. Wang and P.-C. Hsu. Analytical modelling of partial shading and different orientation of photovoltaic modules. *IET Renewable Power Generation*, 4(3):272–282, 2010.
8. J.A. Gow and C.D. Manning. Development of a photovoltaic array model for use in power electronics simulation studies. *IEE Proceedings-Electric Power Applications*, 146(2):193–200, 1999.
9. E. Saloux, A. Teyssedou, and M. Sorin. Explicit model of photovoltaic panels to determine voltages and currents at the maximum power point. *Solar Energy*, 85:713–722, 2011.
10. Weisstein, Eric W. Lambert W-Function. From *MathWorld*—A Wolfram Web Resource. http://mathworld.wolfram.com/LambertW-Function.html.
11. F. Ghani and M. Duke. Numerical determination of parasitic resistances of a solar cell using the lambert w-function. *Solar Energy*, 85(9):2386–2394, 2011.
12. G. Petrone and C.A. Ramos-Paja. Modeling of photovoltaic fields in mismatched conditions for energy yield evaluations. In *Electric Power Systems Research*, 2011, pp. 1003–1013.
13. G. Petrone, G. Spagnuolo, and M. Vitelli. Analytical model of mismatched photovoltaic fields by means of lambert w-function. In *Solar Energy Materials and Solar Cells*, 2007, pp. 1652–1657.

2

Maximum Power Point Tracking

2.1 The Dynamic Optimization Problem

Chapter 1 was extensively dedicated to the discussion of the main aspects concerning the current vs. voltage characteristics of a PV generator and its dependency on the exogenous uncontrollable temperature and sun irradiance variables. The in-depth analysis has shown evidence that the joint variation of temperature and irradiance, as well as the occurrence of drift phenomena that arise along the PV system lifetime, about 25 years, makes the position of the maximum power point (MPP) varying in a wide area. As a consequence, a direct connection of the PV generator to the input port of a power processing system imposing a constant voltage level would be a simple but poor choice from the energy productivity point of view. For instance, a PV battery charger obtained by connecting merely the PV array terminals to the battery would force the PV generator to work at a constant voltage. If this voltage is higher than the PV array open-circuit voltage, then the PV system does not deliver any electric power. Otherwise, the closer the battery voltage to the actual V_{MPP}, the higher the electrical power generated by the PV array. Unfortunately, due to the inherent time variability of V_{MPP} caused by the changes in the operating conditions, the probability that the PV array delivers the maximum power at any time of the day is almost near zero.

The same conclusion holds if the PV array feeds a resistive load, because the intersection between the resistor characteristic and the I-V curve of the PV array cannot always occur in the MPP for the whole day, so that a power lower than the maximum one is delivered to the load. Figures 2.1 and 2.2 show that the operating point resulting from a straightforward connection of a PV generator with a battery or a resistive load cannot be the MPP in any irradiance or temperature condition.

As a consequence, it is mandatory to adopt an intermediate conversion stage, interfacing the PV array and the power system that processes or uses the electrical power produced thereof, which must be capable of adapting its input voltage and current levels to the instantaneous PV source MPP, while keeping its output voltage and current levels compliant with the load requirement. Such a stage must be a *dynamical optimizator*, which means that it must

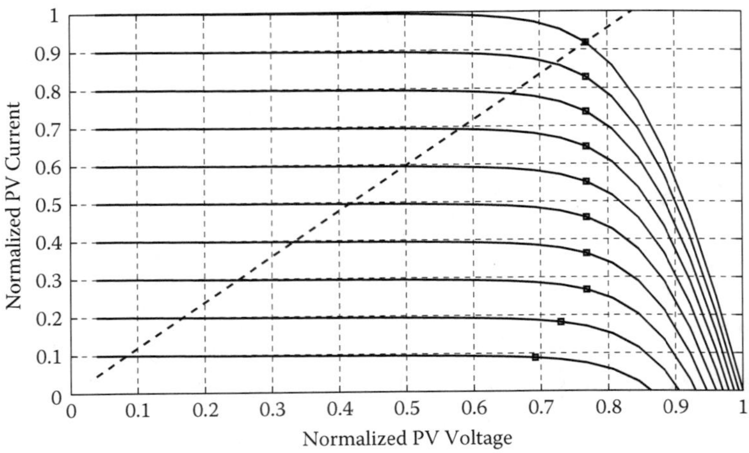

FIGURE 2.1
Operating conditions resulting from a straight connection of a PV generator with a resistive load at varying irradiance levels. Squares are the MPPs.

be able to perform this adaptation in the presence of time-varying operating conditions affecting the PV generator. The adoption of a linear regulator would be ineffective from the efficiency point of view, so that a switching converter is almost always employed. Because of the reduced cost of power devices, the adoption of a switching converter is now in use from very low power levels, e.g., energy scavenging, up to high power applications, e.g.,

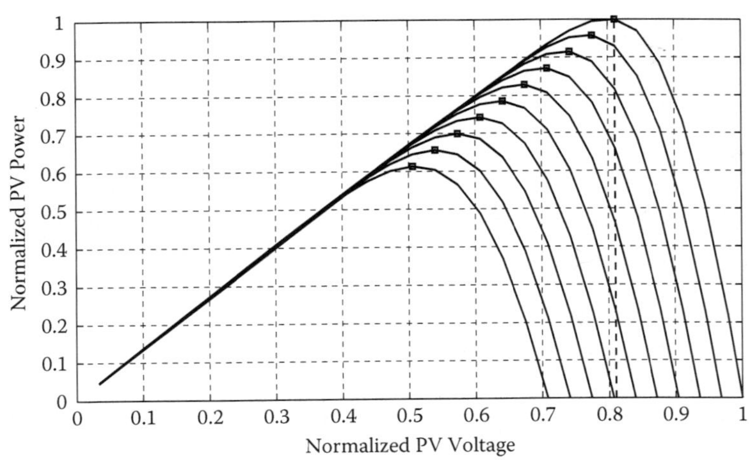

FIGURE 2.2
Operating conditions resulting from a straight connection of a PV generator with a battery at varying temperatures. Squares are the MPPs.

in large solar power plants. Nevertheless, because the additional electronic device increases the system cost, a trade-off between power maximization and system cost must be reached. Indeed, very high-energy efficiency, above 95%, is required to fully exploit the benefits of power electronics in maximizing the energy productivity of PV plants. A careful selection of power converters architecture and power devices is needed to achieve such results. This point is discussed in depth in Chapter 5.

In order to ensure the maximization of the power extracted from the PV source, the interface power converter must be capable of self-adjusting its own parameters at run time, thus changing its input voltage/current levels based on the PV source MPP position. In this book it is assumed, unless specified differently, that the *dynamical optimizator* is a DC/DC converter and that the control parameter is the duty cycle d of the main switch, as shown in Figure 2.3.

Supposing that the DC/DC voltage conversion ratio of the converter is $M(d) = V_o/V_i$, the equivalent resistance and voltage seen at the PV source terminals can be expressed by (2.1):

$$R_{in}(d) = \frac{R_{Load}}{M(d)^2}; \qquad V_i(d) = \frac{V_o}{M(d)} \tag{2.1}$$

Figure 2.4 shows the effect of the duty cycle modification in the PV operating point when it is connected to a DC/DC converter and supplies a resistive load (Figure 2.4a) or a constant voltage load like a battery (Figure 2.4b).

The duty cycle value d must be changed continuously by a controller to ensure that the PV generator always operates at its MPP for whatever irradiance and temperature operating conditions: For this reason it is called maximum power point tracking (MPPT) controller. Based on the instantaneous values of the current and voltage sensed at the PV generator terminals, the

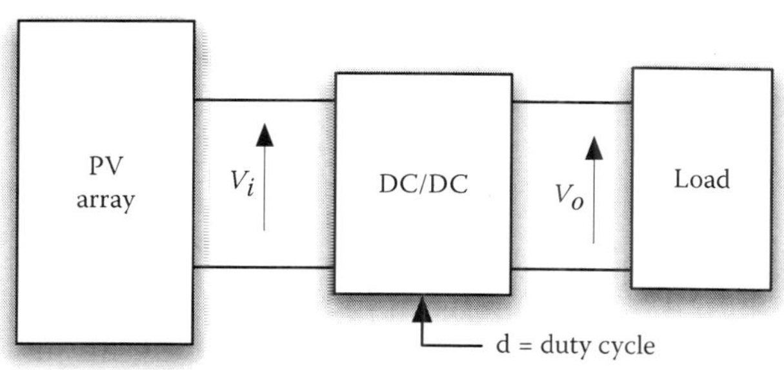

FIGURE 2.3
Connection scheme of a DC/DC converter dedicated to the dynamical optimization of a PV generator.

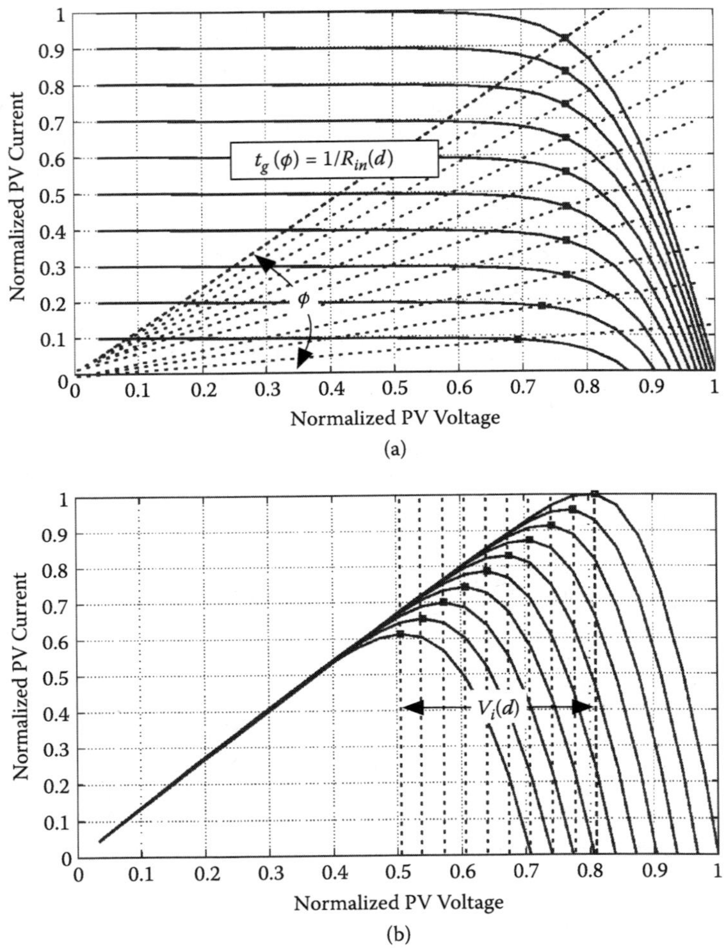

FIGURE 2.4
Operating conditions resulting from a PV generator interfaced to a DC/DC converter. In (a) the PV source has been characterized for different irradiance levels, and in (b) the PV source has been characterized for different temperature levels.

MPPT controller dynamically adjusts the converter duty cycle to follow the MPP, as shown in Figure 2.5.

In most applications, the MPPT controller is conceived to maximize the power extracted from the PV generator. Nevertheless, it might be more convenient to employ the MPPT controller for maximizing the power extracted from the switching converter output terminals, thus transferred to the end user. The difference between the maximization of the input power and that of the output power of the switching converter is in the intrinsic variation of the converter efficiency, which is conditioned by the change of the operating

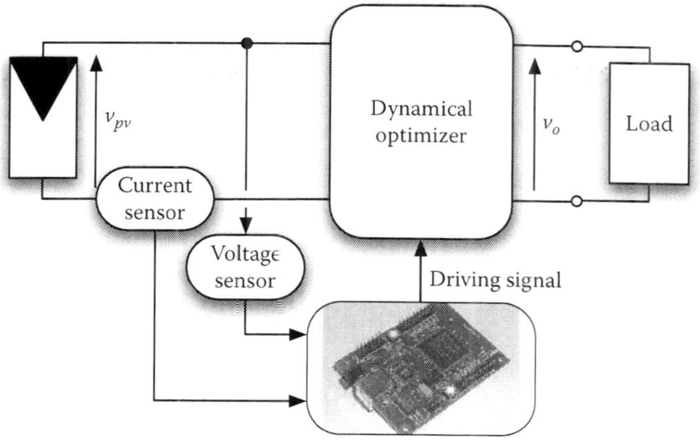

FIGURE 2.5
MPPT implementation.

point during the day. In fact, it should be considered that the irradiation variations cause a significant change of operating current stress on power devices, whose losses are related, in turn, to physical parameters that are heavily conditioned by the temperature. Moreover, in nonuniform irradiance conditions, additional voltage stress is imposed to power devices, as shown in Chapters 4 and 5. All these factors sum up negatively in causing the efficiency of the converter to lower. As a consequence, an increase in the PV power may determine operating conditions for the converter that lead to a lower efficiency, thus potentially resulting in a lower power at the converter output. In literature, the maximization of the power extracted from the PV source is mostly considered the main problem, because the efficiency vs. power rate profile of the power converter is supposed to be designed so that the higher the PV power, the higher the efficiency. However, this point must be carefully considered when a power converter is designed for PV applications, as the power devices must be selected trying to trade off the cost limitations with the need to guarantee a proper shape of the efficiency vs. the power rate according to the power conversion architecture and to the control strategy adopted for the entire system. For this reason, the PV power maximization problem is considered in this book under the perspective of global energy productivity. Chapter 5 provides a detailed discussion of this point.

The MPPT techniques presented in literature and used in most commercial products usually measure both the PV current and voltage values. Direct sensing of the temperature and irradiance is normally avoided, as their measurement requires expensive devices that have to be placed throughout the PV generator, in order to get the values of such variables for each panel or group of them, thus making the measurement quite expensive, especially for large PV plants.

The MPPT controller can be realized based on different methods and algorithms. The most popular methods are known as *perturb and observe* (P&O) and *incremental conductance* (INC). The former is widely used in commercial products and is the basis of the largest part of the most sophisticated algorithms presented in literature. Due to the fact that, as explained in the sequel, the INC approach can be seen as an improved version of the P&O one, this chapter devotes most of its space to the P&O approach.

The practical implementation of MPPT controllers is mostly realized in digital form. The speed of analog-to-digital converters needed to realize MPPT based on digital control is not a critical matter because of the relatively slow variation of the temperature and irradiance to be followed. The computations required by MPPT algorithms allow the designer to use cheap microcontrollers, for basic P&O and INC approaches, or much more expensive devices, like digital signal processing (DSP) and field-programmable gate array (FPGA) systems, when the MPPT approach is based on much more involved and burdensome computations, as those required by neural networks and fuzzy logic techniques. A brief overview of the soft computing approaches to the MPPT problem is given in Section 2.3.

The adoption of a digital controller is not strictly necessary for the MPPT implementation. In fact, it is possible to use analog circuitry to realize the MPPT controller as well [1]. Analog control, however, is not much used in PV MPPT applications because of the difficulty encountered in the design of the controller to take into account tolerances and parametric drifts. The flexibility and the possibility to realize self-adaptive controllers offered by the digital implementation justify the very limited adoption of analog control. However, better performances together with the possibility of ease and cheap realization of distributed MPPT applications are prerogatives of analog MPPT controllers. Brief descriptions of the few techniques that can be implemented by means of analog circuitry only are given in Sections 2.2, 2.6.2, and 2.7.1.

In this chapter, the MPPT problem is discussed with reference to a PV array working in uniform irradiance and temperature operating conditions. Specific distributed MPPT architectures and control algorithms needed for applications where PV generators are subjected to shadowing and other mismatching effects, discussed in the first chapter and for which the P-V curve can exhibit more than one maximum power point, are treated in Chapter 4.

2.2 Fractional Open-Circuit Voltage and Short-Circuit Current

In this section, some typical approaches in which the MPP is estimated by using information extracted by the models of the PV generator are discussed. In some cases, they greatly depend on the system behavior in a particular operating condition, in which some electrical variables can be easily measured.

Such methods are usually defined as *indirect MPPT techniques* because they do not measure the power extracted by the PV source, so that the maximum power point can be only approximately tracked. The main drawback of such techniques is that when the real conditions deviate too much with respect to those supposed in the modeled adopted, e.g., for the effect of parametric drifts, the energy losses are significant.

Alternatively, if the MPPT capability must be assured with high precision and in every environmental condition the PV source is operating in, then *direct MPPT* methods are more appropriate. In such techniques the PV current and voltage are measured continuously, and such measurements are used to realize a proper adjustment of the system operating conditions in order to catch the MPP.

Indirect MPPT methods are based on the estimation or on occasional measurement of one current or one voltage. In [2] it is shown that the MPP is typically located at a voltage close to 76% of the open-circuit voltage of the PV panel. In a more recent study [3], a more reasonable range of 70–82% has been indicated for the percent ratio between V_{MPP} and V_{OC}. A method for tracking the MPP without using an involved algorithm was deduced from this observation based on experimental evaluation on a number of PV panels available on the market. Such a method is quite simple and cheap. A switch connected in series with the PV module is used to disconnect the PV source periodically to allow measurement of its open-circuit voltage and, afterwards, to force the PV system to work at a voltage level that is 76% of the measured voltage. In any case, this method also requires a decoupling unit (e.g., a switching converter) between the source and the load, so that the desired voltage can be settled at the PV source terminals.

This is the weak point of such an approach, which also has a heavy limitation in the fact that the *magic number* 76% does not hold for any operating condition and for *any* PV panel. Moreover, it is easy to understand that, during the time interval used for measuring the open-circuit voltage, the PV generator does not feed energy. The higher the irradiation level at which the measure is done, the larger the missed energy.

In [4] it is stated that the I_{MPP} value is 86% of the short-circuit current; thus the dual approach might be applied by measuring periodically the short-circuit current instead of the open-circuit voltage. The same comments above apply to such a dual approach.

2.3 Soft Computing Methods

Fuzzy and neural network methods can be very useful in affording a hard nonlinear problem, as MPPT is.

Fuzzy logic-based trackers have shown very good performances under varying irradiance conditions without any detailed knowledge of the PV source model. The other side of the coin is that deep knowledge of the system operation and experience are required to choose the relevant computation error and the rule base table. Moreover, neural network strategies require specific training for each type of PV panel or array, and they need new training periods if the long-term parametric drifts are to be accounted for. Such a limitation has been overcome in [5] by using a genetic algorithm to achieve an optimal tuning of the membership functions. Improved performances with respect to the classical MPPT voltage-based approach have also been demonstrated in [6]. The Takagi-Sugeno (T-S) fuzzy approach proposed therein does not require any coordinate transformation, employs simple fuzzy rules, and is able to track rapid irradiance variations.

2.4 The Perturb and Observe Approach

The perturb and observe (P&O) method is the most popular algorithm belonging to the class of the *direct MPPT* techniques; it is characterized by the injection of a small perturbation into the system, whose effects are used to drive the operating point toward the MPP. Other methods derived by the P&O approach are the incremental conductance (INC), the self-oscillation (SO) method, and the extremum seeking (ES) control. Such methods differ from the P&O approach either for the variable observed or for the type of perturbation. All these algorithms have the advantage of being independent of the knowledge of the PV generator characteristics, so that the MPP is tracked regardless of the irradiance level, temperature, degradation, and aging, thus ensuring high robustness and reliability.

The first use of the P&O technique for tracking the MPP in PV systems goes back to the 1970s, when it was used intensively for aerospace applications [7]. At that time a fully analog implementation was adopted; nowadays, with the progress of the performances of low-cost microcontrollers, the digital implementation is preferred because the implementation of the MPPT algorithm results in a simple code [8].

The P&O method is based on the following concept: The PV operating point is perturbed periodically by changing the voltage at PV source terminals, and after each perturbation, the control algorithm compares the values of the power fed by the PV source before and after the perturbation. If after the perturbation the PV power has increased, this means that the operating point has been moved toward the MPP; consequently, the subsequent perturbation imposed to the voltage will have the same sign as the previous one. If after a voltage perturbation the power drawn from the PV array decreases, this means that the operating point has been moved away from the MPP.

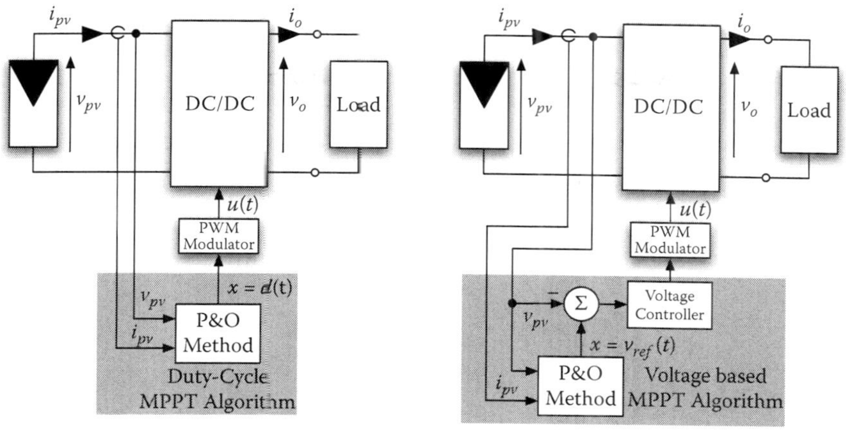

FIGURE 2.6
Basic schemes for implementing the P&O algorithm.

Therefore the sign of the subsequent voltage perturbation is reversed. The switching converter is used to drive the perturbation of the operating voltage of the PV generator.

Two basic P&O configurations can be adopted for controlling the switching converter and realizing the PV source voltage perturbation: The first one involves a direct perturbation of the duty ratio of the power converter. In the second one the perturbation is applied to the reference voltage of an error amplifier that generates the signal controlling the duty cycle. Both solutions are shown in Figure 2.6. In the first case, the converter operates in open loop after each duty cycle perturbation, while in the second case the converter is equipped with a feedback voltage loop.

The general equation describing the P&O algorithm is

$$x_{((k+1)T_p)} = x_{(kT_p)} \pm \Delta x = x_{(kT_p)} + (x_{(kT_p)} - x_{((k-1)T_p)})\cdot sign(P_{(kT_p)} - P_{((k-1)T_p)}) \quad (2.2)$$

where x represents the variable that is perturbed (e.g., the duty cycle or the feedback voltage reference in the schemes shown in Figure 2.6); T_p is the time interval between two perturbations, in the sequel referred to as *perturbation period*; Δx is the amplitude of the perturbation imposed to x, in the sequel referred to as *perturbation amplitude*; and P is the power extracted from the PV array. Figure 2.7 shows the flowchart of the P&O algorithm based on Equation (2.2).

The basic version of the P&O algorithm uses a fixed step amplitude $\Delta x = |x_{(kT_p)} - x_{((k-1)T_p)}|$ that is selected on the basis of performance trade-off between transient rise time and steady-state conditions.

Figure 2.8 shows the operating points of the PV field imposed step-by-step by the P&O algorithm. When the system approaches the MPP, the perturbing nature of the P&O and other similar MPPT algorithms involves oscillations

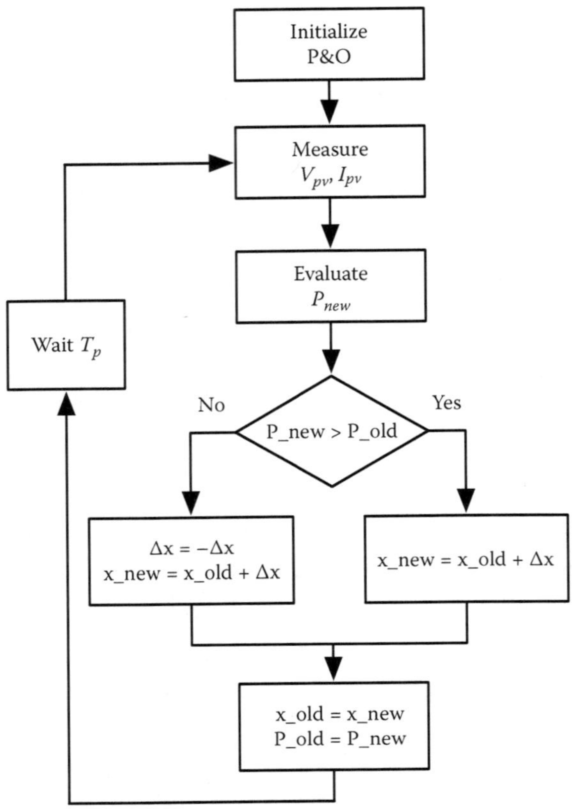

FIGURE 2.7
Basic flowchart for implementing the P&O algorithm.

across the MPP whose characteristics depend on the value of the parameters $\{\Delta x, T_p\}$ in (2.2). In [9] it was shown that the minimum number of steps that ensures a periodic and stable oscillation around the MPP is three, corresponding to a $2\Delta x$ peak-to-peak amplitude and a periodicity of $4T_p$. Such condition is likely the optimal one: Further steps to the left or right side would move the operating point far from the MPP, thus reducing the average power extracted from the PV field. Regardless of the Δx amplitude, a stable three-point behavior is ensured if the time interval T_p is properly selected. If it is settled at a small value, the P&O algorithm can be confused and the operating point can become unstable, entering disordered or chaotic behaviors [10]. On the contrary, a too big value of T_p penalizes the MPPT speed. Guidelines for optimal design of the couple of parameters $\{\Delta x, T_p\}$ are given in the following section.

In literature some variants of the P&O algorithm have been presented. They are especially aimed at reducing the amplitude of the oscillations around the MPP in steady-state irradiance and temperature conditions and at increasing the tracking performances during cloudy days. Although the

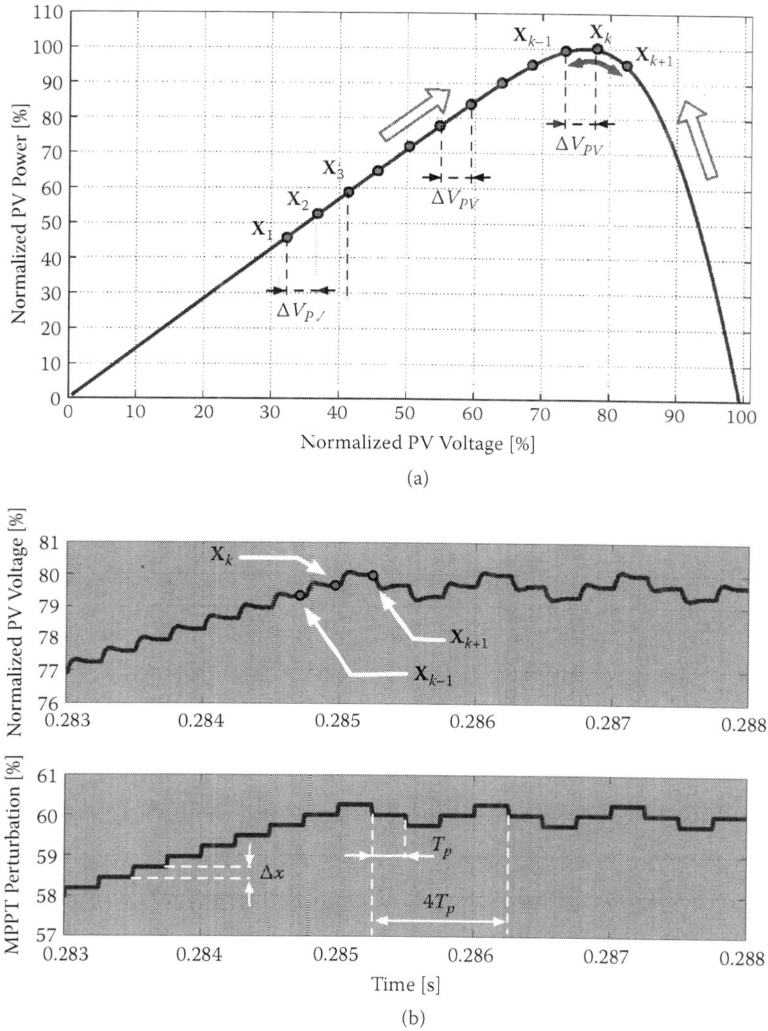

FIGURE 2.8
PV operating points imposed by the P&O algorithm. (a) Power vs. voltage characteristic. (b) Time domain behavior.

design guidelines given in the next section refer to the basic P&O algorithm, they can be easily extended to other perturbative approaches.

2.4.1 Performance Optimization: Steady-State and Dynamic Conditions

Unless sensing input voltage and current of a switching power converter, the MPPT controller works on it as a sampled data feedback system. Due to the nonlinearity of its equation (2.2), it cannot be analyzed/designed by using the classical approaches used for linear regulators, e.g., by analyzing the Bode plots of the loop gain transfer function in the Laplace domain. However, linear open-loop operation between perturbations allows use of time domain analysis. Therefore the design procedure discussed in the following is based on time domain analysis, both in steady-state environmental conditions and in the presence of irradiance variations. Time domain analysis allows optimization of the values of the two P&O parameters $\{\Delta x, T_p\}$ separately.

As shown in [10], the results obtained by using the time domain analysis are in agreement with those achievable by means of the describing function method, mostly used for feedback nonlinear system analysis.

By referring to the schemes of Figure 2.6, the intrinsic dynamic nature of the switching converters introduces a propagation delay between the stimulus Δx and the variables measured for estimating the PV power. If the time delay is greater than T_p, the P&O algorithm has no correct information for deciding the sign of the next perturbation. In order to identify the minimum value to assign to T_p, the behavior of the whole system can be analyzed by considering the system small-signal model evaluated around the MPP. Regarding the PV source, if the oscillations of the operating point (v_{PV}, i_{PV}) are small compared to (V_{MPP}, I_{MPP}), then the relationship among i_{PV}, v_{PV}, G, and T can be linearized as follows:

$$\hat{i}_{PV} = \left.\frac{\partial i_{PV}}{\partial v_{PV}}\right|_{MPP} \cdot \hat{v}_{PV} + \left.\frac{\partial i_{PV}}{\partial G}\right|_{MPP} \cdot \hat{G} + \left.\frac{\partial i_{PV}}{\partial T}\right|_{MPP} \cdot \hat{T} \tag{2.3}$$

where symbols with hats represent small-signal variations around the steady-state values of the corresponding quantities.

At a constant irradiance level it is $\hat{G} = 0$. Moreover, in steady state, due to the relatively high thermal inertia of the PV array [11], the amplitude of variations of the PV array temperature, as a consequence of small oscillations of the operating point, is negligible: Then $\hat{T} = 0$. Therefore (2.3) can be rewritten as

$$\hat{i}_{PV} = \left.\frac{\partial i_{PV}}{\partial v_{PV}}\right|_{MPP} \cdot \hat{v}_{PV} \tag{2.4}$$

where, based on the models introduced in Chapter 1, the derivative of the PV current with respect to the PV voltage can be expressed as follows:

$$\frac{\partial i_{PV}}{\partial v_{PV}} = -\left[R_s + \frac{1}{\frac{I_{sat}}{\eta \cdot V_t} \cdot e^{\frac{v_{PV} + R_s \cdot i_{PV}}{\eta \cdot V_t}} + \frac{1}{R_p}} \right]^{-1} \tag{2.5}$$

R_s, R_p, V_t, and I_{sat} are the values of the parameters of the model of the whole PV field: They have been obtained by scaling the corresponding parameters of a single module for the numbers of modules inside the field (Equations (1.14)–(1.17)).

In any operating point close to the MPP it is

$$v_{PV} = V_{MPP} + \hat{v}_{PV} \tag{2.6}$$

$$i_{PV} = I_{MPP} + \hat{i}_{PV} \tag{2.7}$$

$$p_{PV} = P_{MPP} + \hat{p}_{PV} = V_{MPP} \cdot I_{MPP} + V_{MPP} \cdot \hat{i}_{PV} + I_{MPP} \cdot \hat{v}_{PV} + \hat{v}_{PV} \cdot \hat{i}_{PV} \tag{2.8}$$

Under steady-state conditions in terms of irradiance level and temperature, the MPP occurs where the derivative of the power with respect to the PV voltage is equal to zero:

$$\left.\frac{\partial p_{PV}}{\partial v_{PV}}\right|_{MPP} = 0 \tag{2.9}$$

Writing 2.9 as a function of the PV current and voltage yields

$$\left.\frac{\partial p_{PV}}{\partial v_{PV}}\right|_{MPP} = \left.\frac{\partial (v_{PV} \cdot i_{PV})}{\partial v_{PV}}\right|_{MPP} = \left.\left(i_{PV} + v_{PV} \cdot \frac{\partial i_{PV}}{\partial v_{PV}} \right)\right|_{MPP} = 0 \tag{2.10}$$

Let us then define the differential resistance R_{MPP} of the PV array at the MPP:

$$\frac{1}{R_{MPP}} = -\left.\frac{\partial i_{PV}}{\partial v_{PV}}\right|_{MPP} \tag{2.11}$$

Accordingly, (2.10) can be rewritten as

$$\frac{1}{R_{MPP}} = \left.\frac{i_{PV}}{v_{PV}}\right|_{MPP} = \frac{I_{MPP}}{V_{MPP}} \tag{2.12}$$

From (2.8), (2.4), and (2.12) it results that

$$P_{MPP} = V_{MPP} \cdot I_{MPP} \tag{2.13}$$

$$\hat{p}_{PV} = \hat{v}_{PV} \cdot \left(I_{MPP} - \frac{V_{MPP}}{R_{MPP}} \right) + \hat{v}_{PV} \cdot \hat{i}_{PV} = -\frac{\hat{v}_{PV}^2}{R_{MPP}} \tag{2.14}$$

This is a general result that is not dependent on the DC/DC converter topology adopted for MPPT realization.

Equation (2.14) highlights that in order to investigate the performances of any MPPT algorithm, the study of the dynamic behavior of the PV voltage is mandatory. In fact, in order to allow the MPPT algorithm to make a correct interpretation of the effect of Δx perturbation on the corresponding steady-state variation of the PV power p, it is necessary that the time T_p between two consecutive perturbations be long enough to allow p to reach its *new* steady-state value.

By assuming stationary environmental conditions, if the perturbation Δx introduced by the P&O algorithm is the unique stimulus applied to the system composed of the PV field PV source, the converter, and the load, then it is possible to evaluate the voltage transient of the PV source triggered by the Δx perturbation. As a consequence of (2.14), the transient in the PV power can be obtained by analyzing the step response of the system using the control-to-array voltage transfer function. The block diagram shown in Figure 2.9 provides a simplified representation of the system under study. The dynamic behavior of most switching converters of interest for MPPT applications operating in open loop can be described by means of a second-order model like (2.15).

$G_{vp,x}(s) = \hat{v}_{PV} / \hat{x}$ is the transfer function obtained by using the small-signal linear model of the PV source, described by (2.4), and the small-signal model of the switching converter. This latter can be obtained by applying the state-space averaging modeling technique to the converter [12], which provides:

$$G_{vp,x}(s) = \frac{\hat{v}_{PV}(s)}{\hat{x}(s)} = \frac{\mu \cdot \omega_n^2}{s^2 + 2\zeta \cdot \omega_n \cdot s + \omega_n^2} \tag{2.15}$$

where μ is the static gain, ω_n is the natural frequency, and ζ is the damping factor of a canonical second-order system.

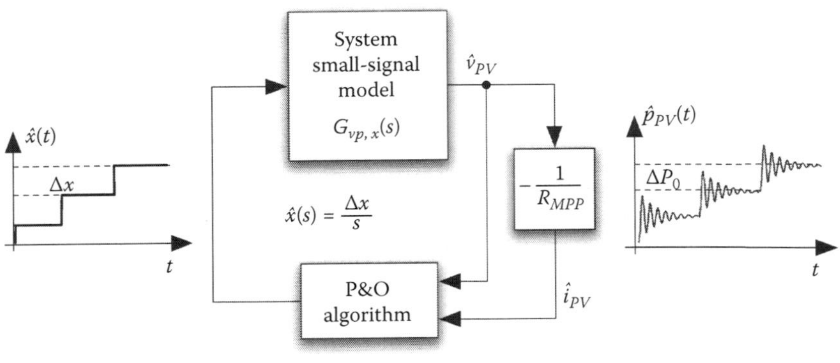

FIGURE 2.9
Model for PV dynamic behavior analysis.

The values of μ, ω_n, ζ can be expressed as function of the converter parameters only when the topology of the DC/DC converter has been selected [12, 13]. In the examples proposed in the next sections the explicit formula of such parameters will be given.

The second-order transfer function $G_{vp,x}(s)$ allows us to obtain design formulas for the P&O parameters Δx and T_p in a closed form. More in general, when the $G_{vp,x}(s)$ transfer function assumes a more complex expression, the values of the parameters of second-order dominant dynamics can be still evaluated numerically, so that the following discussion always remains valid. According to the transfer function (2.15), the response $\hat{v}_{PV}$ to a small step perturbation of amplitude Δx is

$$\hat{v}_{PV}(t) = \mu\Delta x\left(1 - \frac{1}{\sqrt{1-\zeta^2}} \cdot e^{-\zeta\omega_n t} sin\left(\omega_n t\sqrt{1-\zeta^2} + arccos(\zeta)\right)\right) \tag{2.16}$$

On the basis of (2.14) and (2.16), the response $\hat{p}_{PV}(t)$ to a small step perturbation of amplitude Δx can be approximated as shown in Equation (2.17):

$$\hat{p}_{PV}(t) = -\frac{\hat{v}_{PV}^2(t)}{R_{MPP}} \simeq -\frac{\mu^2\Delta x^2}{R_{MPP}}\left(1 - \frac{2}{\sqrt{1-\zeta^2}} \cdot e^{-\zeta\omega_n t} sin\left(\omega_n t\sqrt{1-\zeta^2} + arccos(\zeta)\right)\right) \tag{2.17}$$

where the faster term, whose decay occurs in a very short time, has been neglected.

Figure 2.10 shows the normalized PV power oscillation evaluated numerically (dashed line), the PV power oscillation evaluated by means of (2.17) (black continuous line), and the PV voltage oscillation (gray line). As expected, the figure shows that the power oscillation estimated by using (2.17) is a good approximation of the real power transient only in the final part of the step response where the faster term has extinguished its effect.

The settling time T_ε ensuring that $\hat{p}_{PV}(t)$ is confined within a band of relative amplitude +/−ε around the steady-state value can be evaluated by imposing the condition

$$\hat{p}_{PV}(t) \in [\Delta P_0 \cdot (1-\varepsilon), \Delta P_0 \cdot (1+\varepsilon)] \;\forall\; t > T_\varepsilon \tag{2.18}$$

where ΔP_0 is the final power variation due to the Δx perturbation, given by

$$\Delta P_0 = -\frac{\mu^2 \cdot \Delta x^2}{R_{MPP}} \tag{2.19}$$

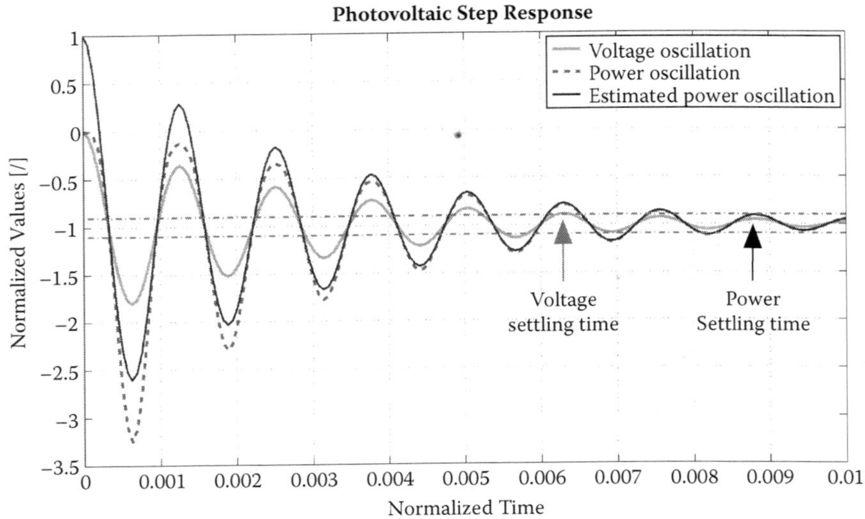

FIGURE 2.10
PV power dynamic behavior.

Evaluating the envelope of (2.17) in the T_ε yields

$$\Delta P_0 \cdot (1+\varepsilon) \simeq \Delta P_0 \left(1 - \frac{2}{\sqrt{1-\zeta^2}} \cdot e^{-\zeta\omega_n T_\varepsilon} \right) \tag{2.20}$$

so that the following expression of the settling time T_ε is obtained:

$$T_\varepsilon \simeq -\frac{1}{\zeta \cdot \omega_n} \cdot ln(\varepsilon / 2) \tag{2.21}$$

It is worth noting that, with the same boundary value ε, the 1/2 factor in (2.21) makes the settling time of the PV power oscillation considerably different with respect to that of the PV voltage oscillation. For the example proposed in Figure 2.10, it is more or less 30% higher than the PV voltage settling time.

Based on the previous modeling results, the MPPT is not affected by mistakes caused by transient oscillations of the PV system caused by its own action if the following condition is fulfilled:

$$T_p \geq T_\varepsilon \simeq -\frac{1}{\zeta \cdot \omega_n} \cdot \ln(\varepsilon / 2) \Big|_{\varepsilon=0.1} \tag{2.22}$$

where the value $\varepsilon = 0.1$ is chosen according to the classical control system theory.

2.4.2 Rapidly Changing Irradiance Conditions

As discussed in Section 2.4.1, the perturbation period T_p should be set not much higher than the system settling time, in order to avoid instability of the MPPT algorithm and limit the oscillations across the MPP in steady state. The value to be assigned to the amplitude of the step perturbation Δx also requires a careful setting: A small value reduces the steady-state losses caused by the oscillation of the PV operating point around the MPP, but it makes the algorithm slower and less efficient in the case of rapidly changing irradiance conditions. The latter circumstance can deceive the P&O algorithm in MPP tracking [14]. Indeed, the possible failure of the P&O algorithm in the presence of varying irradiance may occur if the algorithm is not able to distinguish the variations of the PV power caused by the duty cycle modulation from those caused by the irradiance variation. An adequate choice of Δx can overcome this problem, as discussed in detail in this section.

Let us suppose that the system is working at the MPP in the k-th sampling instant (see Figure 2.11), at an irradiance level equal to G, and that the step perturbation Δx moves the operating point leftward, in the direction of lower array voltage, namely, from the MPP to point A in the absence of irradiation change. Let us then suppose that the irradiance level changes between the k-th and the $(k + 1)$-th sampling instants. In Figure 2.11 it has been assumed that it increases. Then the operating point at the $(k + 1)$-th sampling instant will be point B instead of point A.

Depending on the amplitude of Δx, two possible situations are possible, as shown in Figure 2.11 In the sequel the PV power variation (at a constant irradiance level G) caused by the perturbation of Δx triggered by the P&O algorithm will be named with ΔP_x, while ΔP_G identifies the PV source output power variation caused by the variation ΔG of the irradiance level. The P&O algorithm works properly if

$$|\mathbf{\Delta P_x}| \geq |\mathbf{\Delta P_G}| \tag{2.23}$$

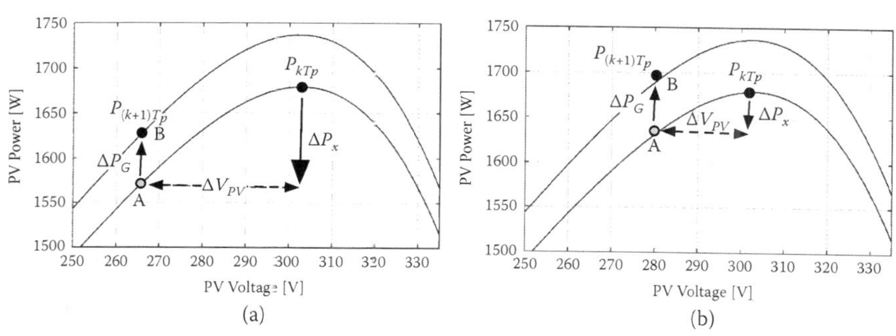

FIGURE 2.11
Operating point of PV field in the presence of an irradiance variation.

Condition 2.23 involves the absolute values because the signs of both power variations are not correlated and cannot be predicted in advance. For the case shown in Figure 2.11a, the inequality (2.23) is satisfied and the algorithm is not confused because it detects $P_{(k+1)T_p} < P_{kT_p}$, and consequently it will change the sign of the next perturbation, so that the operating point moves back to the MPP. In the case shown in Figure 2.11b, instead, the algorithm is confused because $P_{(k+1)T_p} > P_{kT_p}$, so that the next perturbation will have the same sign as the previous one; thus the operating point further moves in the direction of lower array voltage, away from the MPP.

Equation (2.23) can be made explicit by expressing each term as a function of the system parameters and PV field operating condition, so that the PV power variation is related to the Δx perturbation.

If the current and voltage variations with respect to the MPP are indicated as ΔI_{PV} and ΔV_{PV}, respectively, and if $\Delta P_{PV} = P_{(k+1)T_p} - P_{kT_p}$ is the whole power variation when the system moves from the MPP toward B (see Figure 2.11), then the following equations hold:

$$V_B = V_{MPP} + \Delta V_{PV} \tag{2.24}$$

$$I_B = I_{MPP} + \Delta I_{PV} \tag{2.25}$$

$$\Delta P_{PV} = P_B - P_{MPP} = V_{MPP} \cdot \Delta I_{PV} + I_{MPP} \cdot \Delta V_{PV} + \Delta V_{PV} \cdot \Delta I_{PV} \tag{2.26}$$

ΔI_{PV} can be expressed as a function of ΔV_{PV}, but not by means of the linear relation (2.3) since ΔV_{PV} assumes large values. A quadratic expression is much more suitable and allows evaluaton, with an improved accuracy, of whether inequality (2.23) is verified or not.

$$\Delta I_{PV} \simeq \left.\frac{\partial i_{PV}}{\partial v_{PV}}\right|_{MPP} \Delta V_{PV} + \left.\frac{1}{2}\frac{\partial^2 i_{PV}}{\partial^2 v_{PV}}\right|_{MPP} \Delta V_{PV}^2 + \left.\frac{\partial i_{PV}}{\partial G}\right|_{MPP} \Delta G + \left.\frac{\partial i_{PV}}{\partial T}\right|_{MPP} \Delta T_a \tag{2.27}$$

In (2.27) the PV current variation with respect to the irradiance variation ΔG has been considered only with the first derivative term because at the MPP the current variation is almost linear with respect to ΔG, as shown in [11]. Although in (2.27) the effect of the ambient temperature variation is also considered, in the following $\Delta T_a \simeq 0$ will be considered, because in the short time interval between two perturbations (T_p), the PV operating temperature can be assumed constant.

The partial derivative of the PV current can be evaluated as follows:

$$\left.\frac{1}{2}\frac{\partial^2 i_{PV}}{\partial^2 v_{PV}}\right|_{MPP} = -H \tag{2.28}$$

$$H = \frac{1}{2}\frac{1}{\eta V_t}\left(1 - \frac{R_s}{R_{MPP}}\right)^3 \left(\frac{I_{sat}}{\eta V_t} e^{\frac{V_{MPP}+R_s I_{MPP}}{\eta V_t}}\right) \tag{2.29}$$

where R_s, V_t, and I_{sat} are the values of the parameters of the model describing the whole PV generator. The term related to the irradiance variation is given by

$$\frac{\partial i_{PV}}{\partial G} \simeq \frac{\partial i_{ph}}{\partial G} = K_{ph} \tag{2.30}$$

where K_{ph} is a PV material constant [11] that can be evaluated by using Equation (1.2).

By means of (2.11), the PV current variation is expressed as follows:

$$\Delta i_{PV} \simeq -\frac{\Delta V_{PV}}{R_{MPP}} - H\cdot\Delta V_{PV}^2 + K_{ph}\cdot\Delta G \tag{2.31}$$

On the basis of (2.26), (2.31), and (2.12), the PV power variation can be rewritten:

$$\Delta P_{PV} \simeq -\left(H\cdot V_{MPP} + \frac{1}{R_{MPP}}\right)\cdot\Delta V_{PV}^2 + V_{MPP}\cdot K_{ph}\Delta G \tag{2.32}$$

where in (2.32) the higher-order terms have been neglected.

The effect of the Δx step perturbation on the corresponding steady-state variation ΔV_{PV} of the array voltage is evaluated by using the transfer function $G_{vp,x}(s)$ introduced in section 2.4.1:

$$\Delta V_{PV} = G_0\cdot\Delta x \tag{2.33}$$

where G_0 is the DC gain of $G_{vp,x}(s)$.

If Δx is the only stimulus that triggers the PV voltage variation, then

$$\left|\Delta P_x\right| = \left(H\cdot V_{MPP} + \frac{1}{R_{MPP}}\right)\cdot G_0^2\cdot\Delta x^2 \tag{2.34}$$

$$\left|\Delta P_G\right| = V_{MPP}\cdot K_{ph}\cdot\left|\Delta G\right| = V_{MPP}\cdot K_{ph}\cdot\left|\dot{G}\right|\cdot T_p \tag{2.35}$$

where $\dot{G}$ is the average rate of change of the irradiance level inside the period of the MPPT perturbation T_p occurring between the k-th and the $(k + 1)$-th sampling instants.

Under the hypotheses adopted above, the inequality (2.23) is fulfilled if the following condition is fulfilled by the Δx amplitude:

$$\Delta x > \frac{1}{G_0}\sqrt{\frac{V_{MPP}\cdot K_{ph}\cdot\left|\dot{G}\right|\cdot T_p}{\left(H\cdot V_{MPP} + \frac{1}{R_{MPP}}\right)}} = \Delta x_{min} \tag{2.36}$$

Setting the values of T_p and Δx on the basis of Equations (2.21) and (2.36), respectively, ensures that the P&O algorithm is able to track, without errors, irradiance variations characterized by an average rate of change (within T_p) not higher than $\dot{G}$.

It is worth noting that the parameters H, V_{MPP}, and R_{MPP} depend on the irradiance level. Thus when using the (2.21) and (2.36) design equations, the combination of parameters (H, V_{MPP}, R_{MPP}) leading to the highest value of the T_p and Δx parameters must be used. In the next section a design example is proposed in order to show how to apply the proposed MPPT design procedure.

2.4.3 P&O Design Example: A PV Battery Charger

Guidelines for designing the P&O algorithm based on the duty cycle perturbation are analyzed in this section. Figure 2.12 shows the basic scheme of a PV battery charger based on a boost DC/DC converter. An additional control circuitry, providing battery over voltage and over current protections, which must be considered in the practical application, has no effect on the MPPT design procedure.

As explained in Section 2.4.1, the first step is to identify the control-to-PV voltage transfer function: In the case under study the MPPT acts on the duty cycle directly so that $G_{vp,x}(s) = \hat{v}_{PV} / \hat{x}$ corresponds to $G_{vp,d}(s) = \hat{v}_{PV} / \hat{d}$. The scheme of Figure 2.13 shows the whole small-signal model of the PV battery charger that can be used to evaluate $G_{vp,d}(s)$.

As steady-state environmental conditions are supposed, the small-signal model of the PV field is given by its differential resistance obtained by using

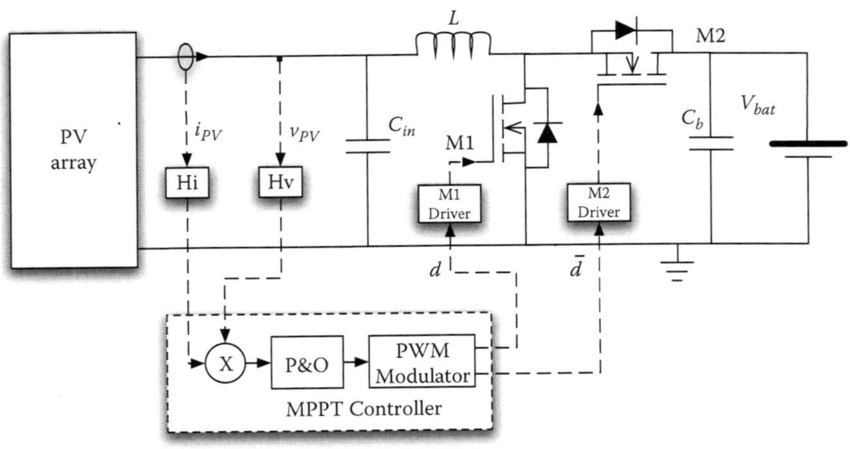

FIGURE 2.12
PV battery charger with a boost converter.

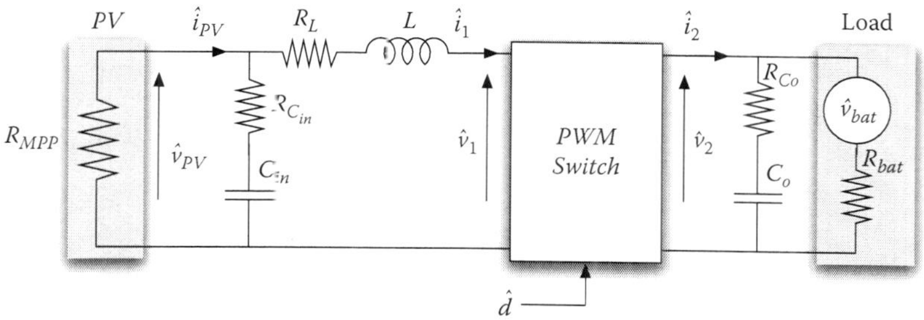

FIGURE 2.13
Small-signal model of PV battery charger.

(2.5) in the MPP. The DC/DC converter, instead, is decoupled in its linear parts and in a linearized small-signal part [12] modeled by the following equations:

$$\begin{cases} \hat{i}_2 = (1-D)\cdot\hat{i}_1 - I_1\cdot\hat{d} \\ \hat{v}_1 = (1-D)\cdot\hat{v}_2 - V_2\cdot\hat{d} \end{cases} \tag{2.37}$$

which completely describe the block *PWM SWITCH* shown in the scheme of Figure 2.13 and deeply analyzed in [12]. The additional small-signal source $\hat{v}_{bat}$ is used to model the low-frequency variations at the converter's output caused by load changes.

The oscillations $\hat{v}_{bat}$ can propagate up to the PV terminals, thus causing a possible deceiving effect on the operation of the MPPT algorithm, because it might be unable to distinguish the PV voltage oscillations caused by $\hat{v}_{bat}$ with respect to the PV voltage oscillations introduced by the perturbation of the duty cycle $\hat{d}$. In this example the battery is considered an ideal voltage source; thus $\hat{v}_{bat} = 0$, $R_{bat} = 0$. The problems related to the disturbances coming from the converter output will be investigated in depth in Chapter 3.

In this example the main parasitic parameters of the DC/DC converter have been also accounted for; nevertheless, $G_{vp,d}(s)$ assumes an expression similar to (2.15):

$$G_{vp,d}(s) = \frac{\mu\cdot\omega_n^2\cdot\left(1+\dfrac{s}{\omega_z}\right)}{s^2 + 2\zeta\cdot\omega_n\cdot s + \omega_n^2} \tag{2.38}$$

where

$$\mu = -V_{bat} \qquad \omega_n = \frac{1}{\sqrt{L\cdot C_{in}}} \qquad \omega_z = \frac{1}{R_{C_{in}}\cdot C_{in}} \tag{2.39}$$

$$\zeta = \frac{1}{2\cdot R_{MPP}}\sqrt{\frac{L}{C_{in}}} + \frac{R_{C_{in}} + R_L}{2}\sqrt{\frac{C_{in}}{L}} \tag{2.40}$$

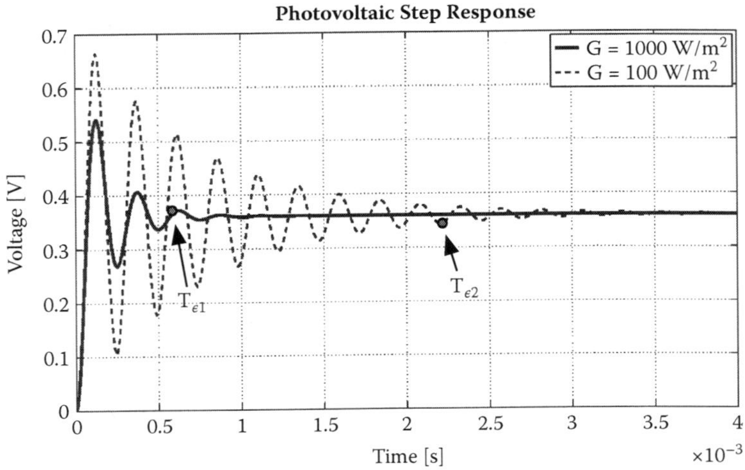

FIGURE 2.14
Settling time evaluation for the PV battery charger.

The high-frequency zero appearing in the $G_{vp,d}(s)$ transfer function, which has been introduced by the resistance $R_{C_{in}}$, does not affect significantly the transient; thus the MPPT analysis can be carried out by using (2.22). If the transfer function assumes a more complex form, it is always possible to estimate the settling time T_ε numerically or graphically by evaluating the time response of the PV voltage for a step duty cycle perturbation. Typical waveforms of the PV voltage oscillation for different irradiance conditions have been shown in Figure 2.14.

The values of the parameters of the boost converter are reported in Table 2.1, and those concerning the PV module have been reported in Table 1.3, where more realistic outdoor environmental conditions have been considered. The damping factor ζ depends on the PV differential resistance R_{MPP}, which is strongly related to the irradiance level; thus in the evaluation of the parameters of the MPPT controller, the worst-case conditions must be considered.

The continuous line in Figure 2.14 has been obtained by considering an irradiance level of $G = 1000$ W/m^2, and the corresponding differential resistance is given by $R_{MPP}|_{G=1000} \simeq 1\Omega$. The dashed line is the PV voltage transient for an irradiance level of $G = 100$ W/m^2, which corresponds to $R_{MPP}|_{G=100} \simeq 11\Omega$. In the analysis and characterization of dynamic systems, $\varepsilon = 0.1$ is usually assumed as a reasonable threshold to consider the transient over; thus the PV power settling times, evaluated by means of (2.21), are, respectively, $T_\varepsilon|_{G=1000} \simeq 0.5ms$ and $T_\varepsilon|_{G=100} \simeq 2.3ms$. Such values are approximately the same marked on the plots of Figure 2.14, where the settling time has been evaluated by analyzing the PV voltage oscillations.

According to (2.22), in order to ensure the right behavior for the P&O algorithm, the minimum value of T_p must be chosen higher than 2.3 ms.

TABLE 2.1

Parameters and Nominal Operating Conditions for the PV Battery Charger

Boost Parameters	Nominal Values
Input capacitance C_{in}	110 μF
Equivalent series resistance of C_{in}	5 mΩ
Output capacitance C_{out}	2 × 22 μF
Inductance L	13.8 μH
Equivalent series resistance of L	20 mΩ
Operating Conditions	**Nominal Values**
Switching frequency f_s	200 kHz
PWM gain modulator (Vm)	1 V
Battery voltage V_{bat}	12 V
Voltage sensor gain H_v	1
Current sensor gain H_i	1 V/A

The behaviors of the PV array voltage and of the duty cycle for different P&O parameters values have been shown in Figure 2.15. Figure 2.15a refers to a simulation of the circuit in Figure 2.12 by using an MPPT perturbation period T_p = 1 ms, while results shown in Figure 2.15b have been obtained with T_p = 2.5 ms. In order to test different environmental conditions, for both simulations, at the time instant t = 30 ms the irradiance changes from 1000 W/m^2 to 100 W/m^2.

For an ideal system operating in steady-state environmental conditions, the P&O step amplitude, which corresponds to $\Delta x = \Delta d$ for the MPPT algorithm acting on the duty cycle, does not affect the behavior of the MPPT algorithm. More in general, as the switching converter intrinsically generates the noise due to the switching ripple on the sensed variable, a minimum value for Δd must be selected in order to overcome such disturbances*:

$$\Delta V_{PV_{\Delta d}} = Go \cdot \Delta d > \Delta V_{PV_{ripple}} \tag{2.41}$$

where $\Delta V_{PV_{\Delta d}}$ is the PV voltage variation due to the perturbation Δd and $\Delta V_{PV_{ripple}}$ is the amplitude of the switching ripple on the PV voltage.

In the boost-based battery charger configuration, if no additional filter is used for removing the switching ripple noise, the minimum Δd can be estimated by using the following condition:

$$\Delta d > \frac{(V_{bat} - V_{MPP}) \cdot V_{MPP}}{16 \cdot L \cdot C_{in} \cdot f_s^2 \cdot V_{bat}^2} \tag{2.42}$$

* The reader can find additional information on the switching ripple attenuation in Section 3.4.

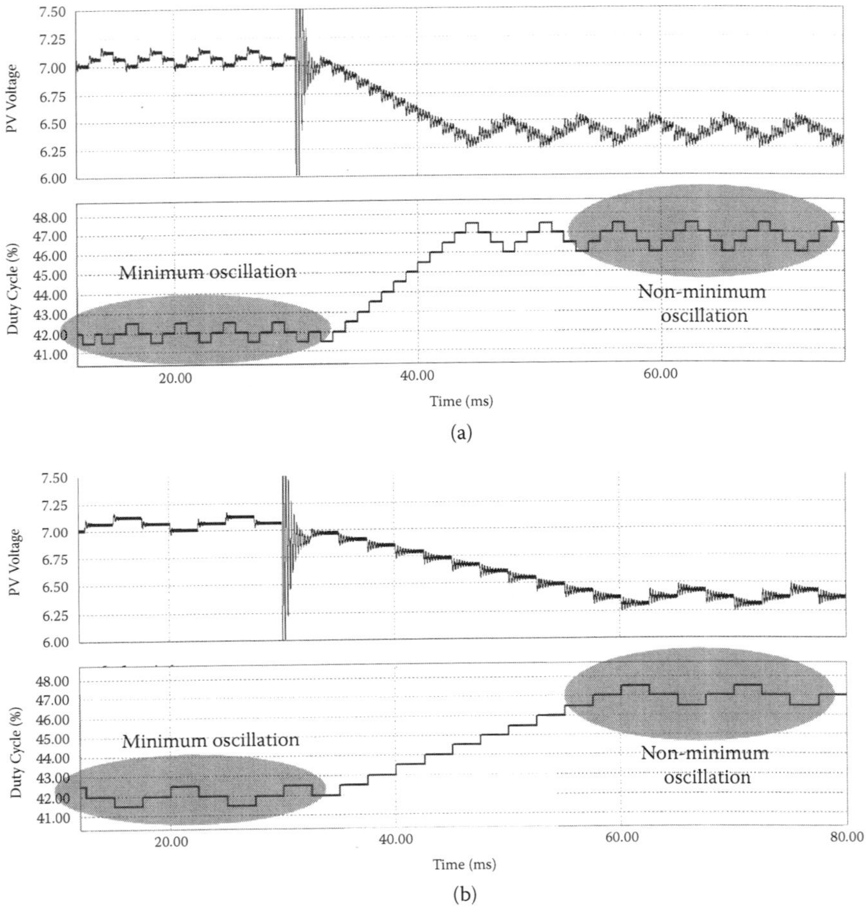

FIGURE 2.15
P&O behavior in steady-state environmental conditions.

For the operating conditions considered in the case under study, applying (2.42) provides $\Delta d > 0.001$. A P&O step amplitude $\Delta d = 0.005$ has been used for the tests whose results are shown in Figure 2.15.

As shown in the two plots, for $t < 30$ ms, the MPPT works correctly because T_p is greater than $T_{\varepsilon|_{G=1000}} = 0.5\,\text{ms}$: In this case, the P&O algorithm is not confused by the transient behavior of the system and the duty cycle oscillates assuming only three different values: $[d_{MPP} - \Delta d, d_{MPP}, d_{MPP} + \Delta d]$.

For $t > 30$ ms, the first simulation highlights that the P&O algorithm is not able to work with only three steps; in fact, in this case $T_p = 1ms < T_{\varepsilon|_{G=100}} = 2.3\,\text{ms}$ and the operating point has a wider swing across the MPP, and then the system is characterized by a lower efficiency with respect to the corresponding case ($T_p = 2.5$ ms) shown in Figure 2.15b. In this second case, after a transient necessary to adapt the system to the new environmental conditions, the P&O continues to work correctly because T_p fulfills the constraint (2.22) for all operating conditions tested in the example.

As discussed in Section 2.4.2, once given the maximum $\dot{G}$ to be tracked without error, the optimal value of Δd can be settled. This is an important aspect conditioning the MPPT performances, especially on a cloudy day, when the PV field is subject to continuous irradiance change and a wrong MPPT configuration can significantly reduce the energy efficiency of the system.

It is very important to adopt a realistic characterization of the performances together with the analysis of the balance of the system related to power components adopted, for a reliable estimation of PV systems energy productivity and economic convenience. Some interesting considerations about such aspects are reported in [15], where an overview of the EN 50530, which is the European standard for measuring the overall efficiency of PV inverters, is provided. The document explains in depth the approach and methodology introduced in the standard for a combined assessment of the conversion as well as the maximum power point tracking efficiency. In particular, the dynamic MPPT efficiency under varying irradiance conditions is identified by using a ramp sequence consisting of irradiance ramps with different gradients as well as irradiance levels.

In detail, ramp gradients ranging from $\dot{G} = 0.5W/m^2/s$ to $\dot{G} = 100W/m^2/s$ are used for the tests with two different irradiance levels, from 100 $W/m^2/s$ to 500 W/m^2 (low to medium irradiance) as well as from 300 $W/m^2/s$ to 1000 W/m^2 (medium to high irradiance). On the basis of the indication given in the standard, for the example under study, a constant rate of change for the irradiance variation $\dot{G} = 100W/m^2/s$ is considered. The worst-case conditions of $G = 100$ W/m^2 will be considered for estimating the parameters that are used in Equation (2.36) for evaluating the right Δd step amplitude.

By means of (2.29), with the data reported in Tables 1.2 and 1.3, the parameter $H|_{G=100} = 0.0836$ A/V^2 has been obtained. Finally, applying (2.36) for the step amplitude Δd provides

$$\Delta d > \frac{1}{V_{bat}} \sqrt{\frac{V_{MPP} \cdot K_{ph} \cdot |\dot{G}| \cdot T_p}{\left(H \cdot V_{MPP} + \dfrac{1}{R_{MPP}}\right)}} \simeq 0.0118 \tag{2.43}$$

where the K_{ph} variable assumes the following value: $K_{ph} = 0.008$ $A{\cdot}m^2/W$.

Figure 2.16 shows the behavior of the PV system in the presence of irradiance variations. The irradiance profile characterized by a positive and negative ramp with a slope of ± 100 $W/m^2/s$, as shown in Figure 2.16a, has been used in the simulation. The P&O algorithm with the optimal values of $T_p = 2.5$ ms and $\Delta d = 0.012$, selected according to Equation (2.43), has been compared with another case in which $\Delta d = 0.0012$ has been used. In Figure 2.16b the duty cycle waveforms of the two cases have been superimposed. The

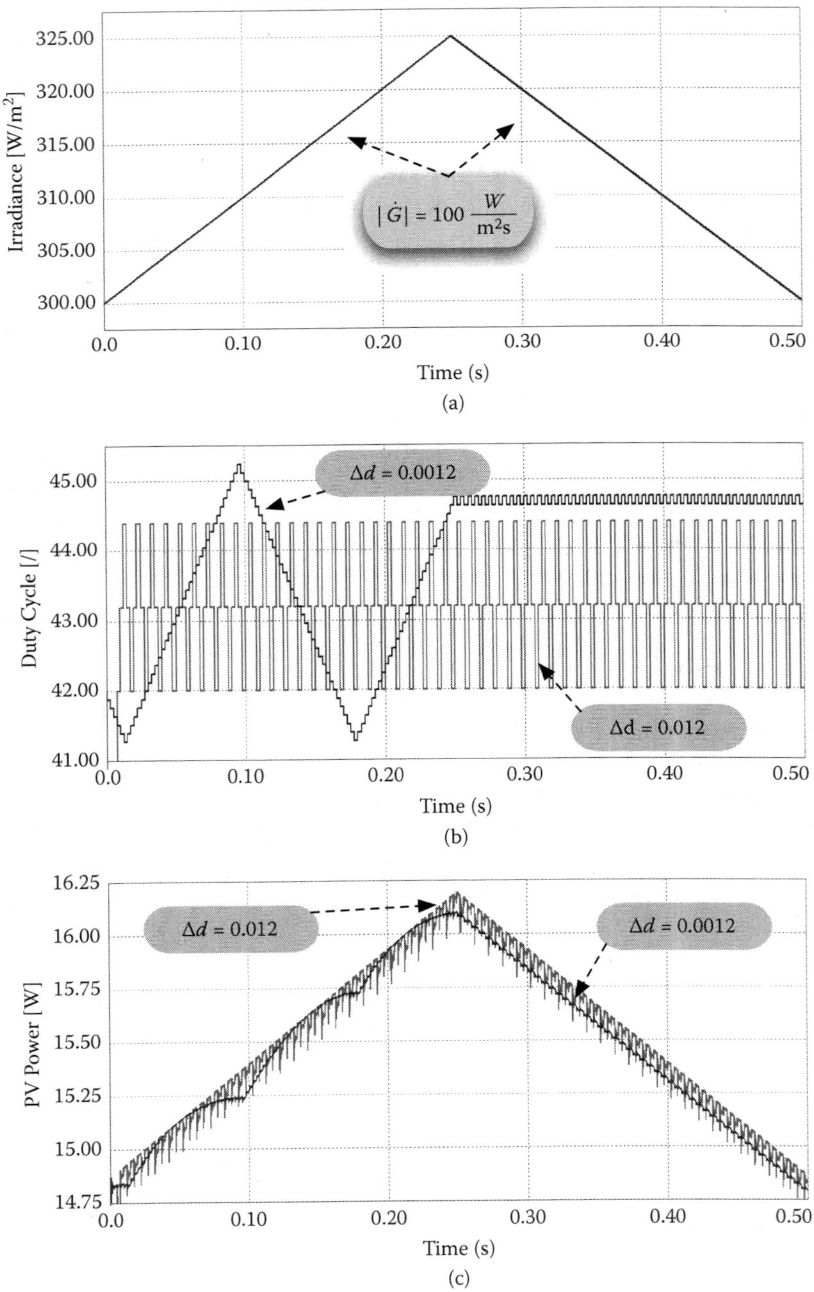

FIGURE 2.16
P&O behavior in dynamic environmental conditions. (a) Irradiance profile. (b) Duty cycle variations. (c) Instantaneous PV power.

simulation shows that if the amplitude of the duty cycle perturbation is well designed, also in the presence of irradiance variation, the duty cycle oscillates assuming only three different values.

In Figure 2.16c the instantaneous power, delivered by the PV source, has been plotted for the two MPPT controllers. The simulation highlights that the optimal Δd, which is one order of magnitude greater than the minimum values required for the steady-state conditions, ensures highest efficiency because the MPPT is not deceived in the tracking process.

For the proposed example, the design setup has also been experimentally validated. In Figure 2.17 the simulation waveforms of the PV battery charger (Figure 2.17a) and the oscilloscope waveforms (Figure 2.17b) of the

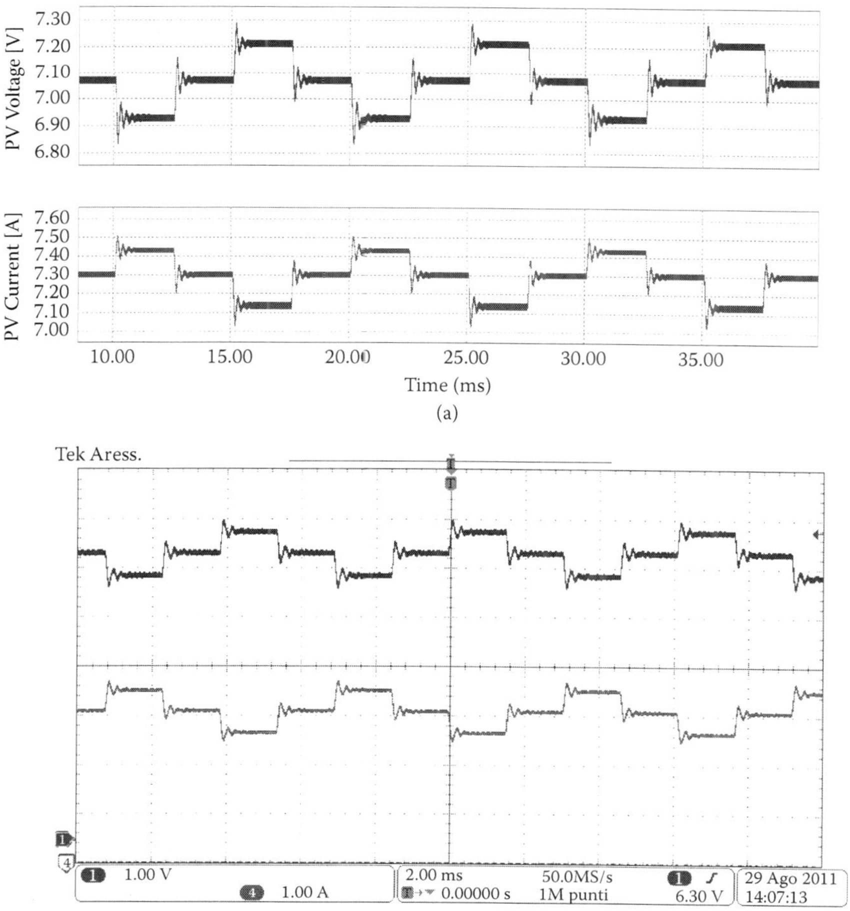

FIGURE 2.17
Experimental test of the MPPT for battery charger. (a) Simulation results. (b) Experimental results.

photovoltaic voltage and current have been shown. The comparison is totally satisfactory and the MPPT P&O algorithm works as predicted by moving the operating point in only three positions around the MPP.

2.5 Improvements of the P&O Algorithm

2.5.1 P&O with Adaptive Step Size

The P&O algorithm is widely used in PV systems due to its simplified control structures and easiness of implementation. In the MPPT algorithm with fixed $\{\Delta x, T_p\}$ parameters a trade-off condition must be achieved in order to choose the controller values for balancing the losses in steady state due to large perturbations around the MPP and the MPPT speed in situations involving quickly changing irradiation conditions or load.

In the previous paragraphs, benefits deriving from the optimization of both the perturbation amplitude Δx and the sampling period T_p have been shown. Nevertheless, overall MPPT performances can be further improved by modifying the basic version of the Hill climbing and P&O algorithms.

For example, in [16], a method in which the duty cycle step size is automatically adjusted according to the derivative of power with respect to the PV voltage (dP/dV) is proposed. The step size becomes tiny as dP/dV becomes very small around the MPP, thus ensuring a very good accuracy at steady state. In such a cases the general equation (2.2) is modified as follows:

$$d_{((k+1)T_p)} = d_{(kT_p)} \pm \Delta d = d_{(kT_p)} \pm N \cdot \frac{|P_{(kT_p)} - P_{((k-1)T_p)}|}{|V_{PV_{(kT_p)}} - V_{PV_{((k-1)T_p)}}|} \tag{2.44}$$

where N is the scaling factor adjusted at the sampling period to regulate the step size. In [17], the variable step size algorithm is implemented according to the slope of the PV power vs. the duty cycle curve on the basis of the following relation:

$$d_{((k+1)T_p)} = d_{(kT_p)} \pm \Delta d = d_{(kT_p)} \pm N \cdot \frac{|P_{(kT_p)} - P_{((k-1)T_p)}|}{|d_{(kT_p)} - d_{((k-1)T_p)}|} \tag{2.45}$$

The performance of the algorithm (2.44) and Equation (2.45) is essentially conditioned by the scaling factor N. The manual adjustment of the value of this parameter is often based on a trial-and-error approach, and the value resulting from this process is just suitable for a given system and specific operating conditions.

In order to ensure the convergence of the MPPT algorithm, the parameter N must meet the following inequality [17]:

$$N \cdot \frac{|\Delta P_{max}|}{|\Delta V_{max}|} < \Delta d_{max} \tag{2.46}$$

where Δd_{max} is the maximum desired step change. In [17] an experimental procedure aimed at identifying the best value for N is proposed. At the start-up, the duty cycle d_{start} is initialized at a low value. Power p_{start} and voltage v_{start} are evaluated with such duty cycle value. Afterwards, the duty cycle value is modified by the maximum desired step change Δd_{max} and corresponding values of ΔP_{max} and ΔV_{max} are evaluated. Equation (2.46), which provides a guide to determine the value of the scaling factor N, can be used in conjunction with (2.36), so that the largest step size of an equivalent fixed step size MPPT algorithm is evaluated by assuming the worst-case operating conditions.

The value of the perturbation period T_p must be designed according to (2.22). The algorithm is explained by means of Figure 2.18: The initial operating point is assumed to be $\mathbf{x}_1$ and the perturbation step size is assumed to be step 1. The sequence $\mathbf{x}_1$–$\mathbf{x}_7$ is obtained by imposing an amplitude of the perturbation proportional to the variation of the PV power.

2.5.2 P&O with Parabolic Approximation

In [18] it is shown that a parabolic interpolation based on the last three sampled voltage-power couples allows us to balance the position of the three

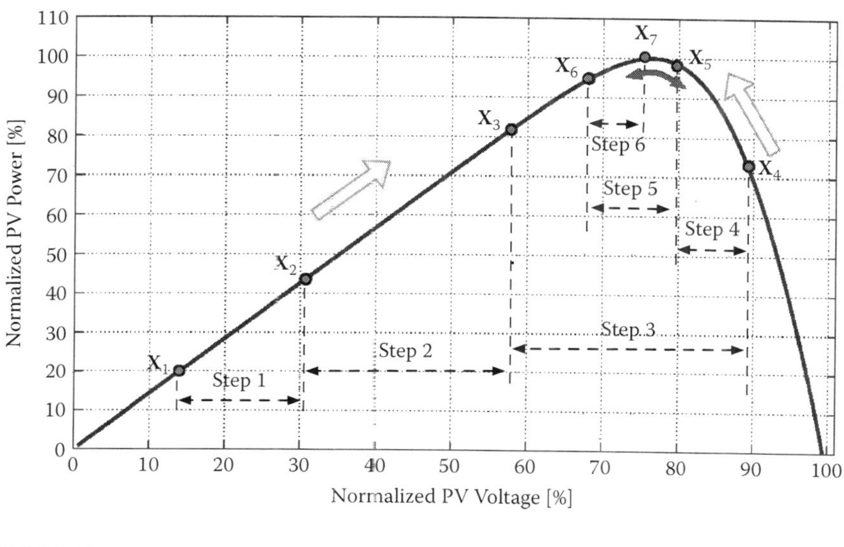

FIGURE 2.18
Variable step MPPT.

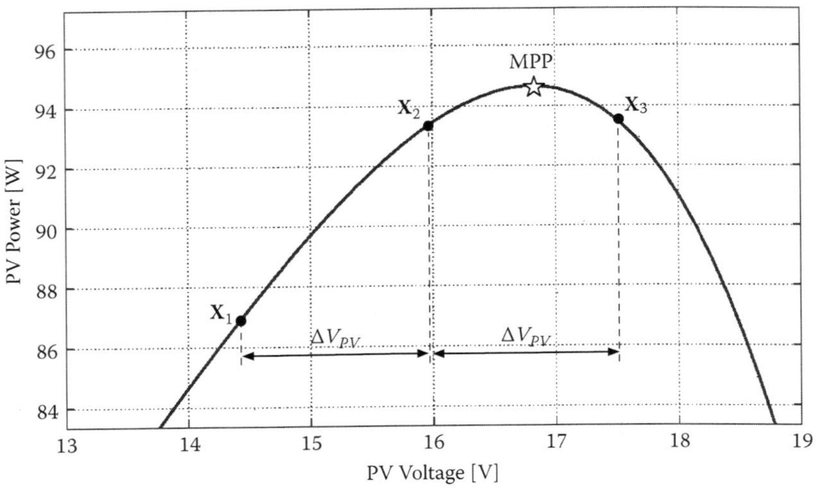

FIGURE 2.19
Operating points location on PV characteristic.

operating points across the MPP. The unbalanced position of the operating points across the MPP, which can occur if any perturbative MPPT approach is used, has a detrimental effect on the MPPT efficiency, and thus it must be corrected properly.

In a fixed step size P&O algorithm, the discretized value of Δx usually does not allow us to have a three-point steady-state behavior with a central point close to the MPP and the other two equally balanced at its sides. A very common situation is that shown in Figure 2.19, where the operating points $\mathbf{x}_1$, $\mathbf{x}_2$, and $\mathbf{x}_3$ are equally spaced in terms of PV voltage, but they are not well balanced with respect to the MPP.

The lower Δx, the lower the distance between $\mathbf{x}_2$ and the MPP. The non-symmetrical position of the three operating points is less efficient than the balanced condition, occurring when x_2 is centered in MPP with $\mathbf{x}_1$ and $\mathbf{x}_3$ equally spaced in terms of PV voltage, and it might also trigger a four-point oscillation, with a further increase of the power loss. A four-point oscillation obtained in an experimental case is shown in Figure 2.20. A sequence of seven operating points around the MPP of the P&O algorithm is plotted. If the position of the operating points is asymmetric with respect to the MPP, the points that are closer to the MPP ($\mathbf{x}_2$ and $\mathbf{x}_3$ in Figure 2.19) have more or less the same power level. In the presence of noise and rounding errors introduced by the measurement system, the action of the MPPT algorithm might be wrong. For instance, when the operating point jumps from position 5 to 6, due to noise or slight variations of the irradiance level, the measured value of the PV power might be different from that evaluated in point 2. As a consequence, ΔP has a sign that causes an additional step in the wrong direction (1 = 7). By comparing the real sequence (1-2-3-4)

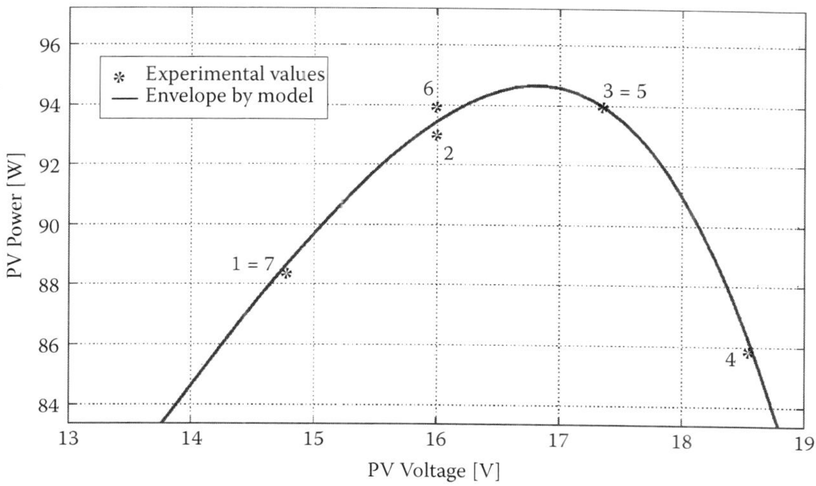

FIGURE 2.20
Steady-state oscillations in a noisy environment.

with respect to the optimal three-point oscillation (1-2-3) or (2-3-4), the difference in terms of power losses is almost evident. It will be quantified in Section 2.8.

If one of the points 2 or 3 is moved much closer to the MPP, the errors are less significant than the power variation due to the jump from one operating point to the following one. This aspect is explained with the support of Figure 2.21 and by considering the relative position of the operating point with respect to the power variation.

In Figure 2.21a the power variation (ΔP) between two possible operating points across the MPP has been put into evidence. In the presence of an uncertainty $\mathbf{u}_{\Delta P}$ due to the measurement process, the result can be affected by an error higher than the nominal value, so that $\mathbf{u}_{\Delta P} > \Delta P$. Due to the concavity of the curve at the MPP, regardless of the step amplitude, the case $\Delta P \simeq 0$ can occur, with a consequent wrong decision of the MPPT algorithm.

Figure 2.21b shows the case of a balanced configuration, obtained by moving forward the operating points without changing the Δx value. After this shift, the system works correctly because it is $\Delta P > \mathbf{u}_{\Delta P}$ for any operating condition.

Some tricks have been proposed in literature for addressing the problem described above. In [19], a memory buffer containing the last three PV voltage-power couples is employed for detecting if the MPP has been reached and if a rapid irradiation variation occurs. The importance of freezing the value of the converter duty cycle when the MPP is reached is also put into evidence: This allows us to draw the maximum power in steady-state conditions without missing any part of it due to the oscillations around the MPP. In [20], and in some other papers reported in the list of references cited

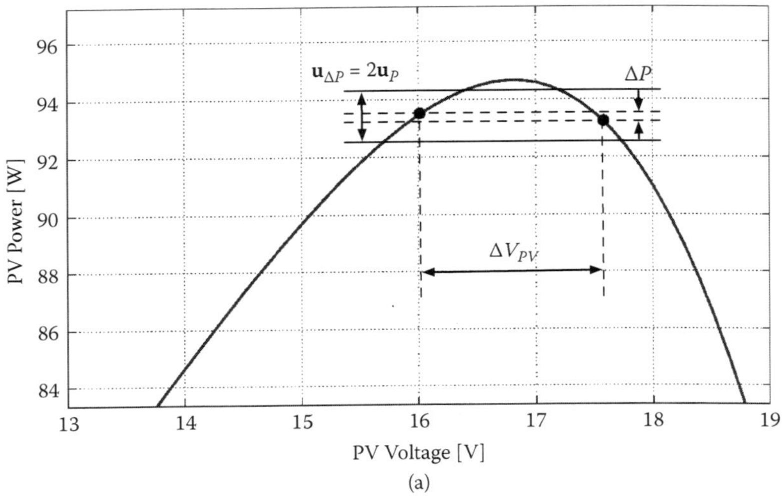

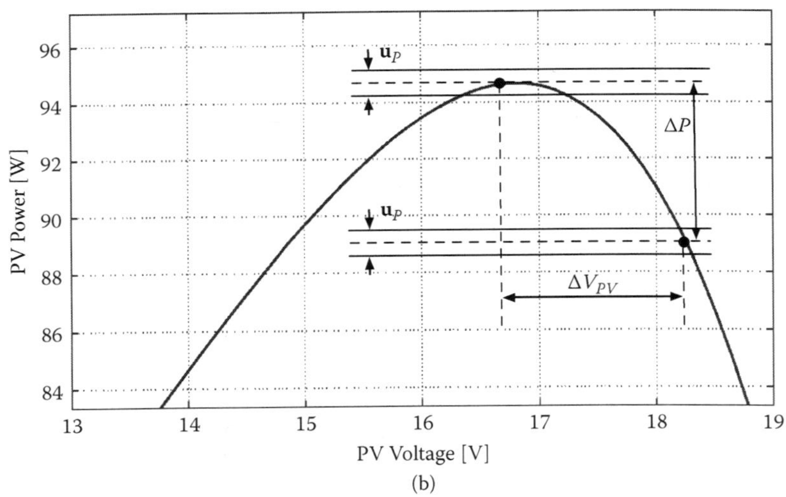

FIGURE 2.21
Balancing the steady-state oscillations for error compensation. (a) Two points with similar power. (b) Two points well displaced on the PV curve.

therein, the basic P&O MPPT technique is compared with some improved versions. Some of them suggest the use of a waiting function aimed at freezing the duty cycle perturbation value if the sign of Δd is reversed several times in a row. This trick reduces the hill climbing of the operating point across the MPP under stationary irradiance conditions, but it deteriorates the system response under changing atmospheric conditions. This emphasizes the P&O erratic behavior under rapidly changing irradiance conditions typical of cloudy days.

Among the techniques proposed in literature, the balancing of the steady-state operating points across the MPP by employing a parabolic interpolation is very effective and relatively simple. Indeed, near the MPP, the parabola is a good approximation of the P-V curve; thus it can be used for estimating the position of the MPP by using the information coming from the last three operating points [18, 21].

A parabola is described in the P-V plane by the following equation:

$$p_{PV} = \alpha \cdot v_{PV}^2 + \beta \cdot v_{PV} + \gamma \tag{2.47}$$

with $\alpha < 0$ and the vertex at $v_{peak} = -\beta/(2 \cdot \alpha)$, with $p_{peak} = \gamma - \beta^2/(4 \cdot \alpha)$.

If the P-V curve and the interpolating parabola have at least three points in common, the coefficients α, β, and γ can be determined by solving the following system of equations:

$$\begin{cases} p_1 = \alpha \cdot v_1^2 + \beta \cdot v_1 + \gamma \\ p_2 = \alpha \cdot v_2^2 + \beta \cdot v_2 + \gamma \\ p_3 = \alpha \cdot v_3^2 + \beta \cdot v_3 + \gamma \end{cases} \tag{2.48}$$

A first-in/first-out array containing the actual operating point and the preceding two, corresponding to the couple of values $[(p_1,v_1) - (p_2,v_2) - (p_3,v_3)]$, is used to find the interpolating parabola. Once such an approximation is obtained, the vertex of the parabola is used as a prediction of the new operating point, as shown in Figure 2.22.

The interpolating parabola is considered misleading in the MPPT process if it exhibits an upward-turned concavity, or if it has a vertex at a voltage

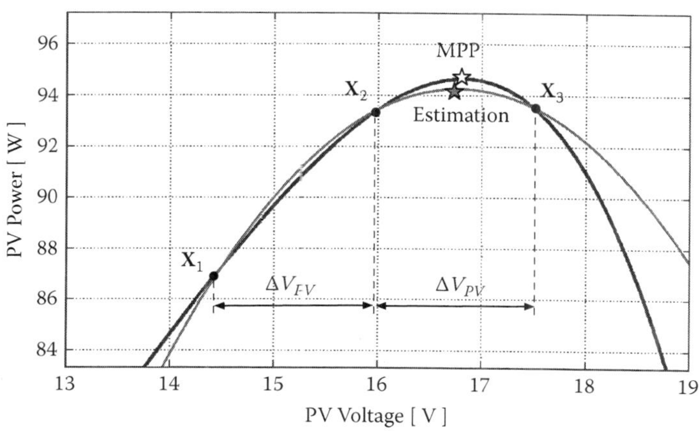

FIGURE 2.22
MPP estimation by means of parabolic (red curve) interpolation.

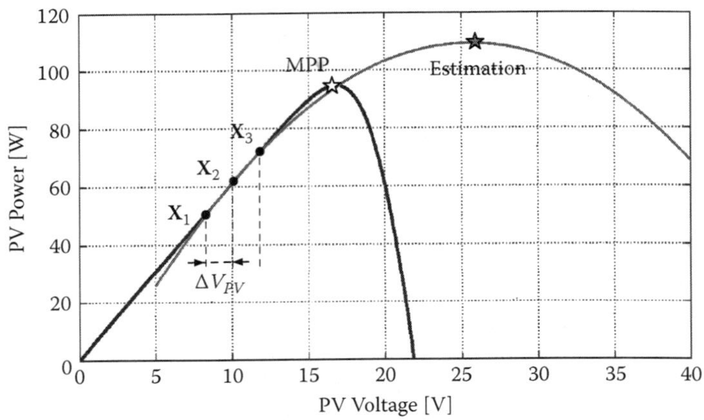

FIGURE 2.23
Wrong estimation of parabolic interpolation.

higher than the open-circuit voltage, as in Figure 2.23. In such cases, as well as if it cannot be found by means of the three points given (e.g., when two of them are superimposed), the perturbation used by the P&O algorithm is that one determined by means of (2.36).

The promptness of the MPPT controller is also improved by exploiting the geometrical properties of the parabolic function. A sequence of the PV operating points is shown in Figure 2.24: By assuming that the MPPT controller starts to work with a fixed step size ΔV_{PV}, after two steps the values x_1, x_2, and x_3 are available for reconstructing the interpolating parabola. If the estimated step amplitude $\Delta V_{parabolic}$ is higher than ΔV_{PV}, the MPPT algorithm can reach the MPP in a few steps. In [18] details concerning the performance in the presence of irradiance variations and more information related to the practical implementation of the parabolic approach are provided.

2.6 Evolution of the Perturbative Method

2.6.1 Particle Swarm Optimization (PSO)

Particle swarm optimization (PSO) has also been proposed as an MPPT approach, especially because it is not computationally burdensome. PSO allows implementation by means of low-cost digital controllers and is characterized by good performance under extreme irradiation conditions. In [22] the authors propose a modification of a classical perturbative approach, directly applied to the duty cycle of a DC/DC converter, based

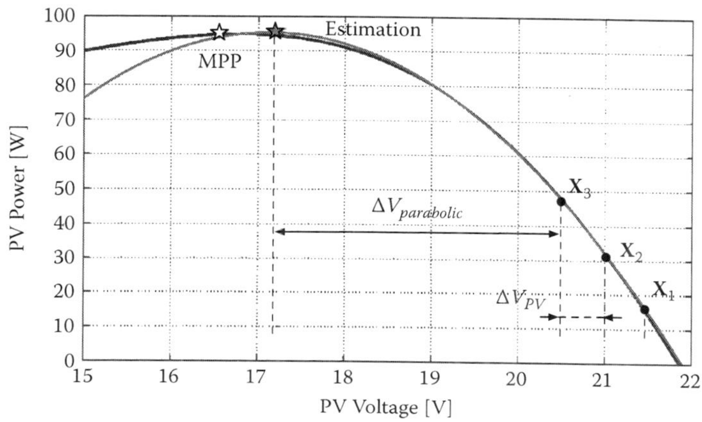

FIGURE 2.24
Promptness MPPT improvement by means of parabolic interpolation.

on a PSO method. The main objective of the study is to achieve a strong reduction, or even the absence, of the steady-state oscillations around the MPP that are a drawback of any perturbative MPPT technique. The PSO algorithm is seen as a perturbative method with an adaptive amplitude of the applied perturbation. The particle velocity has been designed so that its value is close to zero when the system operation approaches the MPP and the value of the DC/DC converter duty cycle is approaching a constant. The application of the method requires a tuning of some parameters that strongly affect performances, some of them being dependent on the specific application. In [22] it has also been demonstrated that PSO is effective in tracking the MPP when the PV generator is affected by partial shadowing phenomena.

In [23] the authors apply a PSO algorithm to the problem of PV arrays working in mismatched conditions. The approach proposed by the authors is not based on the assumption that the mismatched PV panels are still connected in series so that the power vs. voltage curve of the string exhibits more than one maximum power point. In particular, the authors adopt a centralized MPPT controller, running the PSO, acting on the duty cycle of as many DC/DC converters as the PV panels in the string. Such an approach is also proposed in [24], where the authors use the same architecture with a different control algorithm. The multivariable optimization algorithm proposed in [23] adjusts the converter's duty cycle values with the aim of maximizing the total power delivered at the output. Only three agents allow tracking of the absolute MPP; however, some parameters (e.g., the one defining the convergence criterion and the one introduced for detecting an irradiation change) must be designed carefully, according to the real case under analysis, if high tracking performances are required.

2.6.2 Extremum Seeking and Ripple Correlation Techniques

Extremum seeking (ES) control is always confused with the ripple correlation control (RCC) technique. Their applications to the MPPT problem are very similar, the true difference being the source of perturbation used to detect whether the PV array is operating on its MPP or is on the right or left side of the MPP itself. As clearly explained in [25], ES adopts a signal having a low-frequency oscillating component and RCC employs the ripple at the switching frequency generated by the power electronic circuit, so that this is able to fasten the RCC convergence.

In PV MPPT single-phase AC applications, ES can use the oscillating component at 100 or 120 Hz, depending on the frequency of the AC voltage, which can be 50 or 60 Hz, for tracking the MPP. As will be shown in Chapter 3, such an oscillation appears at the DC bus connecting the DC/AC inverter with the DC/DC converter performing the MPPT function in double-stage topologies. Such an oscillation back propagates through the DC/DC converter and, as will be shown in Chapter 3, worsens the performances of the MPPT algorithm. Nevertheless, it can be used for tracking the maximum power point. Alternatively, e.g., in DC applications, a sinusoidal low-frequency disturbance can be even injected for accomplishing the same work, but at the price of additional circuitry. The operating principle is explained in Figure 2.25.

The sinusoidal disturbance affects the PV voltage and has an effect on the PV power. This effect is almost negligible if the operating point voltage is the MPP. If the PV array is working at a voltage lower than the MPP one, the oscillation affecting the PV power is in phase with that imposed on the PV voltage. On the contrary, at any voltage higher than MPP one, the PV power oscillation is displaced by half a period with respect of the voltage oscillation. Figure 2.26 shows that a sinusoidal external dither is superimposed on the PV voltage. The PV power frequency component at the same frequency of the injected signal is multiplied by the latter one, and the average value of the product has a sign that can drive the MPPT toward the peak of the P-V function [26].

Figure 2.27 shows that once the sinusoidal disturbance takes place at the PV terminals due to the DC/AC stage, the demodulated signal to be processed by the low-pass filter is obtained by multiplying the oscillations affecting both the PV power and the PV current [27]. The employment of the inherent oscillations of the instantaneous power in single-phase systems is also proposed in [28], where the authors propose a suitable filtering of the PV current and voltage signals allowing an estimation of the PV power derivative. A suitable control of the DC link voltage allows us to perform the MPPT operation.

Also, in the case of ripple correlation control (RCC), the time derivative of the power is related to the time derivative of the current or of the voltage, but no external oscillation is required. The power gradient is driven to zero by using the natural ripple that occurs due to converter high switching frequency [29].

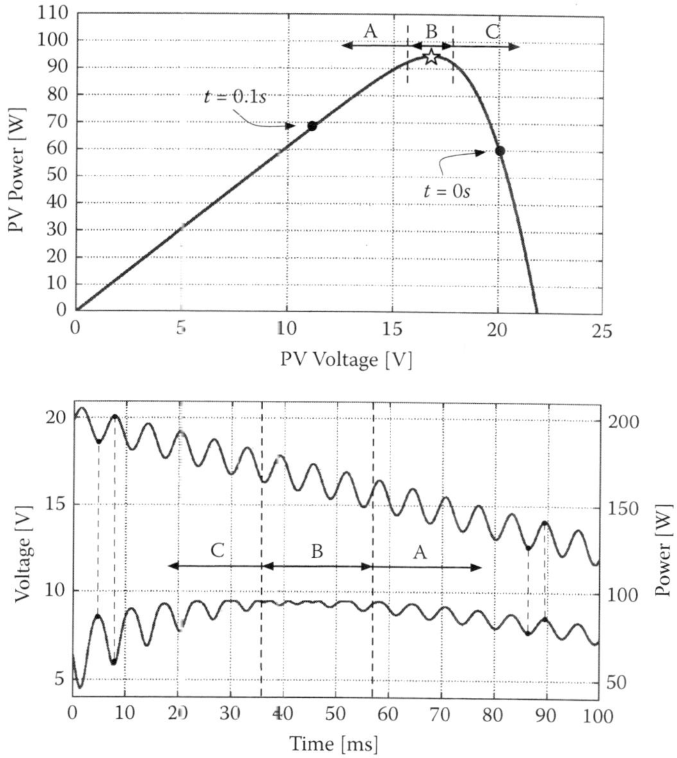

FIGURE 2.25
Principle of operation of the ES control algorithm.

In this way, the RCC method can also be used in PV applications with DC loads or battery chargers as well as three-phase loads or grid connection.

In [30] the RCC technique is improved furthermore, because the study of the wave shapes of the electrical variables in the switching converter has allowed us to reduce the requirements in terms of sampling. In fact, only two samples per switching period, one taken at the minimum value of the PV current, that is, the maximum for the PV voltage, and the second taken at its maximum, which is the minimum for the PV voltage, are needed.

2.6.3 The Incremental Conductance Method

This method was first presented in [31], and it is based on the observation that in the MPP, the following condition occurs:

$$\frac{dP}{dV} = \frac{d(V \cdot I)}{dV} = 0 \tag{2.49}$$

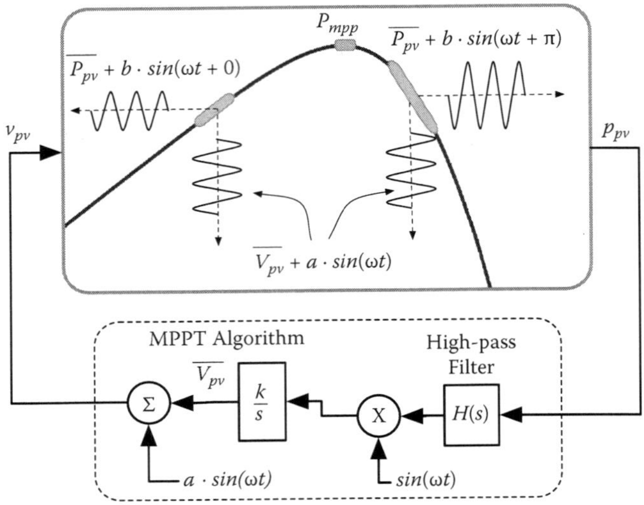

FIGURE 2.26
ES MPPT by an external dither.

By accounting for the dependence of the PV current on the voltage, it is possible to express such a condition as follows:

$$I + V\cdot\frac{dI}{dV} = 0 \tag{2.50}$$

so that the validity of condition (2.49) is equivalent to

$$\frac{I}{V} = -\frac{dI}{dV} \tag{2.51}$$

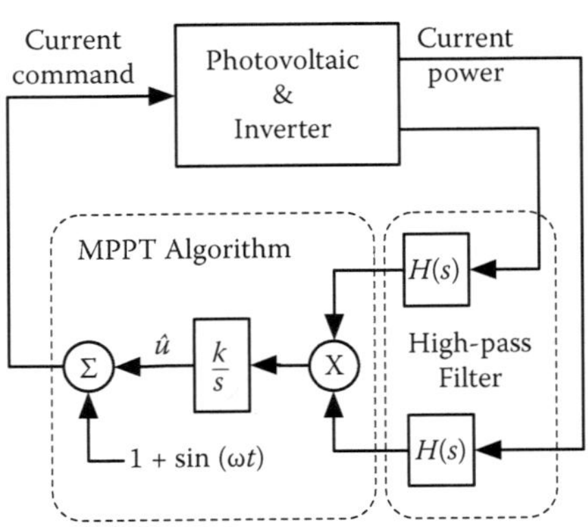

FIGURE 2.27
ES MPPT using the low-frequency oscillation produced by an inverter stage.

which means that, at the MPP, the absolute value of the conductance must be equal to the absolute value of the incremental conductance. Such a condition is the basis of the incremental conductance (INC) MPPT method. Condition (2.51) is verified through a repeated measure of the conductance at two different, yet close enough, values of the PV voltage. As a consequence, the method requires the application of a repeated perturbation of the voltage value, until the following condition occurs:

$$\frac{I_k}{V_k} = -\frac{I_k - I_{k-1}}{V_k - V_{k-1}} \tag{2.52}$$

where the indices k and $k - 1$ refer to two consecutive samples of the PV voltage and current values. In [31] an 8.4% increase in the PV power produced by a PV system equipped with the INC MPPT method is claimed with respect to the P&O method. The reason for this improvement has been mainly ascribed to the fact that the INC method is able to avoid any further oscillation of the operating point when condition (2.52) has been fulfilled. Consequently, the INC method behaves in a way that is similar to P&O during transients, but it is able to avoid any power loss in steady-state conditions because it remains in the MPP unless any exogenous variable makes (2.52) no more fulfilled.

Unfortunately, condition (2.52) holds only for an ideal system, because it is almost never verified because of noises and quantization effects, related to the microcontroller by means of which the INC method is implemented. As a consequence, the method continues to check the validity of (2.52) also in stationary irradiance conditions, so that the theoretical advantage of INC over P&O vanishes.

The evaluation of (2.52) can be useful in order to understand on which side of the P-V curve with respect to the MPP the actual operating point lies. Indeed, by means of (2.49) and (2.50) we get

$$\frac{1}{V}\cdot\frac{dP}{dV} = \frac{I}{V} + \frac{dI}{dV} = G + dG \tag{2.53}$$

where G is the conductance and dG the incremental conductance. It results that, on the left side of the P-V curve with respect to the MPP, $dP/dV > 0$: This means, according to (2.53), that $G + dG > 0$. As a consequence, if the conductance is greater than the absolute value of the incremental conductance, than the operating point is on the left side of the MPP, so that the voltage must be increased in order to move closer to the MPP. Similarly, if $G + dG < 0$, the actual operating point is at a voltage higher than that of the MPP, so that the voltage must be reduced if the MPP has to be approached. Such information is not available if the P&O technique is used, so this is a real advantage ensured by the INC method. The new voltage at which condition (2.51) must be tested is evaluated according to the following iterative formula:

$$V_{PV}^{k+1} = V_{PV}^{k} + sign(G + dG)\cdot\Delta V \tag{2.54}$$

where ΔV is the voltage step chosen for the perturbative phase during which the MPP is searched. The flowchart shown in Figure 2.28 puts into evidence the perturbative nature of the INC algorithm and the use of the information deriving from the comparison between the values of the conductance and the incremental conductance.

Due to fact that the INC method is inherently based on a perturbative approach, the amplitude of the perturbation step needs to be optimized as for the P&O, regardless of whether it acts on the duty cycle or on the reference voltage. In [32] a direct dependency of the step amplitude on the power derivative has been used. The formula proposed therein is referred to as direct control of the converter's duty cycle.

$$\Delta D = \pm N\cdot\left|\frac{dP}{dV}\right| \tag{2.55}$$

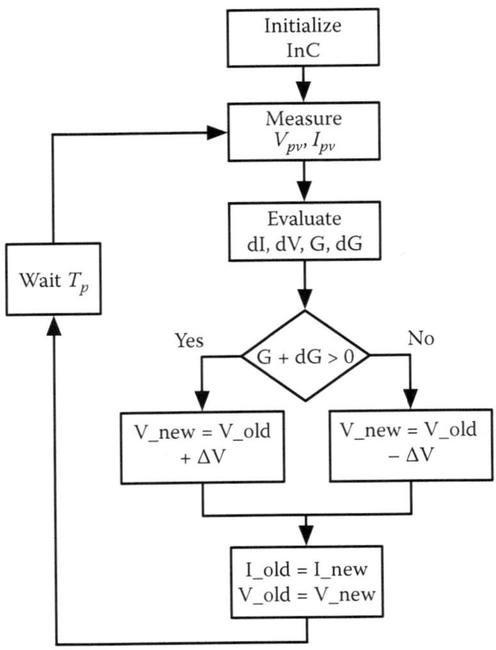

FIGURE 2.28
Flowchart of the incremental conductance algorithm.

In this way, the lower the power derivative, the closer is the MPP and the smaller the duty cycle perturbation that can be settled. The authors also propose an upper bound for the value of the coefficient N, which affects the tracking performances dramatically. The inequality involves the power derivative obtained at the larger ΔD to be used, and the value thereof and its meaning are the same as in (2.46).

$$N < \frac{\Delta D_{max}}{\left|\frac{dP}{dV}\right|_{fixedstep=\Delta D_{max}}} \tag{2.56}$$

Dynamic performances under very fast irradiation variation are claimed to be improved in [33], where a suitable function used for the MPP bounding is introduced. The algorithm step size modes are switched by the extreme values of a threshold function that is the product of the n-th power of the PV array output power and the derivative of the same power:

$$C = P^n \cdot \left|\frac{dP}{dI}\right| \tag{2.57}$$

the parameter n being used for a closer MPP bounding. The function C has two maximum values, for two different current values, one on the left side and one on the right side of the MPP. The proposed MPPT algorithm uses a variable step size mode if the PV array current falls between these two current values. Otherwise a fixed step size mode is used. This method improves both the steady state and the dynamic MPPT response.

2.7 PV MPPT via Output Parameters

The MPPT operation can be achieved by using the DC/DC converter output voltage and current rather than the input ones. The adoption of the output variables simplifies the MPPT algorithm. Moreover, regardless of the power stage and the way the control algorithm is implemented, this approach needs the sensing of only one of the two output variables. Indeed, it has been demonstrated that the MPPT using one output control parameter applies to almost any practical load type, regardless of its nature. Tracking the maximum value of the load current or voltage implies operation of the PV system at its MPP [34]. The controller simplicity translates into a reduction in the hardware involved, in terms of sensors, and in a MPPT algorithm that does not require the multiplication needed for the calculation of the PV power. It should be noted that in the majority of practical PV systems with battery

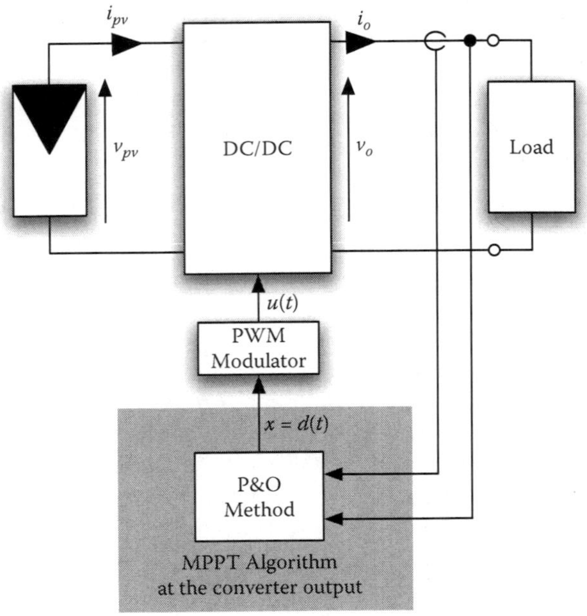

FIGURE 2.29
MPPT scheme via output parameters.

backup, the DC/DC converter output voltage and current are monitored anyway, for the sake of charge control and battery protection. Thus the sensing of the PV array output voltage and current is avoided with benefits in terms of cost, system efficiency, and reliability. In [35] the design guidelines proposed in the previous paragraphs have been applied to the scheme of Figure 2.29.

2.7.1 The TEODI Approach

A new MPPT technique based on the converter output parameters has been recently presented in the literature: It is named TEODI [36–38]. TEODI is the acronym used to synthesize the following concept: **t**echnique based on the **e**qualization of the **o**utput operating points in correspondence of the forced **d**isplacement of the **i**nput operating points of two identical PV systems.

The main advantage offered by this technique is its simplicity because it does not require measurement of the PV power, and this means that it does not require multiplications. Thus its analog implementation is very effective and cheap. Moreover, TEODI is less sensitive than the perturbative MPPT methods to the irradiance variations and noises coming from the converter output and back propagating toward the PV source. As a drawback, in its basic version, it does not work properly in the presence of mismatching between the two PV sections it needs for working.

Even if, in principle, the proposed method is applicable to two arbitrarily large identical subsections of a PV field, due to its intrinsic limitation, TEODI is suitable in the controlling of subsections of the same PV module where the uniform conditions among the PV cells can be ensured with high probability. In such a case the use of TEODI allows us to realize, at a reduced cost, a PV module with its own integrated MPPT controller, which is required in distributed MPPT photovoltaic applications [39].

In the system shown in Figure 2.30, sections A and B represent two identical PV units operating under the same levels of irradiance and ambient temperature. By assuming that the two DC/DC converters are perfectly identical, it is possible to demonstrate that the steady-state operating point of the circuit shown in Figure 2.30 is bounded in a narrow region around the MPP.

The operating principle of TEODI is explained by starting from the sensing section at the converter output. The scheme of Figure 2.30 refers to the parallel configuration of TEODI, which leaves both converters working at the same output voltage and requires the sensing of the currents. It is worth noting that in its dual implementation, obtained by connecting the converters' output ports in series, TEODI only needs the sensing of the output voltages, thus requiring a simpler and cheaper sensing circuitry.

The signal $k \cdot (i_{A2} - i_{B2})$ is preliminarily filtered by means of a low-pass filter (LPF) in order to remove the high-frequency noise. The signal at the LPF output is processed by the proportional-integral controller and the v_c signal is passed to the two pulse-width modulators (PWM).

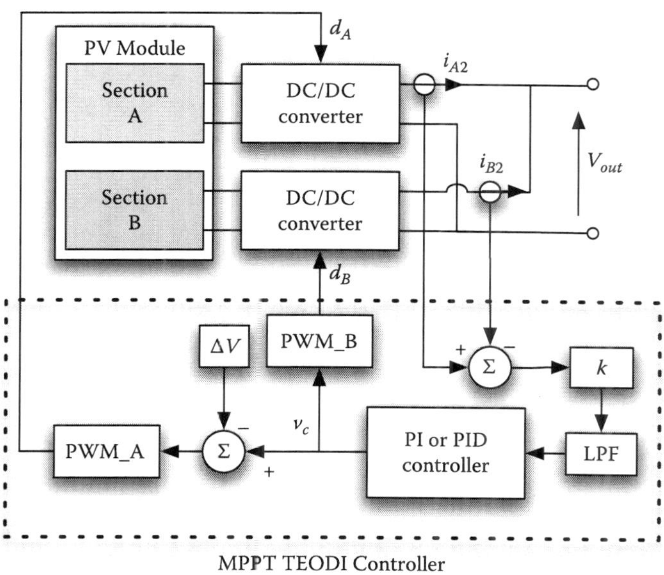

FIGURE 2.30
Block scheme of TEODI for parallel configuration of the power stages.

The duty cycles d_A and d_B of the two DC/DC converters are different because $d_A = d_B - \Delta d$, where $\Delta d = \Delta V/Vs$, Vs is the peak amplitude of the sawtooth carrier waveforms of the two PWM modulators, and ΔV is a constant voltage offset that will be used to settle the TEODI performances. In particular, on the basis of the converter topology, such a parameter defines the voltage displacement (ΔV_{pv}) between the PV operating points of the two identical PV sections.

By assuming that the sign of the displacement voltage ΔV is chosen in order to obtain $V_{PV,A} > V_{PV,B}$, three conditions are possible for the operating points of the two PV sections.

Figure 2.31 shows the P-V curves of both PV sections, the square markers highlighting the operating points of PV section A and the circle markers putting into evidence the operating points of PV section B. In the case shown in Figure 2.31a, the PV power of section A is higher than that delivered by section B, thus corresponding to the condition $i_{A2} > i_{B2}$ at the converter output. In the case shown in Figure 2.31b, the PV power delivered by section A is lower than that produced by section B, thus corresponding to the condition $i_{A2} < i_{B2}$. The signal $k{\cdot}(i_{A2} - i_{B2})$ is at the input of the proportional-integral (PI) controller. The integral action forces the change of the duty cycles d_A and d_B, thus pushing the operating points of the PV sections in the direction that minimizes the difference of the output currents. The sign of the coefficient k must be properly selected in order to associate the right direction to the duty cycles for driving the operating points toward the MPP. Only when the system is in the condition shown in Figure 2.31c does the PV power of section A equal the power of section B, thus corresponding to $i_{A2} = i_{B2}$. In such a case, the input of the PI controller is zero, and consequently the converters' duty cycles remain at a constant value.

As for Δd, in equilibrium conditions, the smaller the value of Δd, the smaller the distance between the operating point of each of the PV modules and the MPP, and thus the higher the MPPT efficiency. In practice, because of the effect of tolerances of the physical components of the two sections and of small, nonetheless unavoidable, effects (due to temperature, humidity, and so on) making the two subsystems not perfectly equal, too small values of Δd could lead to the failure of the TEODI technique. The value of Δd affects both the MPPT efficiency in steady state environmental conditions and the speed of the whole tracking process under environmental dynamic conditions. Indeed, the higher the value of Δd, the higher the MPPT promptness. If $\Delta d = 0$, in the case of two perfectly equal sections, in all three cases shown in Figure 2.31 the operating point of PV section A collapses in the same position of PV section B. The speed would be equal to zero because the error at the input of the PI controller is always zero, and the system would not be able to perform the tracking process. Therefore Δd must settle on the basis of a reasonable compromise between the steady-state MPPT efficiency and required dynamic performances of the system. Considerations similar to those referred to the P&O technique, and discussed in the previous sections, do apply.

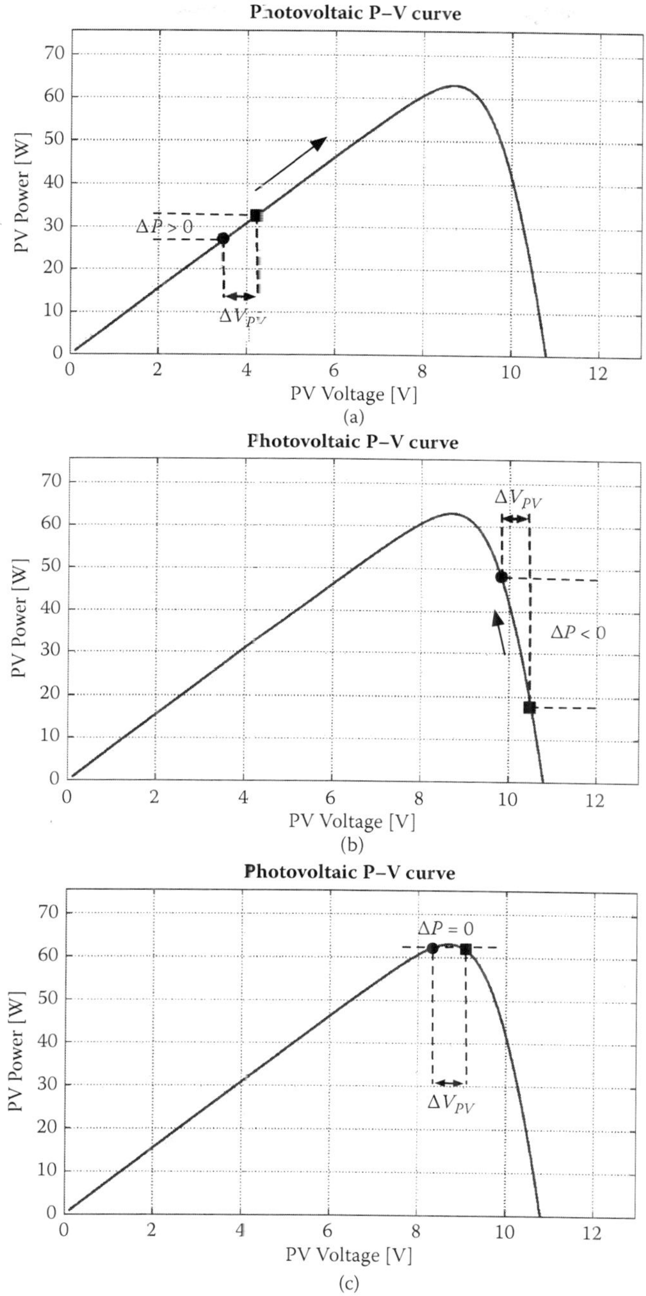

FIGURE 2.31
Power vs. voltage characteristic of photovoltaic modules A and B and the corresponding operating points.

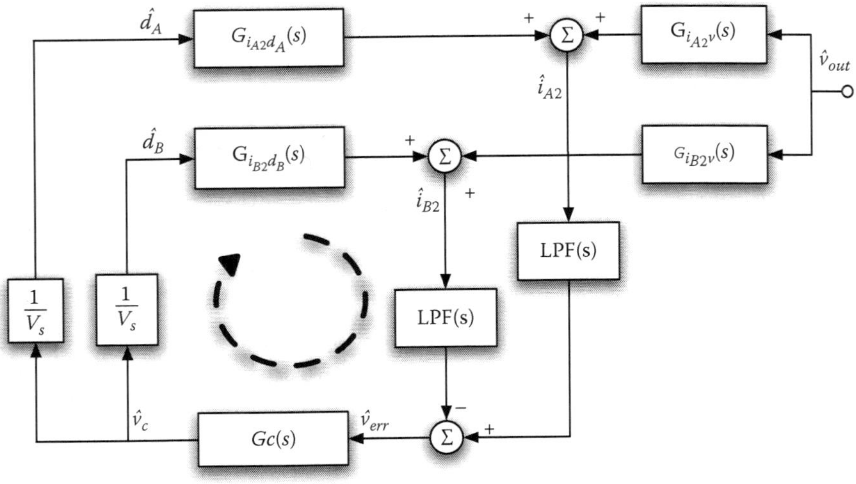

FIGURE 2.32
Equivalent block diagram of the TEODI architecture.

One of the main advantages of TEODI is in the simple design of the MPPT parameters because it consists of the right selection of the compensation network; such design can be performed directly in the Laplace's domain by using the system transfer functions.

Given the TEODI architecture shown in Figure 2.30, the corresponding small-signal model can be obtained by averaging and linearizing the state equations of the whole system made of the converters, the load, and the PV sources.

In the equivalent block diagram shown in Figure 2.32 symbols with hats represent small-signal variations around the quiescent values of the corresponding quantities. The objective of the analysis in the Laplace domain is to identify conditions permitting the design of the transfer function $W(s) = LPF(s) \cdot G_c(s)$ leading to a stable closed-loop system, with an adequate phase margin and a sufficiently high crossover frequency.

By looking at the scheme in Figure 2.32, it results that the equilibrium point is reached when $\hat{i}_{A2} = \hat{i}_{B2}$ and the system loop gain $T_c(s)$ is given by

$$T_c(s) = W(s) \cdot \left(G_{i_{A2}d_A}(s) - G_{i_{B2}d_B}(s)\right) \tag{2.58}$$

The Bode diagram of the $T_c(s)$ loop gain transfer function can be used to evaluate the system stability and dynamic performances of the TEODI algorithm. A particular aspect of the proposed architecture is in the fact that the $T_c(s)$ transfer function has a double loop. The TEODI capability to reject the output voltage oscillations on the PV terminals can be evaluated by means of the $W_{v_{pv,A},v_o(s)}$ and $W_{v_{pv,B},v_o(s)}$ transfer functions:

$$W_{v_{pv,A},v_o(s)} = \frac{\hat{v}_{pv,A}}{\hat{v}_{out}} = \frac{G_{v_{pv,A}v_o}(s)}{1+T_c(s)} \tag{2.59}$$

$$W_{v_{pv,B},v_o(s)} = \frac{\hat{v}_{pv,B}}{\hat{v}_{out}} = \frac{G_{v_{pv,B}v_o}(s)}{1+T_c(s)} \tag{2.60}$$

where $G_{v_{pv,A}v_o}(s)$ and $G_{v_{pv,B}v_o}(s)$ are the output-to-input open-loop transfer functions of the converters used in the corresponding PV sections. Finally the effect of the output voltage oscillation on the TEODI control signal is given by

$$W_{v_{err},v_{out}(s)} = \frac{\hat{v}_{err}}{\hat{v}_{out}} = LPF(s)\cdot\frac{G_{i_{A2}v}(s)-G_{i_{B2}v}(s)}{1+T_c(s)} \tag{2.61}$$

In the frequency range where the condition $T_c(s) >> 1$ holds, the system rejection capability is improved with respect to the open-loop system.

The reader can find further details and numerical examples of the TEODI approach in [36–38].

2.8 MPPT Efficiency

Up to few years ago no guidelines about the benchmarking of different MPPT techniques were available, but only some indication about the reasonable speed of variation of the irradiance to be used in testing an MPPT technique was given. For example, in [40], the slope 30 W/m^2/s is considered a reference value. More recently, a standard for testing DC/AC inverters' efficiency, the EN 50530, has been introduced, and a part of it is devoted to fix the conditions the MPPT algorithm must be subjected to for testing its performances. Steady-state and time-varying irradiation values and slopes are given, and some authors have published the results of their experimental analyses aimed at comparing the most famous MPPT approaches. For instance, in [41], the author assesses that the performances that can be obtained by the two most used algorithms, the P&O and the INC, are approximately the same. The fundamental role of the MPPT efficiency has not been recognized as the conversion efficiency is. In fact, many efforts are usually done by power electronics designers in order to increase the conversion efficiency of the power processing system. In addition to the classical peak efficiency value, the *European efficiency* (η_{ev}) has been proposed as a measure of the performances of the conversion system at different power levels, which means at different irradiation levels along the day. The European efficiency is defined in (2.62) in order to give a weight to the conversion efficiencies in the beginning of the day and approaching sunset, so that a power processing system having a flat and high-efficiency profile has a high η_{EU}.

$$\eta_{EU} = 0.03 \cdot \eta_{5\%} + 0.06 \cdot \eta_{10\%} + 0.13 \cdot \eta_{20\%} + 0.10 \cdot \eta_{30\%} + 0.48 \cdot \eta_{50\%} + 0.20 \cdot \eta_{100\%} \quad (2.62)$$

A similar formula has been also proposed by the Californian Energy Commission: The weights are different but the idea is the same.

$$\begin{aligned} \eta_{CEC} = {} & 0.00 \cdot \eta_{5\%} + 0.04 \cdot \eta_{10\%} + 0.0.05 \cdot \eta_{20\%} \\ & + 0.12 \cdot \eta_{30\%} + 0.21 \cdot \eta_{50\%} + 0.53 \cdot \eta_{75\%} + 0.05 \cdot \eta_{100\%} \end{aligned} \quad (2.63)$$

Nevertheless, the in-depth studies concerning the MPPT efficiency and the factors affecting its value are very few in literature, essentially because of the difficulty in understanding that the efficiency of the PV system is almost the product of the MPPT efficiency and the conversion efficiency.

The MPPT efficiency is defined as follows:

$$\eta_{MPPT} = \frac{\int_{t1}^{t2} P(t)dt}{\int_{t1}^{t2} P_{MPP}(t)dt} \quad (2.64)$$

so that its value is unitary if the operating point remains in the MPP during the whole time interval going from t_1 to t_2. An approximation of the MPPT efficiency can be easily obtained by assuming that the P-V curve of the PV array behaves like a parabola across the MPP (see Figure 2.33). This analysis

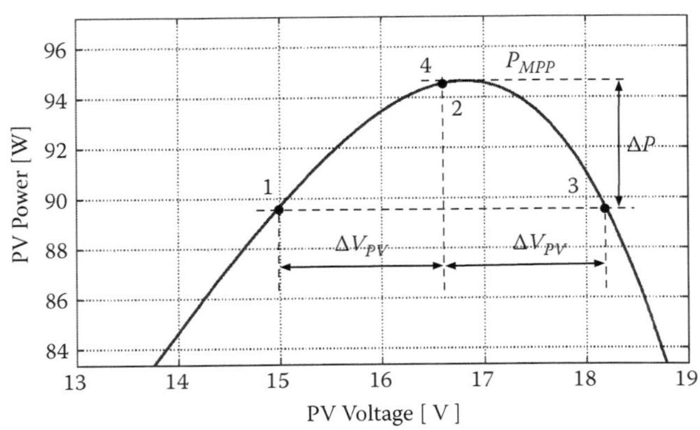

FIGURE 2.33
Three-point behavior in steady state.

allows us to show what is the best steady-state condition that can be obtained by any perturbative MPPT approach. The first analysis is performed by supposing that there are operating points, with the one in the middle placed in the MPP and the side ones at the same power level, because of the parabolic approximation, on the ascending and descending sides of the P-V curve, respectively.

If the MPPT controller leaves the system working in each one of these points for T_p seconds, then the waveform of the power looks as depicted in Figure 2.34.

The period of this waveform is $4 \cdot T_p$, and due to the piecewise constant waveform of the PV power, it is easy to determine the numerator of (2.64), so that the MPPT efficiency is

$$\eta_{MPPT} = 1 - \frac{\Delta P}{2 \cdot P_{MPP}} \tag{2.65}$$

where ΔP depends on the amplitude of the perturbation applied to implement the perturbative MPPT method and on the PV array parameters and operating conditions.

If the steady-state operation consists of four points, two on one side and two on the other side of the MPP, under the same assumption of a parabolic P-V curve with the vertex in the MPP, it results that

$$\eta_{MPPT} = \frac{< P(t) >_{6 \cdot T_p}}{P_{MPP}} = 1 - \frac{\Delta P_1 + \Delta P_2}{2 \cdot P_{MPP}} \tag{2.66}$$

where the condition under analysis is shown in Figures 2.35 and 2.36.

Equations (2.65) and (2.66) reveal that the best condition in terms of MPPT efficiency is achieved when the steady state consists of three points, one in the MPP and two on its sides. The condition consisting of four points, and of course of more than four points, leads to a reduction of the MPPT efficiency and must be avoided as much as possible. In Section 2.4.2 the value of

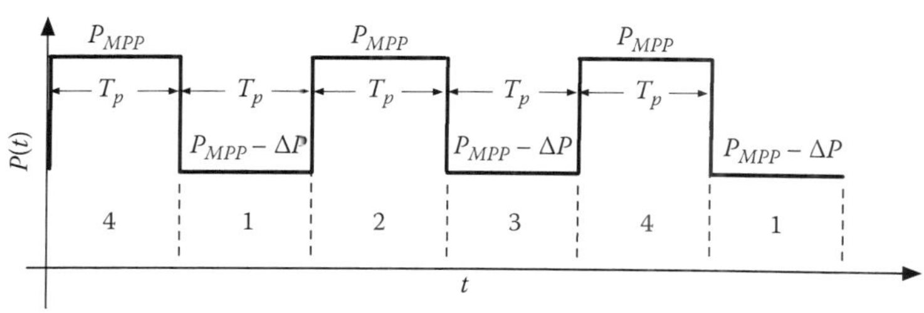

FIGURE 2.34
Power waveform in steady state.

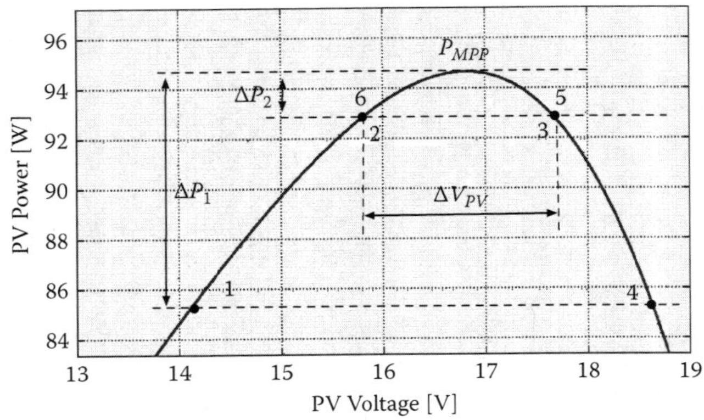

FIGURE 2.35
Four-point behavior in steady state.

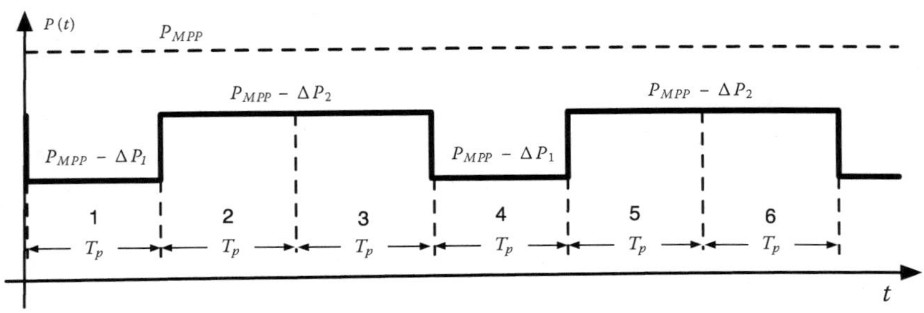

FIGURE 2.36
Power waveform in four points in steady state.

the power drops ΔP used in the expressions above has been calculated as a function of the amplitude of the perturbation signal used in the perturbative MPPT approach and of the PV array parameters.

References

1. S.-K. Ji, D.-H. Jang, S.-K. Han, C.-W. Roh, and S.-S. Hong. Analog control algorithm for maximum power tracker employed in photovoltaic applications. In *2010 Conference Proceedings of IPEC*, October 2010, pp. 99–103.
2. J.H.R. Enslin, M.S. Wolf, D.B. Snyman, and W. Swiegers. Integrated photovoltaic maximum power point tracking converter. *IEEE Transactions on Industrial Electronics*, 44(6):769–773, 1997.

3. W. Xiao, W.G. Dunford, P.R. Palmer, and A. Capel. Application of centered differentiation and steepest descent to maximum power point tracking. *IEEE Transactions on Industrial Electronics*, 54(5):2539–2549, 2007.
4. M.A.S. Masoum, H. Dehbonei, and E.F. Fuchs. Theoretical and experimental analyses of photovoltaic systems with voltage and current-based maximum power-point tracking. *IEEE Transactions on Energy Conversion*, 17(4):514–522, 2002.
5. C. Larbes, S.M. Aït Cheikh, T. Obeidi, and A. Zerguerras. Genetic algorithms optimized fuzzy logic control for the maximum power point tracking in photovoltaic system. *Renewable Energy*, 34(10):2093–2100, 2009.
6. C.-S. Chiu and Y.-L. Ouyang. Robust maximum power tracking control of uncertain photovoltaic systems: A unified t-s fuzzy model-based approach. *IEEE Transactions on Control Systems Technology*, 19(6):1516–1526, 2011.
7. V. Salas, E. Olìas, A. Barrado, and A. Làzaro. Review of the maximum power point tracking algorithms for stand-alone photovoltaic systems. *Solar Energy Materials and Solar Cells*, 90:1555–1576, 2006.
8. E. Koutroulis, K. Kalaitzakis, and N.C. Voulgaris. Development of a microcontroller-based, photovoltaic maximum power point tracking control system. *IEEE Transactions on Power Electronics*, 16(1):46–54, 2001.
9. N. Femia, G. Petrone, G. Spagnuolo, and M. Vitelli. Optimization of perturb and observe maximum power point tracking method. *IEEE Transactions on Power Electronics*, 20(4):963–973, 2005.
10. R.P. Venturini, V.V.R. Scarpa, G. Spiazzi, and S. Buso. Analysis of limit cycle oscillations in maximum power point tracking algorithms. In *IEEE Power Electronics Specialists Conference 2008 (PESC 2008)*, June 2008, pp. 378–384.
11. S. Liu and R.A. Dougal. Dynamic multiphysics model for solar array. *IEEE Transactions on Energy Conversion*, 17(2):285–294, 2002.
12. R.W. Erikson and D. Maksimovic. *Fundamentals of power electronics*. 2nd ed. Kluwer Academic Publishers, Norwell, MA.
13. F.L. Luo and H. Ye. Small signal analysis of energy factor and mathematical modeling for power dc-dc converters. *IEEE Transactions on Power Electronics*, 22(1):69–79, 2007.
14. K.H. Hussein, I. Muta, T. Hoshino, and M. Osakada. Maximum photovoltaic power tracking: An algorithm for rapidly changing atmospheric conditions. *IEE Proceedings on Generation, Transmission and Distribution*, 142(1):59–64, 1995.
15. R. Bründlinger, N. Henze, H. Häberlin, B. Burger, A. Bergmann, and F. Baumgartner. prEN50530—The new European standard for performance characterisation of pv inverters. In *24th European Photovoltaic Solar Energy Conference*, Hamburg, Germany, September 2009, vol. 1, pp. 3105–3109.
16. W. Xiao and W.G. Dunford. A modified adaptive hill climbing MPPT method for photovoltaic power systems. In *2004 IEEE 35th Annual Power Electronics Specialists Conference (PESC 04)*, June 2004. vol. 3, pp. 1957–1963.
17. A. Pandey, N. Dasgupta, and A.K. Mukerjee. Design issues in implementing MPPT for improved tracking and dynamic performance. In *32nd Annual Conference on IEEE Industrial Electronics, IECON 2006*, November 2006, pp. 4387–4391.
18. N. Fermia, D. Granozio, G. Petrone, and M. Vitelli. Predictive adaptive MPPT perturb and observe method. *IEEE Transactions on Aerospace and Electronic Systems*, 43(3):934–950, 2007.

19. Y.-T. Hsiao and C.-H. Chen. Maximum power tracking for photovoltaic power system. In *37th IAS Annual Meeting of Industry Applications Conference*, 2002, vol. 2, pp. 1035–1040.
20. D.P. Hohm and M.E. Ropp. Comparative study of maximum power point tracking algorithms. *Progress in Photovoltaics: Research and Applications*, 11(1):47–62, 2003.
21. F.-S. Pai, R.-M. Chao, S.H. Ko, and T.-S. Lee. Performance evaluation of parabolic prediction to maximum power point tracking for PV array. *IEEE Transactions on Sustainable Energy*, 2(1):60–68, 2011.
22. K. Ishaque, Z. Salam, M. Amjad, and S. Mekhilef. An improved particle swarm optimization (PSO)-based MPPT for PV with reduced steady state oscillation. *IEEE Transactions on Power Electronics*, PP(99):1, 2012.
23. M. Miyatake, M. Veerachary, F. Toriumi, N. Fujii, and H. Ko. Maximum power point tracking of multiple photovoltaic arrays: A PSO approach. *IEEE Transactions on Aerospace and Electronic Systems*, 47(1):367–380, 2011.
24. G. Petrone, G. Spagnuolo, and M. Vitelli. A multivariable perturb-and-observe maximum power point tracking technique applied to a single-stage photovoltaic inverter. *IEEE Transactions on Industrial Electronics*, 58(1):76–84, 2011.
25. A.M. Bazzi and P.T. Krein. Concerning "Maximum power point tracking for photovoltaic optimization using ripple-based extremum seeking control." *IEEE Transactions on Power Electronics*, 26(6):1611–1612, 2011.
26. A.I. Bratcu, I. Munteanu, S. Bacha, D. Picault, and B. Raison. Cascaded DC-DC converter photovoltaic systems: Power optimization issues. *IEEE Transactions on Industrial Electronics*, 58(2):403–411, 2011.
27. S.L. Brunton, C.W. Rowley, S.R. Kulkarni, and C. Clarkson. Maximum power point tracking for photovoltaic optimization using ripple-based extremum seeking control. *IEEE Transactions on Power Electronics*, 25(10):2531–2540, 2010.
28. D. Casadei, G. Grandi, and C. Rossi. Single-phase single-stage photovoltaic generation system based on a ripple correlation control maximum power point tracking. *IEEE Transactions on Energy Conversion*, 21(2):562–568, 2006.
29. T. Esram, J.W. Kimball, P.T. Krein, P.L. Chapman, and P. Midya. Dynamic maximum power point tracking of photovoltaic arrays using ripple correlation control. *IEEE Transactions on Power Electronics*, 21(5):1282–1291, 2006.
30. J.W. Kimball and P.T. Krein. Discrete-time ripple correlation control for maximum power point tracking. *IEEE Transactions on Power Electronics*, 23(5):2353–2362, 2008.
31. K.H. Hussein, I. Muta, T. Hoshino, and M. Osakada. Maximum photovoltaic power tracking: An algorithm for rapidly changing atmospheric conditions. *IEE Proceedings of Generation, Transmission and Distribution*, 142(1):59–64, 1995.
32. F. Liu, S. Duan, F. Liu, B. Liu, and Y. Kang. A variable step size INC MPPT method for PV systems. *IEEE Transactions on Industrial Electronics*, 55(7):2622–2628, 2008.
33. Q. Mei, M. Shan, L. Liu, and J.M. Guerrero. A novel improved variable step-size incremental-resistance MPPT method for PV systems. *IEEE Transactions on Industrial Electronics*, 58(6):2427–2434, 2011.
34. D. Shmilovitz. On the control of photovoltaic maximum power point tracker via output parameters. *IEE Proceedings of Electric Power Applications*, 152(2):239–248, 2005.

35. L. Egiziano, N. Femia, G. Lisi, G. Petrone, G. Spagnuolo, and M. Vitelli. Design and optimization of a maximum power point tracking controller for a PV battery charger. In *IEEE International Symposium on Industrial Electronics, 2007 (ISIE 2007)*, June 2007, pp. 2426–2431.
36. N. Femia, G. Petrone, G. Spagnuolo, and M. Vitelli. A new analog MPPT technique: TEODI. *Progress in Photovoltaics: Research and Applications*, 18(1):28–41, 2010.
37. G. Petrone, G. Spagnuolo, and M. Vitelli. TEODI: A new technique for distributed maximum power point tracking PV applications. In *2010 IEEE International Conference on Industrial Technology (ICIT)*, March 2010, pp. 982–987.
38. G. Petrone, G. Spagnuolo, and M. Vitelli. TEODI: PV MPPT based on the equalization of the output operating points in correspondence of the forced displacement of the input operating points. In *2010 IEEE International Symposium on Industrial Electronics (ISIE)*, July 2010, pp. 3463–3468.
39. N. Femia, G. Lisi, G. Petrone, G. Spagnuolo, and M. Vitelli. Distributed maximum power point tracking of photovoltaic arrays: Novel approach and system analysis. *IEEE Transactions on Industrial Electronics*, 55(7):2610–2621, 2008.
40. B. Bletterie, R. Bruendlinger, and S. Spielauer. Quantifying dynamic MPPT performance under realistic conditions first test results—The way forward. In *Proceedings of 21st European Photovoltaic Solar Energy Conference and Exhibition*, 2006, pp. 2347–2351.
41. T. Andrejasic, M. Jankovec, and M. Topic. Comparison of direct maximum power point tracking algorithms using en 50530 dynamic test procedure. *IET Renewable Power Generation*, 5(4):281–286, 2011.

3

MPPT Efficiency: Noise Sources and Methods for Reducing Their Effects

3.1 Low-Frequency Disturbances in Single-Phase Applications

AC applications of PV systems require a power processing element able to convert the DC power generated by the PV unit into AC power, at 110 or 230 V rms voltage rating by adapting the PV voltage to that needed by the stage performing DC/AC conversion. The adoption of single-stage power processing units is the best choice, especially from the point of view of reliability and efficiency. Some commercial solutions, especially in the power range of interest for residential power plants (e.g., from Fronius and SMA), are based on the topology depicted in Figure 3.1, which consists of only a pulse-width modulation (PWM) inverter controlled in such a way that the maximum power from the PV source is extracted and an almost pure sinusoidal current is injected into the grid or into an AC stand-alone system with energy backup.

Because of the inherent step-down characteristic of the inverting bridge, the adoption of a single-stage inverter requires that the PV voltage be higher than the peak AC voltage value, so that this solution is not suitable for low-power PV applications like microinverters.

Double-stage or multiple-stage solutions are based on a DC/DC converter that controls the PV source and performs the MPPT function, cascaded by a DC/AC conversion stage. Such architecture is shown in Figure 3.2, where the bulk capacitance, placed between the two conversion stages, plays an important role.

The bulk capacitance, indeed, handles the imbalance between the PV power made available at the DC bus by the DC/DC stage and the AC power absorbed by the grid through the DC/AC stage. The difference between the DC power and the AC power is a sinusoidal component at twice the AC grid frequency, namely, either 100 or 120 Hz, as shown in Figures 3.2 and 3.3. The amplitude of the voltage oscillation across the bulk capacitor due to the AC

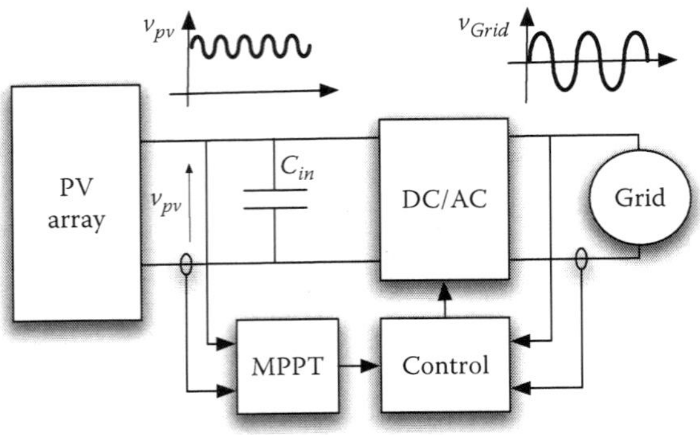

FIGURE 3.1
Single-stage, single-phase PV inverter.

power component can be easily calculated by assuming that the bulk capacitor current has the following expression:

$$i_b(t) = I_b \cdot cos(2 \cdot \omega_{grid} \cdot t) \tag{3.1}$$

where ω_{grid} is the grid frequency. The AC component of the bulk voltage is then given by

$$\hat{v}_b(t) = \frac{I_b}{C_b} \cdot \int_{T_{grid}} \cos\left(2 \cdot \omega_{grid} \cdot t\right) dt = \frac{I_b}{2 \cdot \omega_{grid} \cdot C_b} \cdot \sin\left(2 \cdot \omega_{grid} \cdot t\right) \tag{3.2}$$

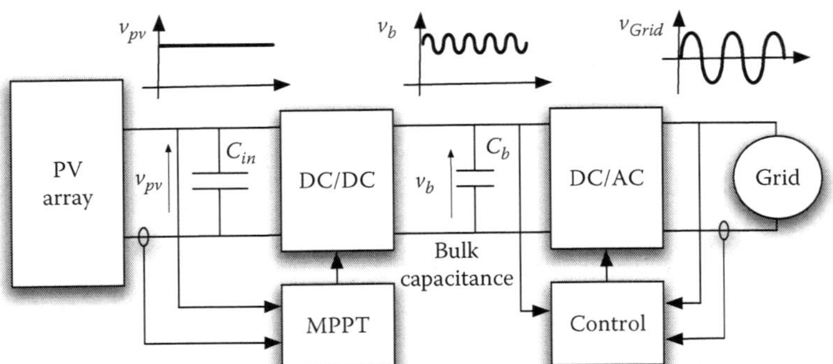

FIGURE 3.2
Double-stage, single-phase PV inverter.

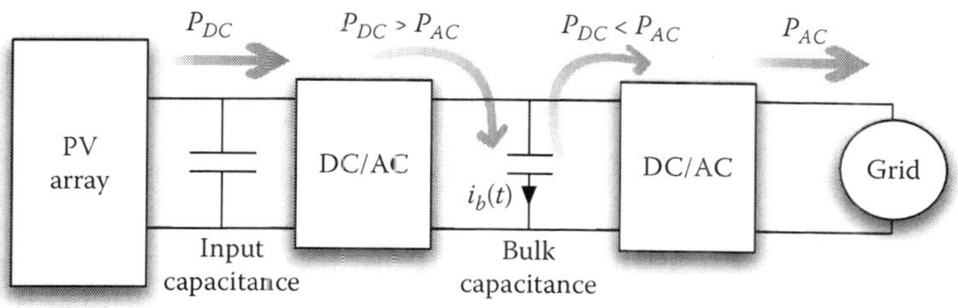

FIGURE 3.3
Role of the bulk capacitance.

and the peak-to-peak value of the voltage oscillation at a frequency $2 \cdot \omega_{grid}$ is

$$\Delta V_t = \frac{P_{PV}}{2 \cdot \omega_{grid} \cdot C_b \cdot V_b} \tag{3.3}$$

where the PV power has been assumed to be $P_{PV} = V_b \cdot I_b$. Equation (3.3) reveals that the voltage oscillation amplitude is not constant along the day, because it depends on the instantaneous PV power, namely on the irradiation level.

The oscillation affecting the voltage of the bulk capacitor has detrimental effects on both the DC and the AC part of the power processing system, so that a large electrolytic capacitor is mostly used at the DC link. In fact, as (3.3) highlights, the larger the bulk capacitance, the smaller the voltage oscillation. Unfortunately, bulk electrolytic capacitors are a weak point of the conversion chain, because of the effects that the operation temperature has on their lifetime. A significant effort is done by PV inverter manufacturers in order to keep the working temperature of the bulk capacitor as close as possible to that at which the capacitor manufacturer has tested the component for some thousands of hours, so that the mean time between failures (MTBF) is increased.

A reduced value of the bulk capacitance might allow use of film capacitors instead of electrolytic ones, but the resulting increased low-frequency oscillation given by (3.2) would have an impact on the quality of the current at the inverter output, so that suitable control techniques must be adopted in order to reduce such an effect.

The oscillation back propagates through the DC/DC converter up to the PV terminals, thus degrading the quality of the MPPT operation performed by the DC/DC converter itself. A simplified expression of the nonlinear current vs. voltage characteristic allows us to obtain a relationship between the amplitude (3.3) and the MPPT efficiency [1]. Neglecting the series and parallel resistances in the PV array model provides the following PV MPP current and relevant derivatives:

$$I_{MPP} = I_{ph} - I_0 \cdot \left(e^{\frac{V_{MPP}}{\eta \cdot V_T \cdot n_s}} - 1 \right) \tag{3.4}$$

$$\frac{\partial I_{MPP}}{\partial V_{MPP}} = -\frac{I_0}{\eta \cdot V_T \cdot n_s} \cdot e^{\frac{V_{MPP}}{\eta \cdot V_T \cdot n_s}} \tag{3.5}$$

$$\frac{\partial^2 I_{MPP}}{\partial V_{MPP}^2} = -\frac{I_0}{(\eta \cdot V_T \cdot n_s)^2} \cdot e^{\frac{V_{MPP}}{\eta \cdot V_T \cdot n_s}} \tag{3.6}$$

These expressions allow us to approximate the PV current vs. voltage curve across the MPP by means of the second-order Taylor series:

$$i_{PV} = I_{MPP} + \frac{\partial I_{MPP}}{\partial V_{MPP}} \cdot (v_{PV} - V_{MPP}) + \frac{\partial^2 I_{MPP}}{\partial V_{MPP}^2} \cdot (v_{PV} - V_{MPP})^2 \tag{3.7}$$

that is,

$$i_{PV} = \gamma + \beta \cdot v_{PV} + \alpha \cdot v_{PV}^2 \tag{3.8}$$

where

$$\gamma = I_{MPP} - V_{MPP} \cdot \frac{\partial I_{MPP}}{\partial V_{MPP}} + \frac{1}{2} \cdot \frac{\partial^2 I_{MPP}}{\partial V_{MPP}^2} \cdot V_{MPP}^2 \tag{3.9}$$

$$\beta = \frac{\partial I_{MPP}}{\partial V_{MPP}} - V_{MPP} \cdot \frac{\partial^2 I_{MPP}}{\partial V_{MPP}^2} \tag{3.10}$$

$$\alpha = \frac{1}{2} \cdot \frac{\partial^2 I_{MPP}}{\partial V_{MPP}^2} \tag{3.11}$$

The oscillation of the bulk voltage causes an oscillation of the PV voltage across the MPP: If $M(D) = V_b/V_{PV}$ is the voltage conversion ratio of the DC/DC converter, then $\Delta V_{PV} = \Delta V_b/M(D)$. The PV voltage is thus given by

$$v_{PV}(t) = V_{MPP} + \hat{v}(t) = V_{MPP} + \Delta V_{PV} \cdot sin(2\omega_{grid} \cdot t) \tag{3.12}$$

so that the instantaneous power and average power over one period of the grid voltage can be calculated:

$$
\begin{aligned}
p_{PV}(t) &= v_{PV}(t)\, i_{PV}(t) \\
&= \gamma\cdot(V_{MPP} + \hat{v}(t)) + \beta\cdot(V_{MPP} + \hat{v}(t))^2 + \alpha\cdot(V_{MPP} + \hat{v}(t))^3 \\
P &= \frac{1}{T_{grid}} \int_{T_{grid}} p(t)dt \\
&= \gamma\cdot V_{MPP} + \beta\cdot V_{MPP}^2 + \alpha\cdot V_{MPP}^3 + \frac{3\cdot\alpha\cdot V_{MPP} + \beta}{2}\cdot\Delta V_{PV}^2 \\
&= P_{MPP} + \frac{3\cdot\alpha\cdot V_{MPP} + \beta}{2}\cdot\Delta V_{PV}^2
\end{aligned}
\tag{3.13}
$$

Equation (3.13) can be used to obtain an expression of the MPPT efficiency as a function of the bulk voltage oscillation amplitude:

$$
\eta_{MPPT} = \frac{P}{P_{MPP}} = 1 + \frac{3\cdot\alpha\cdot V_{MPP} + \beta}{2\cdot P_{MPP}\cdot M^2(D)}\cdot\Delta V_b^2 \tag{3.14}
$$

The voltage oscillation amplitude that ensures a desired MPPT efficiency [1] is then given by

$$
\Delta V_b = M(D)\cdot\sqrt{\frac{2\cdot P_{MPP}\cdot(\eta_{MPPT} - 1)}{3\cdot\alpha\cdot V_{MPP} + \beta}} \tag{3.15}
$$

Equation (3.14) reveals that the MPPT efficiency is as high as the oscillation amplitude is small. It must be highlighted that α and β are negative; thus the term $3\cdot\alpha\cdot V_{MPP} + \beta$ appearing in (3.13), (3.14), and (3.15) is negative.

Equation (3.15) provides the constraint to be fulfilled by the oscillation amplitude on the bulk voltage if a desired minimum MPPT efficiency is required. The value of ΔV_b given by (3.15) can be put into (3.3) in order to obtain the value of C_b. Equation (3.15) also puts into evidence that a step-up DC/DC converter, characterized by $M(D) > 1$, allows use of a smaller bulk capacitance for a given MPPT efficiency: The higher $M(D)$, the higher η_{MPPT}.

The previous discussion highlights the weak point of the single-stage PV inverter topologies. In particular, they include only one storage element, that is, a large electrolytic capacitor, at the PV source terminals, so that a reduction of that capacitance aimed at using more reliable components would cause an increase of the PV voltage oscillation amplitude, with a consequent reduction of the MPPT efficiency. On the contrary, the presence of $M(D)$ in

(3.14) and (3.15) highlights that the additional DC/DC converter is very helpful in reducing the effects of the oscillations coming from the DC link, not only at the frequency of the bulk voltage oscillation but also in a wider range of frequencies.

The strategy of using the DC/DC converter as an active filter was presented in [2] and is described in the next section. Some other approaches have also been presented in literature [3]. In some cases they are referred to applications involving fuel cells (FCs). FCs are electrochemical systems that are able to convert a chemical energy vector, e.g., hydrogen or methanol, into electrical power by means of the action of a catalyst and of the contribution of oxygen at the cathode. Similar to PV systems, FCs are low-voltage devices that, in the power range of a few kilowatts, are able to deliver hundreds of amperes in DC. While any disturbance affecting the PV current has only a detrimental effect in terms of MPPT efficiency, current ripple significantly affects the fuel consumption and life span of FCs [4]. In other words, while in PV systems the ripple can be tolerated at the price of a reduced MPPT efficiency, in FC systems the current ripple must be kept below 10% of the average value in order to preserve the stack functionality. As a further difference between the PV and FC systems in this aspect, the effect of the ripple frequency must be pointed out. In PV systems, any oscillation at whichever frequency is detrimental for the MPPT efficiency. In FC applications, instead, the high-frequency ripple, mainly due to the switching operation of the converter, is easily filtered by a small capacitance put in parallel to the stack terminals. Instead, the low-frequency component at 100/120 Hz cannot be passively filtered in an effective way and has a strong impact on the stack aging because of the stresses it causes on the FC membrane.

In [5] a DC/DC converter controlled by means of a double-feedback loop allows us to compensate the 100/120 Hz harmonic component. The latter is caused by a full bridge inverter, connected in cascade with the DC/DC one, thus forming a classical double-stage power processing system that allows injection of the power produced by a fuel cell into the grid. The approach proposed in [5] requires a careful design of the controller, because the interaction between the dynamics of the two loops deeply affects the attenuation capabilities. Low-frequency disturbances can be injected by the outer control loop if its bandwidth is close to the grid frequency.

In [6] a power zero ripple filter (ZRF) has been proposed. The authors point out that the high FC current flowing in the ZRF dramatically increases the conduction losses, so that they suggest a practical implementation involving four ZRFs, with a significant increase of topology and control strategy complexity. In [7] a novel topology called pulse-link DC/AC converter is proposed. An inductor and a capacitor connected in series are inserted between the DC/DC and the DC/AC stages, which increase the complexity and the number of components used in the power processing system. A similar approach is proposed in [8], where the active filter is inserted in the DC/DC converter by using a center tap isolation transformer. Another interesting approach, based on a feed-forward control structure, is proposed in [9]. This technique requires a perfect phase lock of the sinusoidal oscillation at twice the grid frequency, so that the use of an additional phase-locked loop (PLL) circuit is mandatory. Furthermore, as in any feed-forward approach, the proposed technique suffers uncertainties related to unmodeled phenomena and disturbances. Another limitation of the approach proposed in [9] is represented by the fact that the attenuation is obtained only at twice the grid frequency. In practice, the disturbance produced by the AC connection at 100 or 120 Hz is prevailing, but also additional harmonics are generated by the system nonlinearity.

A feed-forward technique is also used in [10]. The method is explained in more detail in [11], where a digital implementation is also proposed. Supposing that the DC/DC converter always operates in a given mode, e.g., continuous conduction mode, the availability of the formula expressing the voltage conversion ratio allows calculation of the correction to be instantaneously given to the duty cycle value in order to compensate the oscillation of the DC/DC converter output voltage and to keep the input voltage constant. Such a correction is at twice the grid frequency, and is superimposed to the stationary duty cycle values driven by the MPPT control loop. The main limitation of this simple and intuitive technique is in the fact that the voltage conversion ratio has to be known in explicit form, which does not change during the system operation. Unfortunately, at low irradiation levels, e.g., at sunrise and sunset, any DC/DC converter based on a classical boost, buck, or buck-boost topology and using diode rectifiers may get close, and even enter, discontinuous conduction mode. In this case the method gives a wrong compensation of the disturbance, unless the discontinuous current operating mode is detected and the right formula is used to predict the correct compensating duty cycle amplitude.

An effective technique for reducing the effects of the noise on the MPPT efficiency consists in employing closed-loop switching converters instead of open-loop ones. Such an approach is described in the next section by referring to the P&O technique.

3.1.1 The Perturb and Observe Approach Applied to Closed-Loop Switching Converters

In Section 2.4.2 the guidelines for settling the parameters T_p and Δx were given. Therein, the assumption that Δx is the only stimulus applied to the system has been adopted in order to explain the basic concept. Unfortunately, it is not fully realistic, except for a PV battery charger where the battery is approximated with a constant voltage generator.

In this section the advantages of using the P&O algorithm on a feedback-compensated switching converter, and based on the scheme of Figure 2.6, will be shown. This approach is particularly suitable for controlling PV systems in the presence of noise at the converter output, which is detrimental for the P&O

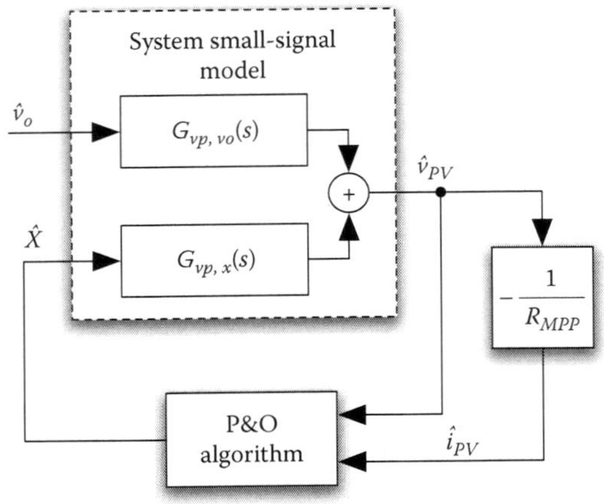

FIGURE 3.4
Effect of output voltage noise on the PV voltage.

performances. The benefits of the feedback compensation in the P&O operation will be analyzed, and an example of how to apply the P&O design equation to the closed-loop converters will be proposed.

The effect of the noise on the PV voltage is described by using the output-to-input $G_{vp,vo}(s)$ transfer function, as shown in Figure 3.4. In this scheme the converter transfer functions depend on the architecture selected for performing the MPPT.

Both the P&O perturbation ($\hat{x}$) and the output voltage noise ($\hat{v}_o$) affect the PV voltage. An estimation of the maximum ΔV_{PV} variation is

$$\Delta V_{PV} = |G_o \cdot \Delta x| + |G_{vp,vo}(\omega) \cdot \Delta V_o| \tag{3.16}$$

where the first term has already been introduced in (2.33) and represents the P&O effect, $G_{vp,vo}(\omega)$ is the magnitude of the output-to-input transfer function evaluated at the noise frequency ω, and ΔV_o is the amplitude of the output voltage noise (V_o). Equation (3.16) involves the absolute values because the signs of both voltage variations are not correlated and cannot be predicted in advance. The use of the absolute quantities ensures that we take into account the worst case for the ΔV_{PV} variation. As a consequence, by substituting eq. (3.16) in (2.32) we get

$$\Delta P_{PV} \simeq -\left(H \cdot V_{MPP} + \frac{1}{R_{MPP}} \right) \cdot \left(|G_o \cdot \Delta x| + |G_{vp,vo}(\omega) \cdot \Delta V_o| \right)^2 + V_{MPP} \cdot K_{ph} \Delta G \tag{3.17}$$

where the symbols have the same meaning described in Section 2.4.2.

As explained in Section 2.4.2, in order to avoid that the P&O is confused by the effect of additional noise $\hat{v}_o$, the PV power variation produced by a perturbation Δx must be greater than the sum of all the other PV power variations. In this case the P&O step amplitude must fulfill the following inequality:

$$\Delta x > \frac{1}{|G_0|}\sqrt{\frac{V_{MPP}\cdot K_{ph}\cdot|\dot{G}|\cdot T_p}{\left(H\cdot V_{MPP}+\dfrac{1}{R_{MPP}}\right)}}+\frac{|G_{vp,vo}(\omega)|}{|G_o|}\cdot\Delta V_o \tag{3.18}$$

In a more compact form,

$$\Delta x > \Delta x_{\dot{G}} + \Delta x_{\Delta V_0} \tag{3.19}$$

where $\Delta x_{\dot{G}}$ is the step amplitude needed for compensating the irradiance variation and $\Delta x_{\Delta V_0}$ is the step amplitude needed for compensating the output noise. According to the assumption done on (3.16), Equation (3.19) considers the worst case in the evaluation of Δx. It is worth noting that even under a constant irradiance level ($\Delta x_{\dot{G}} = 0$), the amplitude of the duty cycle perturbation must be large enough in order to overcome the negative effects of the oscillations at the converter output.

Under steady-state atmospheric conditions, it is mandatory to keep Δx as low as possible in order to bound the amplitude of the oscillations of the PV array operating point around the MPP to guarantee a high MPPT efficiency.

A technique for reducing $\Delta x_{\Delta V_0}$ is based on the adoption of a compensation network, with a suitable frequency bandwidth B, able to remove from the PV array voltage all the oscillations due to noises appearing at a frequency lower than B.

The PV voltage v_{PV} is sensed and compared with the output signal of the MPPT controller. As shown in Figure 2.6b, the latter is used to perturb the voltage reference of a closed-loop DC/DC converter. In Figure 3.5 the corresponding small-signal model is shown: It includes the small-signal model of the power stage (Figure 3.4) and of the PV voltage compensation loops.

The objective is to design the transfer function $G_c(s)$ leading to a stable closed-loop system, with an adequate phase margin, with the additional constraint that the closed-loop transfer function $W_{vp,vo}(s)$ between the converter output voltage and the PV array voltage is sufficiently small at the frequency of the noise to be suppressed. From Figure 3.5, the expression of $W_{vp,vo}(s)$ is obtained:

$$W_{vp.vo}(s) = \frac{\hat{v}_{PV}(s)}{\hat{v}_o(s)} = \frac{G_{vp,vo}(s)}{1+T_c(s)} \tag{3.20}$$

where $T_c(s) = G_{vp,d}(s)\cdot G_{PWM}\cdot H_v\cdot G_c(s)$ is the loop gain of the system under study. H_v is the gain of the PV voltage sensor and G_{pwm} is the gain of the PWM modulator, as shown in Figure 3.5.

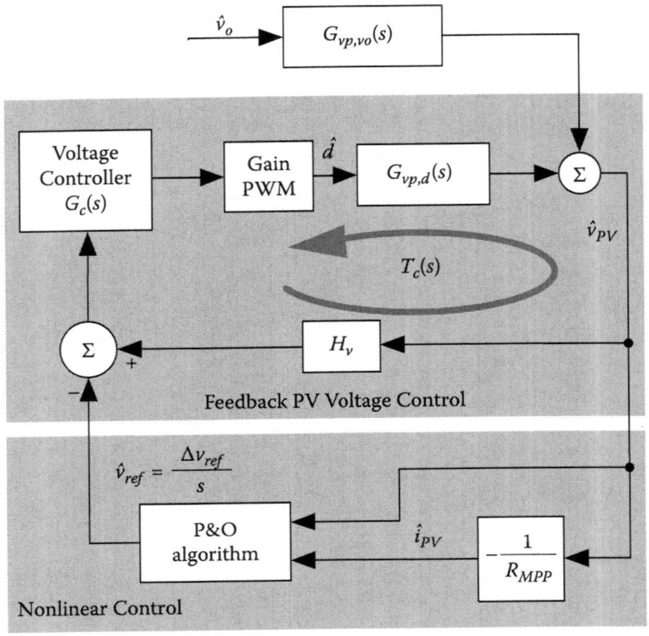

FIGURE 3.5
Small-signal model with a PV voltage compensation network.

The closed-loop transfer function $W_{vp,v_{ref}}(s)$, between $\hat{v}_{PV}(s)$ and $\hat{v}_{ref}(s)$, is

$$W_{vp.v_{ref}}(s) = \frac{\hat{v}_{PV}(s)}{\hat{v}_{ref}(s)} = \frac{1}{H_v}\frac{T_c(s)}{1+T_c(s)} \simeq \frac{1}{H_v} \tag{3.21}$$

with the approximation in (3.21) holding only in the frequencies range where $T_c(s) >> 1$. Such a condition must be fulfilled within bandwidth B to comply with the wish of suppressing noise whose frequency spectrum falls within B.

Regarding the MPPT algorithm, the closed-loop transfer functions (3.20) and (3.21) have, respectively, the same roles of $G_{vp,vo}(s)$ and $G_{vp,x}(s)$ for the open-loop system shown in Figure 3.4. This analogy allows us to fix the right values of the P&O parameters straightforwardly by using the results of the previous discussion. The perturbation period T_p is chosen by analyzing the behavior of $\hat{p}(t) = -\hat{v}_{PV}^2 / R_{MPP}$ and by using (3.21) in order to calculate the response $\hat{v}_{PV}(t)$ to a small step perturbation $\hat{v}_{ref}(t)$. The amplitude Δv_{ref} is instead given by

$$\Delta v_{ref} > \Delta v_{ref,\dot{G}} + \Delta v_{ref,\Delta V_o} \tag{3.22}$$

namely:

$$\Delta v_{ref} > H_v \sqrt{\frac{V_{MPP} \cdot K_{ph} \cdot |\dot{G}| \cdot T_p}{\left(H \cdot V_{MPP} + \dfrac{1}{R_{MPP}}\right)}} + H_v \cdot |W_{vp,vo}(\omega)| \cdot \Delta V_o \tag{3.23}$$

The effect of ΔV_o can be minimized by reducing the value of $|W_{vp,vo}(\omega)|$ as much as possible. Further details concerning the benefits of the voltage-compensated converter for PV applications are given in [2].

3.1.2 Example of P&O Design for a Closed-Loop Boost Converter

Figures 3.6 and 2.12 show the same circuit, but the former one includes a 100 Hz voltage oscillation at the converter output and also the $G_c(s)$ compensation network.

The small-signal model shown in Figure 2.13 allows us to determine the following transfer function $G_{vp,vo}(s)$:

$$G_{vp,vo}(s) = \frac{\mu \cdot \omega_n^2 \cdot \left(1 + \dfrac{s}{\omega_z}\right)}{s^2 + 2\zeta \cdot \omega_n \cdot s + \omega_n^2} \tag{3.24}$$

with $\mu = 1 - D$ and all the other coefficients given in (2.39) and (2.40).

The linear compensation network is designed on the basis of the desired dynamic performances: Given the desired crossover frequency and the

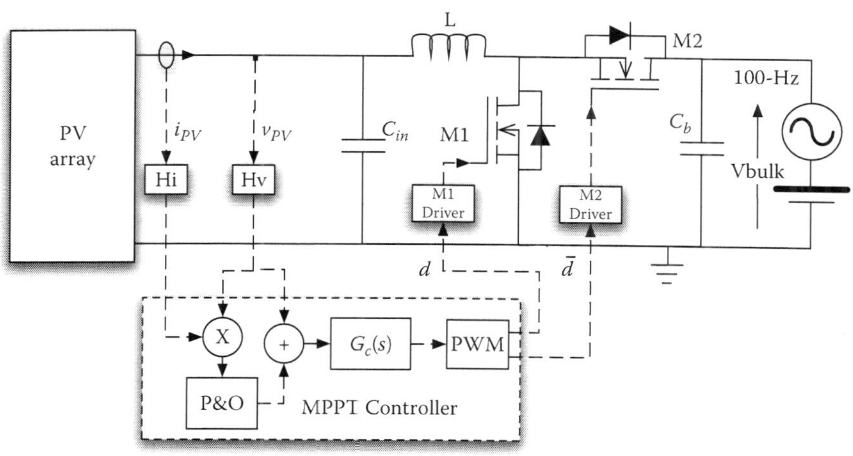

FIGURE 3.6
PV battery charger with a boost converter.

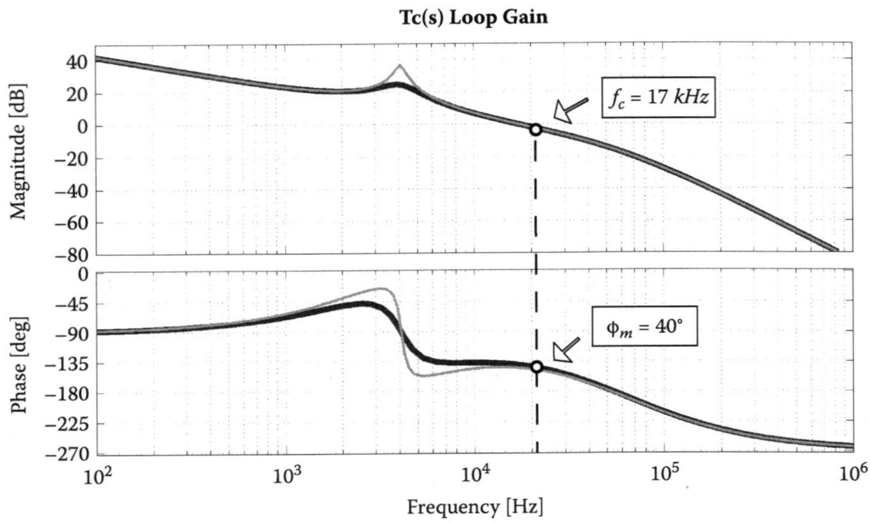

FIGURE 3.7
Bode plots of the converter loop gain $T_c(s)$. Grey curve $T_c(s)$ at $G = 100$ W/m^2, black curve $T_c(s)$ *at* $G = 1000$ W/m^2.

phase margin of the system loop gain, the values of the parameters in $G_c(s)$ can be easily determined [12]. In order to get a crossover frequency lower than $f_{sw}/10$ and a phase margin $\phi_m > 35°$, a PID compensator is used:

$$G_c(s) = k \cdot \frac{(s+\omega_{z1}) \cdot (s+\omega_{z2})}{s \cdot (s+\omega_{p1}) \cdot (s+\omega_{p2})} \tag{3.25}$$

The data listed in Table 2.1 lead to the following values of the parameters in $G_c(s): k = -6 \cdot 10^6$, $\omega_{z1} = 2 \cdot 10^4\, rad/s$, $\omega_{z2} = 2.5 \cdot 10^4\, rad/s$, $\omega_{p1} = 3 \cdot 10^5\, rad/s$, and $\omega_{p2} = 4.5 \cdot 10^5\, rad/s$. With this value the open-loop transfer function guarantees a crossover frequency of $f_c \simeq 17$ kHz and a phase margin of $\phi_m \simeq 40°$.

The transfer function $G_{vp,d}(s)$ depends on the irradiance level G through the value of the parameter R_{MPP} appearing in the denominator of (2.38). In particular, the value of ζ and the dynamic characteristics (f_c, ϕ_m) of $T_c(s)$ must be designed in the worst case. Figure 3.7 shows the converter loop gain $T_c(s) = G_c(s) \cdot G_{PWM} \cdot G_{vp,d}(s)$ at a low irradiance level, $G = 100$ W/m^2, and at a high one, $G = 1000$ W/m^2. It is assumed that $H_v = 1$.

The Bode diagrams show that the compensated system preserves almost the same dynamic performance in any operating condition.

The optimal values of the MPPT controller parameters T_p and $\Delta x = \Delta v_{ref}$ are designed as follows. T_p is fixed by taking into account the relation (2.22) and by evaluating the step response of the compensated closed-loop transfer function $W_{vp,vref}(s)$, which is shown in Figure 3.8 together with the corresponding

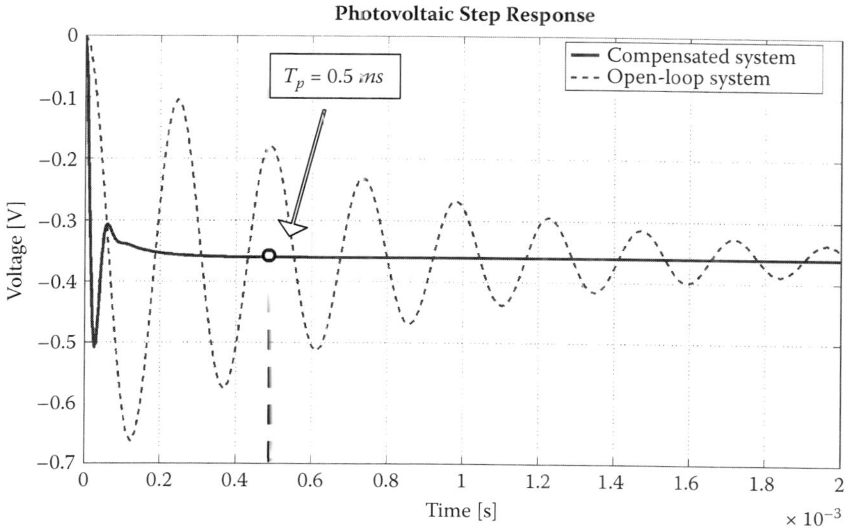

FIGURE 3.8
Step responses of the PV system for low irradiance G = 100 W/m². Black curve: closed-loop $W_{vp,x}(s)$ step response; dashed curve: open-loop $G_{vp,x}(s)$ step response.

open-loop transfer function $G_{vp,d}(s)$. Both functions have been evaluated in the worst case in terms of irradiance (G = 100 W/m²). The feedback loop introduces a significant benefit on the speed and damping of the closed-loop system, so that T_p = 0.5 ms can be adopted. This value is five times smaller than that used for T_p if the same boost converter operates in open loop.

The step perturbation Δv_{ref} is obtained by means of (3.23). $\Delta v_{ref,\dot{G}}$ is due to the irradiance variation and is evaluated by using the same PV parameters considered in the previous example, namely, $R_{MPP}|_{G=100} \simeq 11\Omega$, $H|_{G=100} = 0.0836$ A/V², and $K_{ph} = 8mA \cdot m^2 / W$. The step amplitude required to compensate an irradiance variation $\dot{G} = 100\, W / m^2 / s$ is $\Delta v_{ref,\dot{G}} = 0.0646\text{V}$.

Regarding the term $\Delta v_{ref,\Delta V_o}$ on the right side of (3.22), a value $\Delta V_o = 3$ V has been assumed for the noise at 100 Hz, while the DC component of the bulk voltage has been fixed at 12 V.

The value of $|W_{vp,vo}(\omega)|$ at the noise frequency can be deduced from the Bode diagram shown in Figure 3.9. At f = 100 Hz the system introduces an attenuation of 50 dB, corresponding to $|W_{vp,vo}|$ = 0.0032, so that the second contribution to the step amplitude needed for compensating the noise is $\Delta v_{ref,\Delta V_o} = |W_{vp,vo}| \cdot \Delta V_o = 0.0095\text{V}$.

Finally, from (3.22), a value of $\Delta v_{ref} > 0.0741\text{V}$ must be chosen; in the simulation Δv_{ref} = 100 mV is used.

The comparison between $\Delta v_{ref,\dot{G}}$ and $\Delta v_{ref,\Delta V_o}$ reveals that in the feedback-compensated system the second contribution is almost negligible.

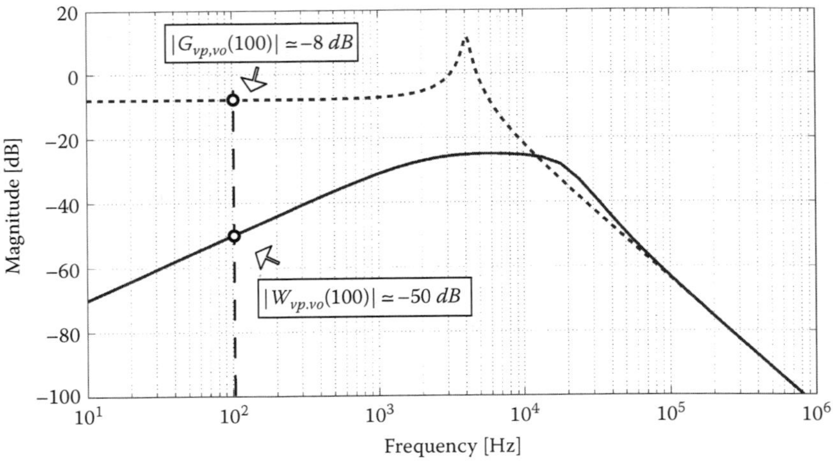

FIGURE 3.9
Bode diagrams (magnitude) of output-to input transfer functions. Dashed line: $G_{vp,vo}(s)$; continuous line: $W_{vp,vo}(s)$. Both diagrams have been evaluated at a low irradiance level, $G = 100\text{W/m}^2$.

The Bode diagram of $G_{vp,vo}(s)$ reported in Figure 3.9 highlights the modification of Δd due to the 100 Hz noise in the scheme analyzed in Section 2.4.3. At 100 Hz the attenuation is 8 dB, corresponding to $|G_{vp,vo}| = 0.4$.

Settling the minimum value for $T_p = 2.5$ ms and applying (2.36) provide the new Δd. The value $\Delta d_{\dot{G}} = 0.012$ comes from the analysis carried out in Section 2.4.3, while the noise term is $\Delta d_{\Delta V_o} = 0.1$. The latter value reveals that the configuration proposed in Section 2.4.3 cannot be used in the presence of

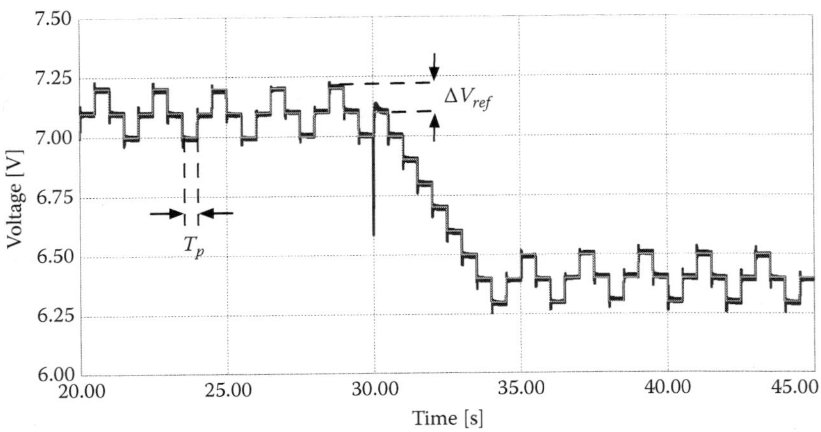

FIGURE 3.10
Steady-state operating conditions of the PV system. Black curve: PV voltage; grey curve: reference voltage V_{ref} imposed by the P&O algorithm with the relative parameters T_p and ΔV_{ref}.

the 100 Hz noise because it would require a step perturbation that is too high and not feasible by a practical point of view.

The circuit in Figure 3.6 has been simulated at different irradiance levels: G = 1000 W/m^2 for t < 25 ms, and G = 100 W/m^2 for t > 30 ms. Figure 3.10 shows the behavior of the PV voltage as a function of the voltage reference given by the P&O controller. As explained in Section 2.4.3, the MPPT algorithm, in both stationary irradiance conditions, imposes a three-level oscillation that is confined around the MPP. The transient after the irradiance variation is due to the fact that the MPPT algorithm adjusts the operating point by driving it on a different PV characteristic.

In order to appreciate the benefit of the compensation network, the behavior of the proposed closed-loop system has been compared with the open-loop system designed in Section 2.4.3. In both systems a 100 Hz voltage oscillation of amplitude ΔV_c = 3 V has been superimposed on the converter output voltage. Figure 3.11 shows the effect of such an oscillation

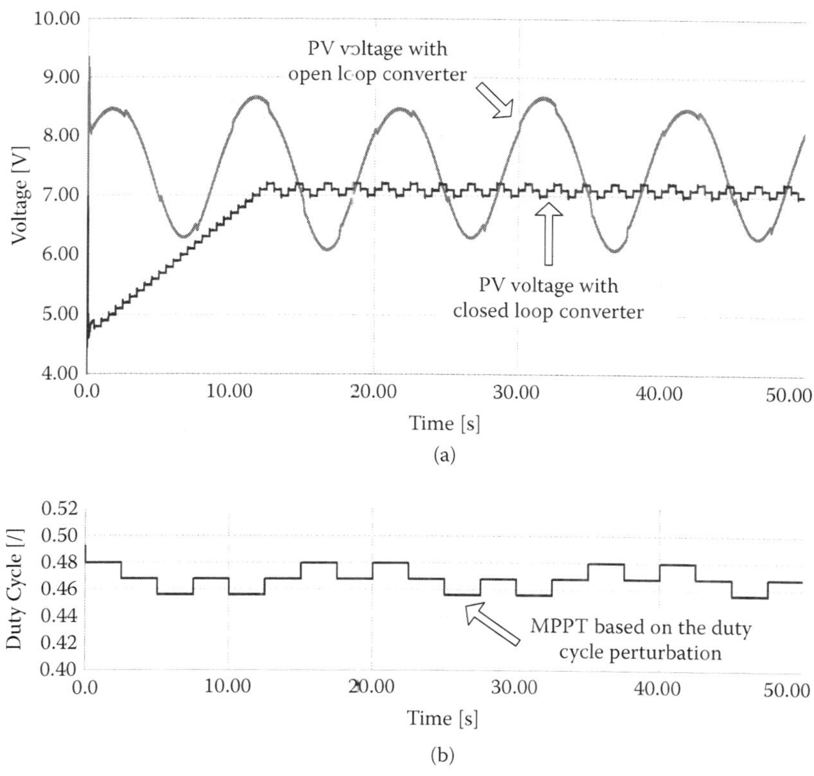

FIGURE 3.11
Steady-state operating conditions with an irradiance level of G = 1000 W/m^2. (a) PV voltage in the closed-loop and open-loop systems. (b) Output of the P&O algorithm acting on the duty cycle of the open-loop system.

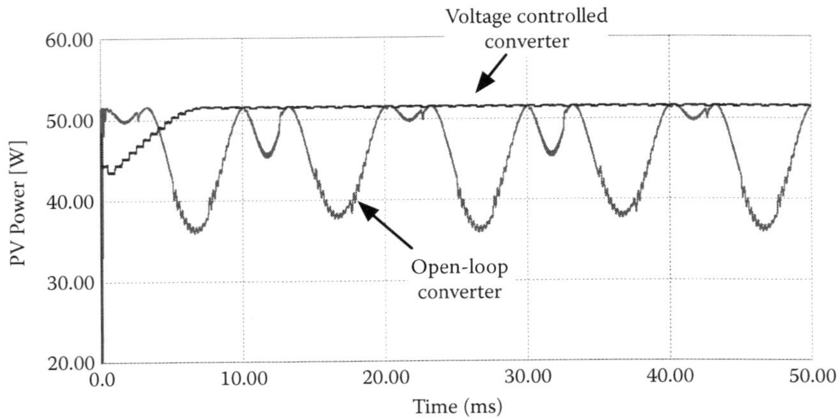

FIGURE 3.12
Instantaneous PV power for a fixed irradiance value of $G = 1000$ W/m^2.

on the PV voltage. While the closed-loop system does not exhibit a significant PV voltage variation due to the 100 Hz oscillation, the open-loop system is affected by two drawbacks. The first is the poor rejection capability of the converter, which transfers almost totally the output oscillation to the PV terminals. The second is the wrong behavior of the P&O algorithm: The duty cycle is updated with a sequence that does not look like a three-step stairs, as shown in Figure 3.11b. Finally, Figure 3.12 shows that the power generated by the PV source in the close-loop system is very close to the maximum.

Moreover, the noise rejection capability of the closed-loop system is effective not limited to 100 Hz only, as it extends to the whole range of frequencies below the crossover frequency. As shown in Figure 3.9, $W_{vp,vo}(s)$ is always much lower than 0 dB. The open-loop system, instead, shows a $G_{vp,vo}(s)$ that, in some cases, might be even higher than 0 dB, thus producing an amplification of the noise coming from the converter output, with a detrimental effect on the PV power production and on the MPPT operating conditions.

3.2 Instability of the Current-Based MPPT Algorithms

The largest part of the MPPT algorithms presented in literature and used in commercial products is based on the voltage-mode feedback control, or at least on a combination of control loops in which the final objective is to regulate the PV voltage. Indeed, the logarithmic dependency of the PV voltage on the irradiation level makes the MPPT algorithm based on the regulation of the PV voltage less sensible to the irradiance variation [13]. On the

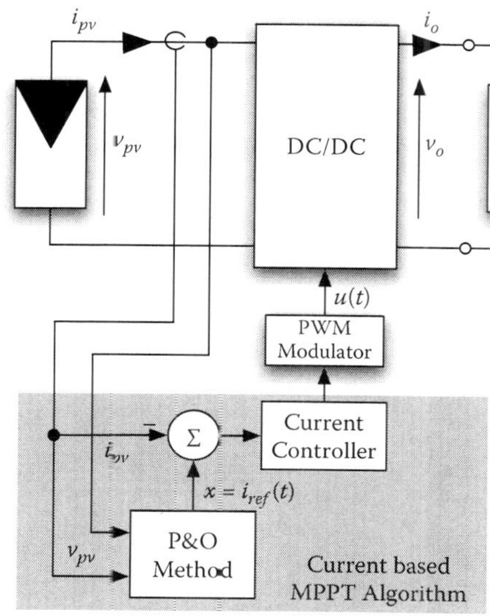

FIGURE 3.13
Control scheme of the current-based MPPT.

other side, the linear dependency of the PV current on the irradiance level would be very useful for a fast current-based MPPT, but the occurrence of irradiance drops might lead to the failure of configurations based on the direct regulation of the PV current. Figure 3.13 shows the basic scheme of a PV current regulator: In terms of MPPT control, it is the dual configuration of Figure 2.6b.

The difference in robustness between the voltage-mode-based and current-mode-based MPPT with respect to an irradiance drop is explained in Figures 3.14 and 3.15.

The former shows the I-V and P-V curves at three different irradiance levels. The voltage-based MPPT controller settles the best operating condition (the point marked with the star in Figure 3.14) at a given irradiance level by means of the condition $v_{ref} = V_{MPP(G_2)}$. In the presence of an irradiance variation ($\Delta G = \pm 100$ W/m^2), at a constant value of the reference voltage, the new steady-state PV operating point is given by the intersection of the vertical line of equation $v_{ref} = V_{MPP(G_2)}$ with the new I-V and P-V curves. Figure 3.14 shows that regardless of the sign of the irradiance variation, the new operating point is not so far from the new MPP. This means that during the transient, if the MPPT algorithm is not able to change v_{ref} promptly, the MPPT efficiency changes, but the system remains under control.

Figure 3.15 shows the same I-V and P-V curves, but with inverted axes: The PV voltage and the PV power are plotted vs. the PV current at three different irradiance levels. The optimal operating condition, marked with the

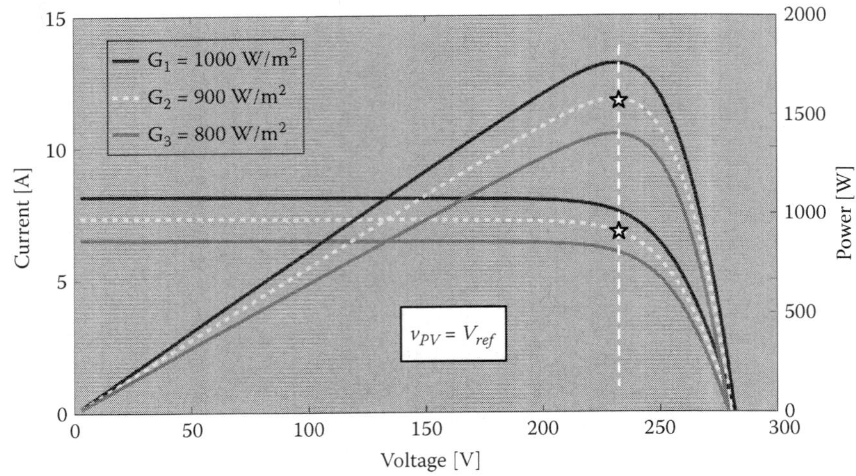

FIGURE 3.14
P-V and I-V curves at different irradiance levels: voltage-based MPPT.

star in Figure 3.14, is fixed at the same irradiance level of the previous case by the current-controlled MPPT technique by imposing $i_{ref} = I_{MPP(G_2)}$. Where $I_{MPP(G2)}$ corresponds to the current delivered by the PV field when it operates in MPP at the irradiance level G2. At this value of the reference current, in the presence of an irradiance variation, the PV operating point moves vertically. Figure 3.15 shows that an increase of irradiance moves the operating

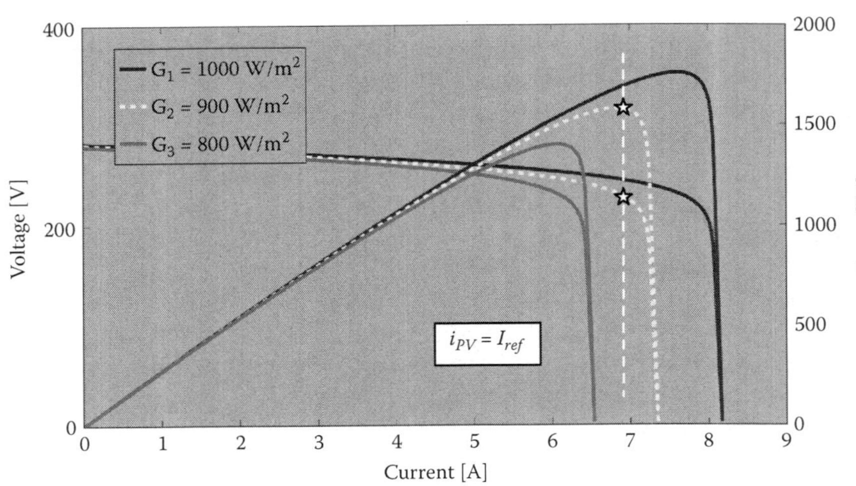

FIGURE 3.15
P-I and V-I curves at different irradiance levels: current-based MPPT.

condition far from the new MPP, without compromising the system stability. In fact, the PV field works in the region where the voltage increases slightly and the current is fixed at i_{ref}. The wrong behavior occurs in the presence of an irradiance drop: In this case the current-based control can cause a system instability because the PV curves, characterized by the irradiance G_3, are on the left side of the vertical line corresponding to the control law $i_{ref} = I_{MPP(G_2)}$; thus the only possible operating point is at zero voltage. The only way to avoid this voltage drop is to employ an extremely prompt MPPT controller that must be able to change the i_{ref} value very quickly.

The maximum irradiance variation that does not trigger this phenomenon in current-based MPPT is estimated as follows. If the optimal operating condition is $i_{ref} = I_{MPP(G_1)}$ at the irradiance level G_1, the minimum allowable irradiance $G_2 < G_1$ is given by the condition $i_{ref} = I_{MPP(G_1)} = I_{sc(G_2)}$, where $I_{sc(G_2)}$ is the PV short-circuit current at G_2.

The relationship between the short-circuit current (I_{SC}) and the irradiance level of a PV cell is given by $I_{sc(G)} = K_{ph} \cdot G$, and the ratio between the PV current in the MPP and the short-circuit current is almost constant for the different irradiance value. In [13, 14] it is shown that

$$\frac{I_{MPP(G)}}{I_{sc(G)}} \simeq \alpha \qquad \forall G \in [0,1000] \tag{3.26}$$

where the value of α usually belongs to the interval $[0.85 \div 0.95]$. Imposing

$$I_{MPP(G_1)} = i_{ref} = I_{sc(G_2)} \tag{3.27}$$

$$\alpha \cdot K_{ph} G_1 = i_{ref} = K_{pk} \cdot G_2 \tag{3.28}$$

and rewriting (3.28) in terms of relative irradiance variation lead to

$$\frac{G_1 - G_2}{G_1} = \frac{\Delta G}{G} = 1 - \alpha \tag{3.29}$$

By considering the typical value of α, the equation (3.29) reveals that a 5% ÷ 15% drop of the irradiance level can compromise the system stability of an MPPT algorithm performing a direct regulation of the reference current i_{ref}. This is the reason why very few examples of current-based MPPT approaches can be found in literature.

Some attempts have been done in order to improve the performance of the current-based approach by suitably modifying the basic structure of the control loop, so that the i_{ref} signal is promptly changed in the presence of a rapid irradiance variation. In [15] an inner current loop is coupled with an MPPT-oriented voltage loop to the aim of controlling a grid-connected inverter. In [16, 17] a hysteretic capacitor voltage control is used for the MPPT function in an energy harvesting system, with the main aim of reducing the power consumption of the control circuitry. A robust MPPT current-based architecture adopting the sliding mode control is described in much deeper detail in Section 3.3.

3.3 Sliding Mode in PV System

In the previous paragraphs it has been shown that in grid-connected PV systems, like the one depicted in Figure 3.16, the inverter operation generates an oscillation of the voltage of the bulk capacitor C_b at a frequency equal to twice the grid frequency, which propagates through the MPPT DC/DC converter and affects the photovoltaic voltage, thus degrading the MPPT efficiency.

In literature, solutions avoiding large DC-link capacitance C_b have been proposed. They are mostly based on complex architectures that penalize the efficiency and reliability [18]. The linear voltage-mode-based controllers used in [2] and in Section 3.1.1 specifically to reject the disturbances coming from the grid in some cases suffer a possibly not easy design of the compensation network to comply with worst-case conditions and limited robustness whenever both continuous conduction mode (CCM) and discontinuous conduction mode (DCM) operating conditions occur, e.g., during irradiance transitions from high to low levels.

This paragraph shows how the low-frequency ripple mitigation can be achieved by means of the sliding mode control (SMC). Such a nonlinear control technique can be very effective in ensuring extreme robustness and fast response, not only in applications involving switching converters [19–21], but also in more complex power electronic systems [22, 23]. SMC is based on the

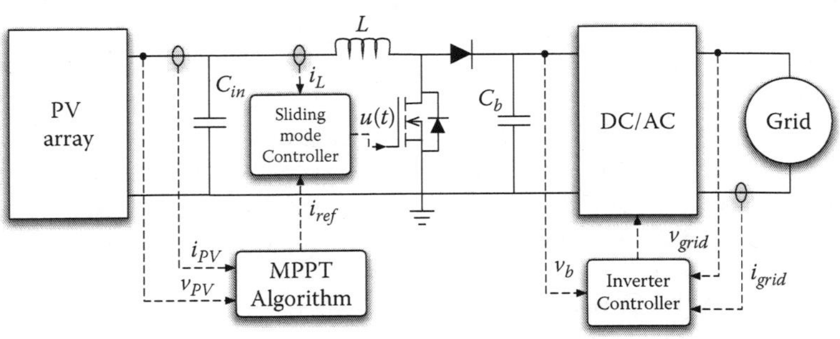

FIGURE 3.16
MPPT P&O with direct perturbation of the reference current.

implementation of a control equation, which forces the system variables to stay on a selected surface, called sliding surface.

Many publications available in the literature discuss the SMC. Those ones focused on photovoltaic applications are few and mainly devoted to the control of the DC/AC stage for regulating the current injected into the grid [24, 25].

In this section, the SMC will be used with the specific objective to regulate the PV source in order to remove the 100 Hz disturbance coming from the converter output.

Without loss of generality, the proposed analysis will be carried out by considering the scheme of Figure 3.16. It has been assumed that the first stage of a grid-connected photovoltaic system is a step-up converter used in order to increase the PV voltage to a suitable value that allows the operability of the DC/AC stage. The choice of boost-derived topologies is very common because they provide the additional benefit to sink a continuous current at its input, thus allowing us to minimize the input capacitance.

For the proposed topology, the regulation of the PV field can be easily performed if the SMC is used to control the average input current of the boost converter by imposing the following sliding equation:

$$\Psi = i_L - i_{ref} = 0 \tag{3.30}$$

where the reference current i_{ref} can be provided by a classical perturbative MPPT controller. Figure 3.17 shows a possible implementation of the SMC with two comparators and a flip-flop. Such a simple scheme can be directly used to generate the signals driving the MOSFET gates:

$$\begin{cases} i_L < i_{ref} - \dfrac{H(t)}{2} & turn\ ON\ MOSFET \\ i_L > i_{ref} + \dfrac{H(t)}{2} & turn\ OFF\ MOSFET \end{cases} \tag{3.31}$$

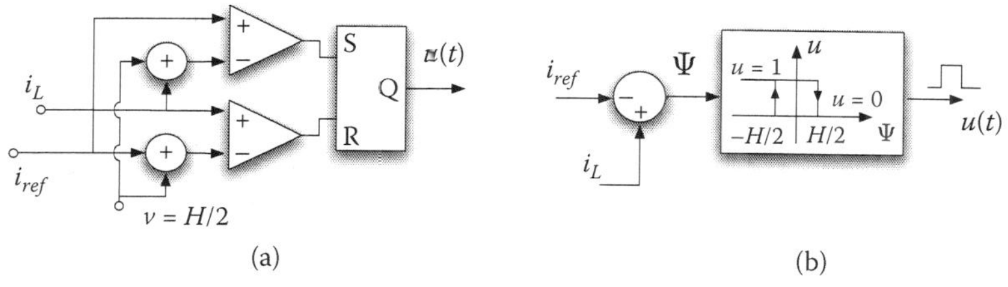

FIGURE 3.17
Sliding mode controller. (a) Practical implementation. (b) Logic scheme.

In steady-state conditions, $H(t)$ fixes the peak-to-peak amplitude of the inductor current ripple and represents the SMC hysteresis band, while i_{ref} fixes its average value so that such a control will be identified as sliding mode current control (SMCC).

The analysis of the boost converter leads to the following differential equations:

$$\begin{cases} L\dfrac{di_L}{dt} = v_{PV} - v_b \cdot (1-u) \\ C\dfrac{dv_{PV}}{dt} = i_{PV} - i_L \end{cases} \tag{3.32}$$

where $u=1$ means MOSFET turned ON and $u=0$ means MOSFET turned OFF.

The necessary and sufficient conditions for local surface reachability are [19–21]

$$\begin{cases} lim_{\Psi\to 0^-}\dfrac{d\Psi}{dt} > 0 \quad u=1 \\ lim_{\Psi\to 0^+}\dfrac{d\Psi}{dt} < 0 \quad u=0 \end{cases} \tag{3.33}$$

From Equation (3.30), in steady-state condition (i_{ref} = constant) the time derivative of the sliding equation is

$$\frac{d\Psi}{dt} = \frac{di_L}{dt} \tag{3.34}$$

Merging Equation (3.33) with (3.34) and using the first row of (3.32) provides

$$\begin{cases} lim_{\Psi\to 0^-}\dfrac{d\Psi}{dt} = \dfrac{v_{PV}}{L} > 0 \\ lim_{\Psi\to 0^+}\dfrac{d\Psi}{dt} = \dfrac{v_{PV} - v_b}{L} < 0 \end{cases} \tag{3.35}$$

As in the boost converter, it is $vb > v_{PV} > 0$; both the previous inequalities are fulfilled. The local stability can be verified by using the following conditions [19]:

$$\frac{d\Psi}{dt} = 0 \to 0 < u_{eq} < 1 \tag{3.36}$$

where u_{eq} represents an equivalent continuous control input that maintains the system evolution on the sliding surface.

Substituting the control input u with u_{eq} in Equations (3.30) and (3.32) allows us to write the condition (3.36) in the following way:

$$0 < u_{eq} = \frac{v_b - v_{PV}}{v_b} + \frac{L}{v_b}\frac{di_{ref}}{dt} < 1 \tag{3.37}$$

that is,

$$-\frac{v_b - v_{PV}}{L} < \frac{di_{ref}}{dt} < \frac{v_{PV}}{L} \tag{3.38}$$

where $-(v_b - v_{PV})/L$ corresponds to the inductor current slope when the MOSFET is OFF and v_{PV}/L corresponds to the inductor current slope when the MOSFET is ON. Therefore if the current reference slope is constrained to the inductor current slope, then correct operation of the SMCC is ensured.

In order to analyze the intrinsic capability of the SMCC to reject the noises coming from the output of the converter, the low-frequency equations of the switching converters will be used. These equations are the same used for analyzing the switching converters operating with the PWM modulation; such an approach is still valid because for each signal its average value has been evaluated on a switching period, so that the effect of the low-frequency variation can be analyzed. It is also worth noting that even if the analysis is carried out by focusing the attention on the 100 Hz oscillation of the output voltage, the results can be extended to any disturbances occurring at frequencies lower than 1/10 of the switching frequency, which is the range of validity of the average model [12].

In the following, we will label with $f_{sw}(t)$ the switching frequency (which is time varying in SMCC), with v_{b0} the average output voltage of the boost converter, and with $\Delta v_b(t)$ the instantaneous deviation of the average output voltage of the boost from v_{b0} due to the energy storage/release process performed by the bulk capacitor.

The relation among the above quantities is

$$L = \frac{v_{PV} \cdot D(t)}{H(t) \cdot f_{sw}(t)} \tag{3.39}$$

In the sequel, we will label with D_0 the following quantity:

$$D_0 = 1 - \frac{v_{PV}}{v_{b0}} \tag{3.40}$$

The SMC action imposed by Equation (3.31) introduces a natural correction on the duty cycle $D(t)$, which depends on the instantaneous conversion ratio of the converter only. Thus in order to avoid the propagation of the $\Delta v_b(t)$ oscillation on v_{PV}, the duty cycle changes according to the following equation:

$$D(t) = 1 - \frac{v_{PV}}{v_{b0} + \Delta v_b(t)} \tag{3.41}$$

In a PWM control system without any feedback loop, the duty cycle is fixed, and consequently in Equation (3.41), the PV voltage cannot remain constant in the presence of a variation on v_b.

The duty cycle fraction $\Delta D(t)$ needed to compensate the bulk capacitor voltage oscillation is obtained as follows:

$$D(t) = D_0 + \Delta D(t) = 1 - \frac{v_{PV}}{v_{b0} + \Delta v_b(t)} \tag{3.42}$$

$$1 - \frac{v_{PV}}{v_{b0}} + \Delta D(t) = 1 - \frac{v_{PV}}{v_{b0} + \Delta v_b(t)} \tag{3.43}$$

$$\Delta D(t) = \frac{v_{PV}}{v_{b0}} \cdot \frac{\Delta v_b(t)}{v_{b0} + \Delta v_b(t)} \tag{3.44}$$

If the waveform $\Delta v_b(t)$ is equal to

$$\Delta v_b(t) = \Delta v_b \cdot \sin(4\pi \cdot f_{grid} \cdot t) \tag{3.45}$$

where the amplitude Δv_b is given by Equation (3.3), then from (3.44) and (3.45) it is possible to get

$$\Delta D(t) = \frac{v_{PV}}{v_{b0}} \cdot \frac{\Delta v_b \cdot \sin(4\pi \cdot f_{grid} \cdot t)}{v_{b0} + \Delta v_b \cdot \sin(4\pi \cdot f_{grid} \cdot t)} \tag{3.46}$$

If the inductor current ripple is constant and equal to H_0, the switching frequency is given by

$$f_{sw}(t) = \frac{v_{PV} \cdot D(t)}{L \cdot H_0} \tag{3.47}$$

In the sequel we will label with f_{sw0} the nominal value of f_{sw}:

$$f_{sw0} = \frac{v_{PV} \cdot D_0}{L \cdot H_0} \tag{3.48}$$

The switching frequency oscillation $\Delta f_{sw}(t)$ can be found by considering that

$$f_{sw0} + \Delta f_{sw}(t) = \frac{v_{PV} \cdot (D_0 + \Delta D(t))}{L \cdot H_0} \tag{3.49}$$

$$\Delta f_{sw}(t) = \frac{v_{PV}}{L \cdot H_0} \cdot \Delta D(t) \tag{3.50}$$

The previous equation shows that the relative oscillation $\Delta D(t)/D_0$ of the duty cycle corresponds to the relative oscillation $\Delta f_{sw}(t)/f_{sw0}$ of the switching frequency. From Equation (3.46), it is also evident that the duty cycle oscillation does not depend on converter parameters, while the switching frequency oscillations due to the bulk oscillation are related to the inductance and the hysteresis band, as shown in Equation (3.50).

Figure 3.18 shows the waveforms of $D(t)$ and $f_{sw}(t)$ evaluated by means of Equation (3.46) and by using the values reported in Table 3.1.

The main drawback of SMCC is represented by the variability of the switching frequency, which is due to the variation introduced by the DC-link voltage oscillations, which can be estimated with (3.50) and limited with an appropriate selection of the converter parameters, as to the occurrence of DCM operation. In fact, when the current reference is lower than $H/2$ the lower boundary saturates to zero, increasing in this way the switching frequency to nonsustainable limits. To solve this problem, it is possible to adopt the burst mode (BM) operation [26, 27] when $i_{ref} < H/2$. In this way, it is pos-

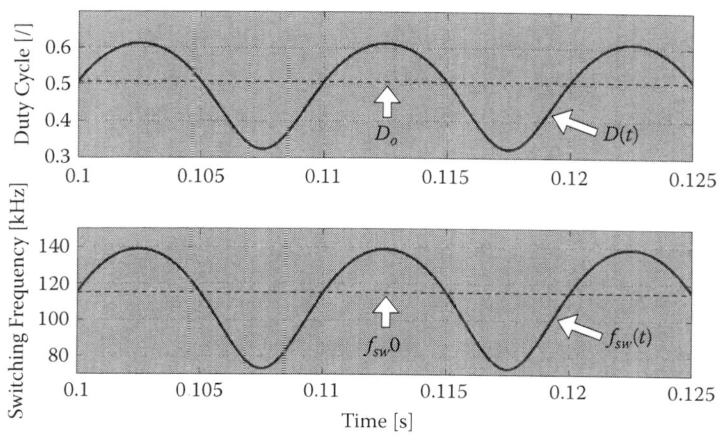

FIGURE 3.18
Low-frequency variation of duty cycle and switching frequency required to compensate the voltage oscillation of v_b.

TABLE 3.1

List of Parameters Used in the Simulation

Parameter	Value
v_{PV}	225 V
v_{b0}	460 V
f_{grid}	50 Hz
f_{sw0}	120 kHz
$H_0(HysteresisBand)$	2 A
$P_{pv} \simeq P_{MPP}$	≃1800 W
Δv_b (DC-link voltage oscillation)	≃127 V
L	500 µH
C_{in}	50 µF
C_b	30 µF

sible to get the desired average input current by decreasing the switching frequency rather than increasing it.

Of course, the drawbacks associated to DCM can be overcome by using a synchronous version of the boost converter. In such a case, the boost converter never enters DCM. Another advantage of the synchronous version is represented by its higher efficiency; on the other hand, it is characterized by a higher cost, and also by the need of more complex high-side gate drivers [12].

3.3.1 Noise Rejection by Sliding Mode: Numerical Example

The SMCC has been tested by implementing in the PSIM® circuit-based simulator the scheme of Figure 3.16 [28]. The results have been obtained by considering an irradiance value of $G = 1000$ W/m^2 and ambient temperature of 25°C. The parameters of the DC/DC converter are shown in Table 3.1.

Figure 3.19 reveals the capability of the SMCC to reject the oscillations of v_b; although the SMCC is implemented by means of a very simple circuit (Figure 3.17a), the large oscillation appearing on the bulk voltage is almost totally removed from the PV voltage, which is characterized by a DC value only.

Figure 3.20 put in evidence that the duty cycle oscillation (Figure 3.20a) and the switching frequency oscillation (Figure 3.20b) match the analytically predicted behavior of Figure 3.18, thus confirming that the bulk voltage oscillation is translated in a switching frequency oscillation and in an equivalent duty cycle variation, which ensure an almost constant PV voltage.

Unfortunately the pure SMCC suffers stability problems in the presence of fast and large variation in the irradiance conditions. Figures 3.21 and 3.22 show the simulation of the system in Figure 3.16, in which the P&O MPPT algorithm has been used to perturb the reference current i_{ref}. The P&O parameters have been settled to the following values: $T_p = 15$ ms and $\Delta i_{ref} = 0.1$ A.

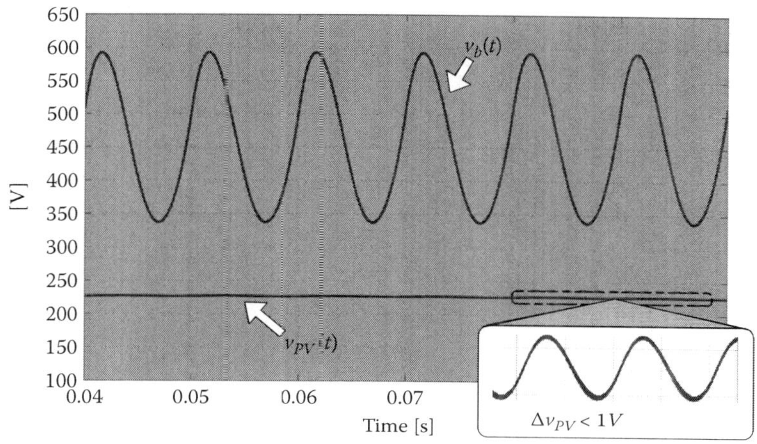

FIGURE 3.19
Waveforms of a PV voltage and bulk voltage.

Two different amplitudes of the irradiance drop have been considered: $\Delta G = 50$ W/m^2 and $\Delta G = 100$ W/m^2. In the first case, shown in Figure 3.21, at $t = 0.3$ s the irradiance changes from 1000 to 950 W/m^2, and the system experiences a transient after which it is able to reach a new steady-state condition. In the second case, at $t = 0.3$ s, the irradiance changes from 1000 W/m^2 to 900 W/m^2, and the system crashes because, after the step irradiance variation, the reference current provided by the MPPT controller is higher than the short-circuit current associated to the new irradiance condition, as shown in Figure 3.22. In this case the higher-threshold $i_{ref} + H/2$ is never reached, creating a sort

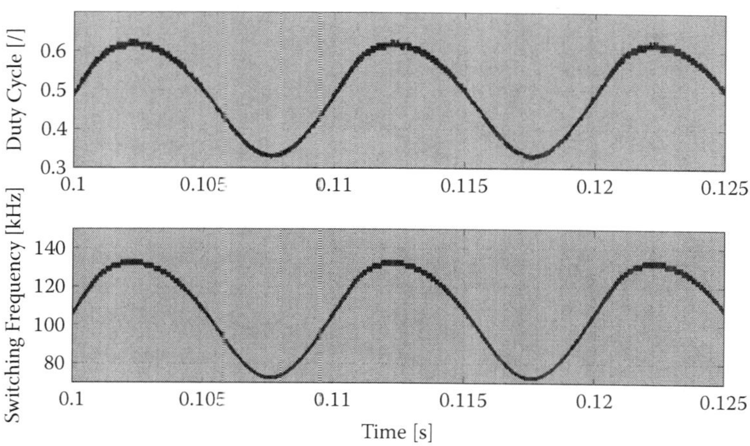

FIGURE 3.20
Duty cycle and switching frequency waveforms in the PSM simulation.

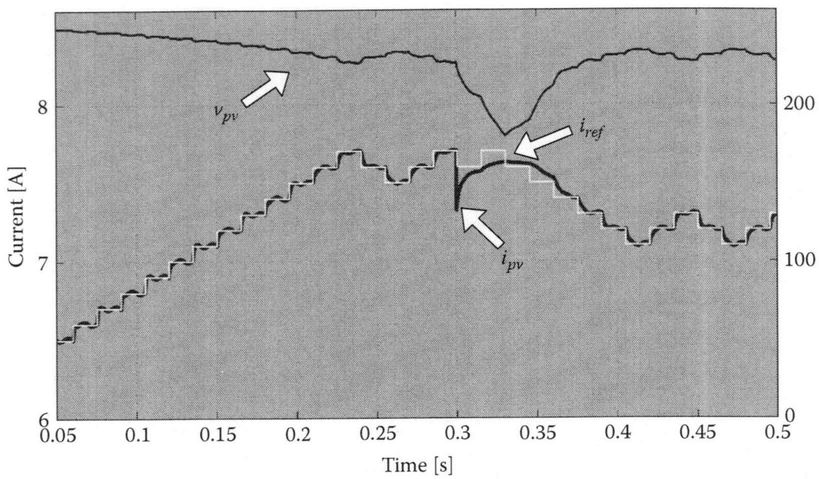

FIGURE 3.21
PV voltage and PV current in the presence of a small irradiance drop.

of stall in the SMCC control. This means that the "hitting condition," which represents one of the basic constraints in order to achieve SMC, has not been fulfilled [20].

The wrong behavior in the presence of negative step variations of the irradiance not only affects the SMCC but also occurs in other direct current control strategies without the voltage loop, and it is due to the linear dependency of the PV current with respect to the irradiance. Thus no stability condition is

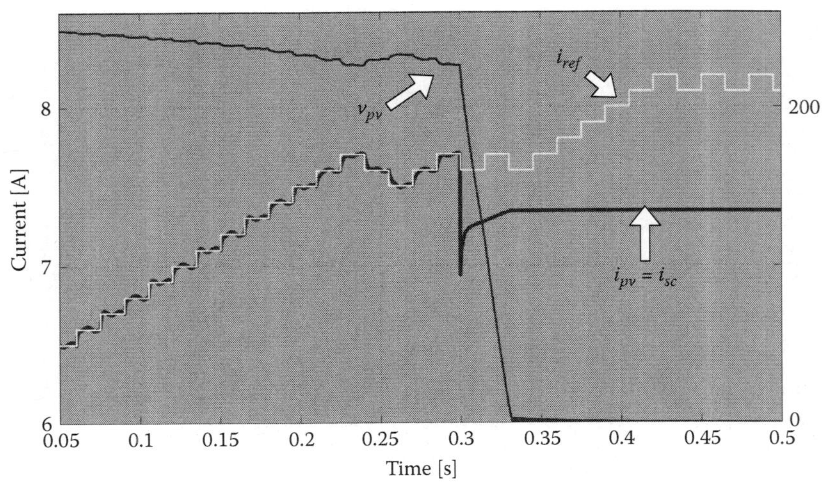

FIGURE 3.22
PV voltage and PV current in the presence of a large irradiance drop.

ensured when the PV field works in the region where its behavior is similar to that of a current generator, as explained in Section 3.2.

3.3.2 MPPT Current Control by Sliding Mode

A sliding mode current-based P&O MPPT strategy is described in this section, able to track very promptly the variations in the irradiance value and, additionally, to guarantee the rejection of the low-frequency voltage oscillations affecting the PV array due to the inverter operation in grid-connected systems. The main peculiar aspect of the approach discussed in this section lies in the fact that it is a current-based technique, joining the benefits of a sliding mode inner current loop with the well-known advantages of an outer voltage-based control loop. The inner loop is based on the sensing of the current flowing into the capacitor, which is usually put in parallel with the PV generator. Although the main role of the input capacitor C_{in} consists in bypassing the switching ripple of the inductor current, in the SMC application herein discussed the PV current control can be obtained by regulating the instantaneous value of the capacitor current. Further details concerning this architecture have been shown in [29–31].

3.3.2.1 Basic Configuration of Sliding Mode with Voltage Controller

Figure 3.23 shows a possible control strategy operating on the inductor current of the DC/DC converter in order to regulate the PV current value. The current reference i_{Lref} of the inductor current controller is given by

$$i_{Lref} = i_{PV} + i_{vr} \tag{3.51}$$

where i_{PV} is the actual PV current and i_{vr} is the output of the voltage controller. The feed-forward action on i_{PV} considered in (3.51) helps in promptly detecting the irradiation variations and avoid the crashing problems described in the previous paragraph. In steady-state conditions, it is $i_{PV} = i_{Lref}$ and $i_{vr} = 0$.

According to the classical sliding mode DC/DC converter current control theory, the signal $u(t)$ driving the MOSFET is a function of the sliding equation $\Psi = i_{Lref}(t) - i_L(t) = 0$, so that the inductor current i_L is regulated according to the current reference signal i_{Lref}.

At the circuit node connecting the PV array, the boost converter input inductor L and capacitor C_{in} (see Figure 3.23), it is

$$i_{PV} = i_{C_{in}} + i_L \tag{3.52}$$

From (3.52) and (3.51), the current reference i_{Lref} can be expressed as

$$i_{Lref} = i_{C_{in}} + i_L + i_{vr} \tag{3.53}$$

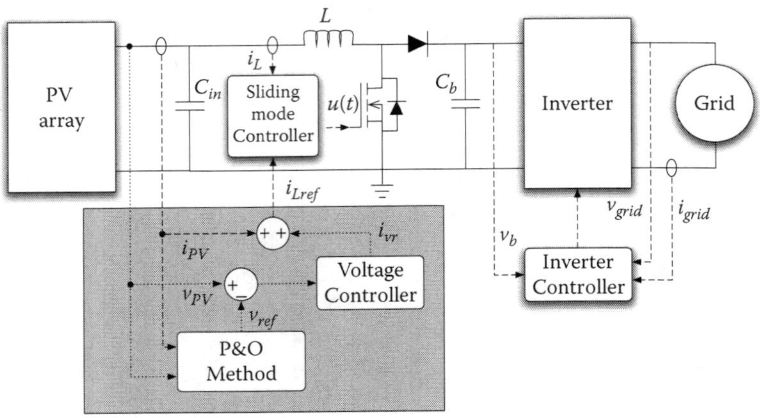

FIGURE 3.23
Sliding mode architecture for compensating the fast PV current variation.

Using the previous expression in the sliding control equation, $\Psi = i_{Lref}(t) - i_L(t) = 0$, leads to the following simplification:

$$\Psi = i_{C_{in}} + i_L + i_{vr} - i_L = 0 \tag{3.54}$$

$$\Psi = i_{C_{in}} + i_{vr} = 0 \tag{3.55}$$

The simplified control objective (3.55) reveals that the control structure of Figure 3.23 can be simplified as in Figure 3.24, with the inner control loop, which is now aimed at regulating the input capacitance current $i_{C_{in}}$. This simplification is important because the practical implementation of the scheme shown in Figure 3.23 would require current sensors with too wide bandwidth for i_{PV} and i_L, both affecting the value of the MOSFET control signal $u(t)$. Instead, in the scheme of Figure 3.24, $u(t)$ only depends on the instantaneous value of $i_{C_{in}}$. Consequently, only one wide-bandwidth current sensor is required because the other one, dedicated to i_{PV}, can be narrow bandwidth. Additionally, sensing the capacitor current is easier than for the inductor current due to the fact that the former has a zero DC component. Another advantage of the system shown in Figure 3.24 consists in its easier analysis with respect to that depicted in Figure 3.23.

The definition of the sliding surface Ψ as in (3.55) suggests that in sliding mode operation, the capacitor current $i_{C_{in}}$ changes in order to reject the perturbations on the bulk capacitor voltage v_b and to track the perturbations in the irradiance level. Thus the fact that Ψ does not depend on those variables

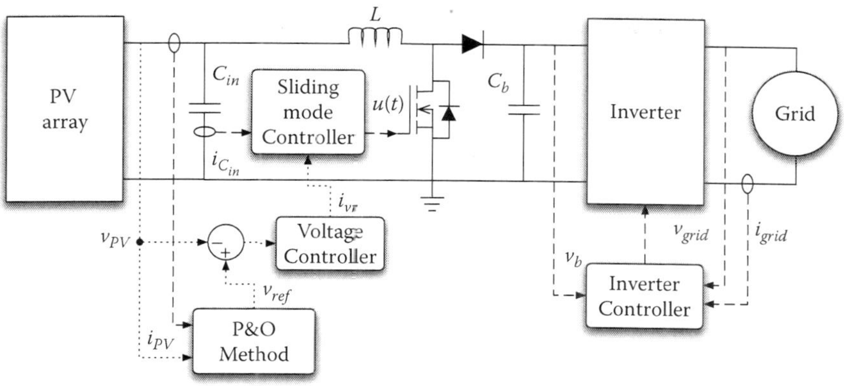

FIGURE 3.24
System scheme based on input capacitor current control.

ensures the MPPT operation and the rejection of the low-frequency disturbances coming from the output of the DC/DC converter.

Two conditions must be fulfilled in order to ensure the sliding mode operation [21]:

$$\Psi = 0 \tag{3.56}$$

$$\frac{d\Psi}{dt} = 0 \tag{3.57}$$

From the first condition (3.56), also accounting for the characteristic equation of the input capacitor,

$$i_{C_{in}} = C_{in} \cdot \frac{dv_{PV}}{dt} \tag{3.58}$$

the following condition is obtained:

$$\frac{dv_{PV}}{dt} = -\frac{i_{vr}}{C_{in}} \tag{3.59}$$

From the second sliding mode condition (3.57), also considering that $i_{C_{in}} = i_{PV} - i_L$, we get

$$\frac{d\Psi}{dt} = \frac{di_L}{dt} - \frac{di_{PV}}{dt} - \frac{di_{vr}}{dt} = 0 \tag{3.60}$$

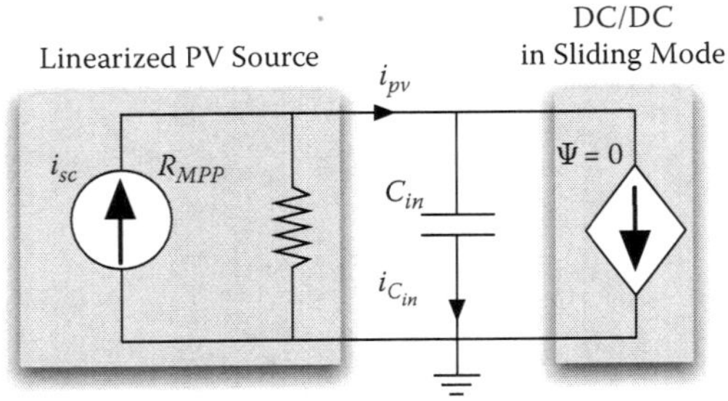

FIGURE 3.25
Small-signal model of the open-loop system.

The analysis of Equation (3.60) can be done by using the linearized model shown in Figure 3.25, wherein the switching converter is represented by a controlled current source whose value is imposed by the SM equation, and the PV generator is given as a Norton model including the photo-induced current source and the differential resistance R_{MPP}, which is calculated in the generator's operating point. By looking at Figure 3.25, the result is

$$\frac{di_{PV}}{dt} = -\frac{1}{R_{MPP}} \cdot \frac{dv_{PV}}{dt} + \frac{di_{sc}}{dt} \tag{3.61}$$

As a result, the sliding mode condition given in (3.60) can be thus rewritten as

$$\frac{di_L}{dt} + \frac{1}{R_{MPP}} \cdot \frac{dv_{PV}}{dt} - \frac{di_{sc}}{dt} - \frac{di_{vr}}{dt} = 0 \tag{3.62}$$

In addition, the boost converter model provides the following relationship:

$$\frac{di_L}{dt} = \frac{v_{PV}}{L} - \frac{v_b \cdot (1-u)}{L} \tag{3.63}$$

where v_b is the DC/DC converter output voltage and the value $u = 1$ is used in the MOSFET ON state and the value $u = 0$ in the OFF state, corresponding to the following driving conditions:

$$u = 1 \text{ if } \Psi < 0 \tag{3.64}$$

$$u = 0 \text{ if } \Psi > 0 \tag{3.65}$$

In order to obtain the constraints that must be fulfilled to ensure the SM operation, the equivalent control technique [19] can be referred to.

The constraints on the inputs and state variables are defined by ensuring that the average control signal u_{eq} fulfills the inequality $0 < u_{eq} < 1$. From Equations (3.59)–(3.63) the following equivalent control equation is obtained:

$$\frac{v_{PV}}{L} - \frac{v_b \cdot (1 - u_{eq})}{L} + \frac{1}{R_{MPP}} \cdot \frac{dv_{PV}}{dt} - \frac{di_{sc}}{dt} - \frac{di_{vr}}{dt} = 0 \tag{3.66}$$

From (3.55) and (3.60) it can be deduced that the sliding mode equilibrium point is defined by $\{i_{PV} = i_L, i_{vr} = 0\}$. In (3.66), at the equilibrium point, it is $\frac{dv_{PV}}{dt} = 0$; thus the constraints in terms of maximum slope values $\frac{di_{vr}}{dt}$ and $\frac{di_{sc}}{dt}$ are given by

$$\frac{v_{PV} - v_b}{L} < \frac{di_{vr}}{dt} + \frac{di_{sc}}{dt} < \frac{v_{PV}}{L} \tag{3.67}$$

According to the superposition principle $\left(\frac{di_{vr}}{dt} = 0\right)$ and the linear relation among the short-circuit current and the irradiance $i_{sc} = K_{ph} \cdot G$, Equation (3.67) gives the constraints for the maximum irradiance variation rate ensuring the sliding mode operation:

$$\frac{v_{PV} - v_b}{L \cdot K_{ph}} < \frac{dG}{dt} < \frac{v_{PV}}{L \cdot K_{ph}} \tag{3.68}$$

It is worth noting that the lower bound in (3.68) is negative because of the adoption of a boost converter. This means that the larger the converter voltage boosting factor, the faster the negative irradiance variation can be tracked without losing the sliding mode behavior. This might be the case of a step-up DC/DC converter used in PV module-dedicated microinverter applications, in which the low PV module voltage must be boosted up very much, so that the quantity $\frac{v_{PV} - v_b}{L}$ is deeply negative. Inequality (3.68) reveals that the maximum irradiance variation that can be tracked without losing the sliding mode control depends also on the inductance value; thus L should be properly selected in order to follow the expected irradiance profile and variations, according to the specific applications. For instance, stationary PV power plants will be subjected to slow irradiance variations, while in PV applications dedicated to sustainable mobility fast irradiance variations need to be tracked.

Moreover, putting $\frac{di_{sc}}{dt} = 0$ in (3.67) allows us to take into account the effect of variations on i_{vr}; in particular, the following constraint on $\frac{di_{vr}}{dt}$ is obtained:

$$\frac{v_{PV} - v_b}{L} < \frac{di_{vr}}{dt} < \frac{v_{PV}}{L} \tag{3.69}$$

which again shows that the maximum i_{vr} slope value that can be tracked without missing the sliding mode control depends on the inductor current derivatives in the OFF and ON MOSFET states. In this case, the voltage controller can be designed to fulfill this dynamic constraint.

3.3.2.2 Voltage Controller Design

In sliding mode operation, Equations (3.55) and (3.59) give the time domain relation between the PV voltage and the reference current; thus it allows us to design the voltage compensator. The corresponding transfer function $G_{v/i}(s)$ between the input capacitor voltage, which is the PV voltage, and the current reference i_{vr}, is

$$G_{vp,i_{vr}}(s) = \frac{v_{PV}(s)}{i_{vr}(s)} = -\frac{1}{C_{in} \cdot s} \tag{3.70}$$

This expression allows us to point out one of the main features of the proposed control technique. Indeed, it is worth noting that the transfer function (3.70) is not dependent on any PV generator's parameter, so that the control approach can have the same performances regardless of the environmental conditions and of the PV array type/size connected at the DC/DC converter's input terminals. This feature is not achievable through classical feedback control. Indeed, as shown in the example of Section 3.1.2 and remarked in [2], the closed-loop system must be designed by considering the worst-case conditions. Furthermore, in the range of validity of the sliding mode conditions, Equation (3.70) is valid without the assumption of the small-signal approximation. Of course the equivalent series resistance (ESR) of the input capacitor changes Equation (3.70) by introducing an additional zero in the loop gain transfer function. In order to avoid that such a zero affects the controller design, capacitors with low ESR, e.g., ceramic capacitor, are recommended for practical implementation. Anyway, the presence of the ESR does not affect the features commented on above.

A PI compensator can be used in the voltage controller loop of Figure 3.24. Given the PI transfer function $G_v(s)$ in (3.71) and the voltage error $e_v(s)$ definition (3.72) that compensates the negative sign appearing in (3.70), the closed-loop transfer function $T_{vp}(s)$ of the system can be expressed by (3.73)

$$G_v(s) = k_p + \frac{k_i}{s} \tag{3.71}$$

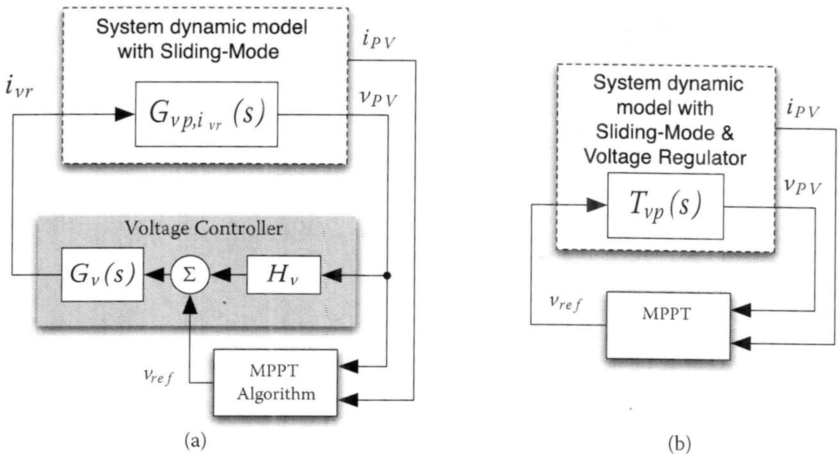

FIGURE 3.26
Dynamic model of the PV system based on input capacitor current SM control. (a) Double-loop representation. (b) Equivalent single-loop representation.

$$e_v(s) = -(v_{ref}(s) - H_v \cdot v_{PV}(s)) \tag{3.72}$$

$$T_{vp}(s) = \frac{v_{PV}(s)}{v_{ref}(s)} = \frac{k_p s + k_i}{C_{in}s^2 + H_v \cdot (k_p s + k_i)} \tag{3.73}$$

Figure 3.26 shows the dynamic models of the system under study. The transfer function $T_{vp}(s)$ (3.73) allows analysis of the behavior of the whole PV system in the presence of a step perturbation on the PV voltage reference. Thus as shown in the previous chapter, it can be designed by accounting for a desired MPPT speed.

3.3.3 Sliding Mode MPPT Controller: Numerical Example

The SMC on the input capacitor current has been tested by means of PSIM simulation, based on the scheme of Figure 3.24. The parameters of the DC/DC converter are the same as in the previous example and are reported in Table 3.1. The parameters of the PI voltage controller, $G_v(s)$, can be selected by designing the settling time t_ϵ of the transfer function $T_{vp}(s)$ in (3.73) according to Equation (2.22). Given the relationships between the settling time t_ϵ of the closed-loop voltage, the damping (ζ), and the natural frequency (ω_n) of the whole PV system, we get

$$T_{vp}(s) = \frac{as + b}{s^2 + 2\zeta\omega_n s + \omega_n^2} \tag{3.74}$$

The comparison of Equation (3.74) with Equation (3.73) gives the following relationships for the parameters of the transfer function $G_v(s)$:

$$k_p = \frac{2C_{in}\zeta\omega_n}{H_v} \tag{3.75}$$

$$k_i = \frac{C_{in}\omega_n^2}{H_v} \tag{3.76}$$

From Table 3.1, the nominal switching period is $T_{sw} \simeq 1/f_{sw0} = 8.3\ \mu s$. Choosing $t_\varepsilon / T_{sw} = 20$, which means imposing a system settling time 20 times greater than the nominal switching period, provides $t_\varepsilon = 166\mu s$. The second design consideration is related to the damping of the system, which can be settled at $\zeta = 0.7$.

According to the definition of the equivalent time constant, which is $\tau = \frac{1}{\zeta\omega_n}$, the settling time of the closed-loop system can be approximated by $t_\varepsilon = 4\tau$ when it is assumed that the settling error step response is smaller than 2%. Finally, if the gain of the voltage sensor is $H_v = 0.1$, then the PI controller transfer function ensuring the desired t_ε and ζ becomes the following:

$$G_v(s) = 24 \cdot \frac{s + 24625}{s} \tag{3.77}$$

In order to verify the stable behavior of the SM on the input capacitor current, the circuit of Figure 3.24 has been tested in the same irradiance condition of the circuit in Figure 3.17 and with the same MPPT parameters, although in this case the P&O algorithm is acting on the voltage reference and not directly on the current reference.

In Figure 3.27 the PV voltage and PV current are shown; an irradiance change from 1000 W/m^2 to 900 W/m^2 occurs at the time instant t = 0.3 s. The sudden irradiance variations, having an instantaneous effect on the PV short-circuit current, have been correctly tracked, with the system permanently tracking the maximum power point even at the very high rate of irradiance variation that has been considered. Indeed, comparing the voltage behavior with the simulation proposed in Figure 3.22, now no sensible variation in the PV voltage is visible and the system does not crash. Moreover, it is worth noting that also in this case, a bulk voltage oscillation of 127 V has been considered in the scheme of Figure 3.24; the simulation results put into evidence that the large low-frequency oscillations affecting the bulk voltage are almost totally rejected at the PV terminals. Moreover, the upper and lower limits of the irradiance variations $\frac{dG}{dt}$ have been calculated by using Equation (3.68) for the numerical example considered in this section. In the considered case, if $K_{ph} = 0.008\ \text{A}\cdot\text{m}^2/\text{W}$, it must be

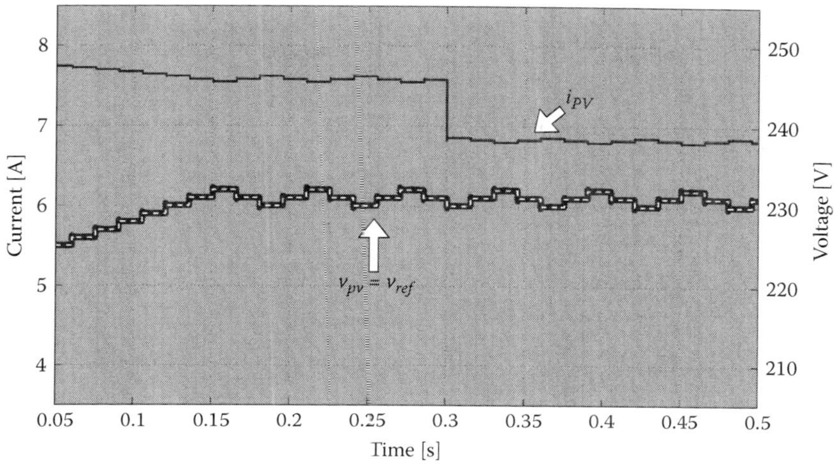

FIGURE 3.27
Simulation of the system of Figure 3.24 using the sliding mode and G_v controllers. MPPT parameters: $\Delta v_{ref} = 1$ V, $T_p = 15$ ms.

$$-58.75\frac{\mathrm{W}}{\mathrm{m^2ms}} < \frac{dG}{dt} < 56.26\frac{\mathrm{W}}{\mathrm{m^2ms}} \tag{3.78}$$

In Figure 3.28 an irradiance change from 1000 to 500 W/m^2 with a slope of $-50\frac{W}{m^2\mu s}$ has been applied at $t = 0.3$ s: clearly the system maintains an optimal stable behavior.

Figure 3.29 shows the magnification of the PV voltage in time interval (0.255 s, 0.325 s) in comparison with the bulk voltage. The simulation shows that the PV voltage waveform is free of the 100 Hz oscillations caused by the inverter operation, thus demonstrating that the sliding mode applied to the current of the input capacitor has been able to reject the back propagation of the large oscillations of amplitude Δv_b affecting the converter output voltage.

Finally, it is worth noting that in the simulation of Figures 3.27 and 3.28 a nonoptimal T_p MPPT parameter has been used; this is because the intent was to verify the stability of the SMC and not to test the MPPT dynamic performances. Anyway, the t_ε specification also allows us to define the optimal MPPT perturbation period T_p of the P&O algorithms, that is, approximately $1.5 \cdot t_\varepsilon$. Such a value of T_p ensures that the PV power has reached its steady state when the MPPT controller measures it, thus avoiding the MPPT deception as explained in Section 2.4.1 and in [2, 32]. Thus the case shown in Figure 3.28 has been resimulated by using $T_p = 250$ µs.

Figure 3.30 puts into evidence the proper design of the P&O parameters leading to a three-point behavior of the PV voltage with an MPPT speed that is about one order of magnitude higher than the case shown in Figure 3.28.

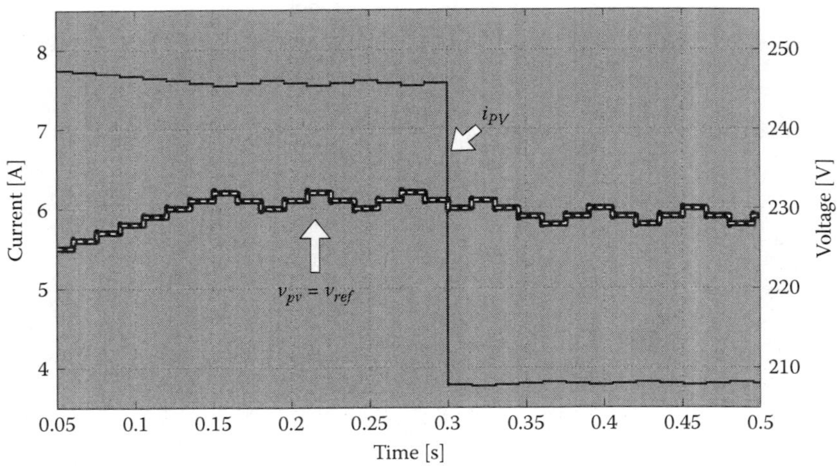

FIGURE 3.28
Simulation of the system of Figure 3.24 using the sliding mode control. MPPT parameters: Δv_{ref} = 1 V, T_p = 15 ms.

The same figure shows the waveform of the closed-loop voltage reference v_{ref} and the accurate tracking performed by both $G_v(s)$ and the sliding mode controller.

The features listed above make the control strategy shown in Figure 3.24 suitable for all PV applications in which the adoption of electrolytic capacitors at the DC bus must be avoided or wherever an excellent MPPT performance is required.

Moreover, the fast MPPT capability makes the proposed architecture suitable for a large range of applications for which the sudden irradiation changes are very common, e.g., in sustainable mobility and for the PV integration on cars, trucks, buses, ships, and so on.

3.4 Analysis of the MPPT Performances in a Noisy Environment

It is well known that the result of a measurement is only an approximation or estimation of the value of the specific quantity subject to measurement, that is, the *measurand*, and thus the measure is complete only when it is accompanied by a quantitative statement of the uncertainty affecting it.

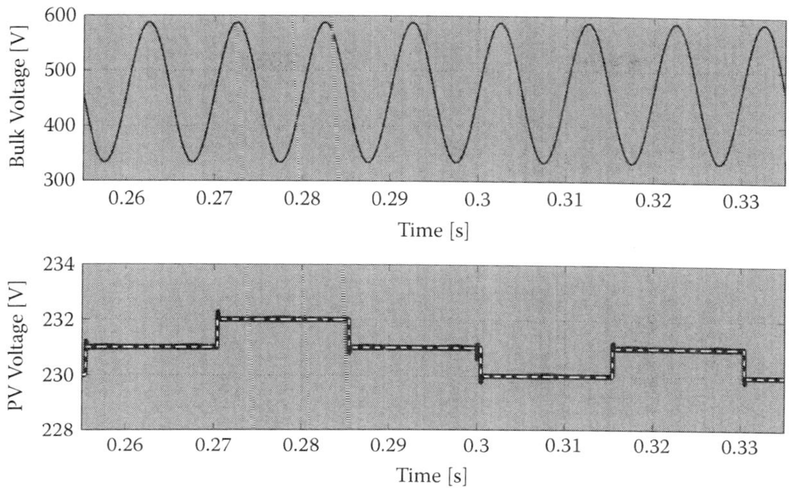

FIGURE 3.29
Bulk voltage and PV voltage.

The uncertainty is the entity that identifies the maximum deviation of a measure with respect to the true value, including stability, precision, resolution, and other factors involved in the measurement and elaboration process.

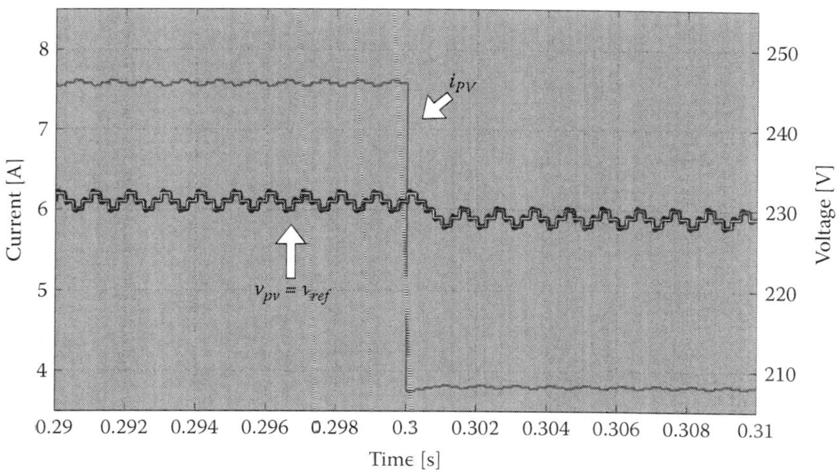

FIGURE 3.30
Simulation of the system of Figure 3.24 using the sliding mode control. MPPT parameters: Δv_{ref} = 1 V, T_p = 250 μs.

In whatever MPPT algorithms based on the measurement of the electrical or environmental parameters, the noises and the errors contribute to increase the uncertainty associated to the variable involved in the MPPT, and as a consequence, the decision process can be compromised.

Nonidealities of sensors and sensing amplifiers, analog-to-digital converter (ADC) resolution, and internal arithmetic quantization of the adopted microprocessor (representation of data) can result in measurement uncertainty and bias, which tricks the MPPT algorithm to settle away from the MPP, penalizing both the tracking capability and the steady-state settling point of the PV system. Moreover, the use of switching converters for controlling the operating point of the solar array increases the system noise due to the presence of high-frequency switching ripples.

It is evident that all the above sources of noise and uncertainty should be taken into account in the evaluation of the MPPT performances in order to identify possible solutions to mitigate their effect.

To date, there is not so much in scientific literature that specifically addresses the effect of noise on the MPPT algorithm; in [33] a comparison of convergence characteristics of different search strategies in the presence of noise is discussed in detail. In [34] the behavior of MPPT in a noisy environment has been analyzed by using a probabilistic model based on the statistical characteristics of the noise signal. Analysis presented therein highlights that noise in the voltage measurement shifts the settling point of the MPPT to the right-hand side of the characteristics curve, while noise in the current measurement reduces the MPPT tracking speed. Enhanced signal filtering and larger perturbations are found to be effective in building the system immunity to noise.

In this section, the behavior of the MPPT algorithms in a noisy environment will be analyzed by means of a similar approach. The method adopted exploits the concept of the uncertainty of a measure to characterize different types of errors. Indeed, being that uncertainty is a cumulative entity, it can be used for evaluating the global effect of errors introduced in the measurement and elaboration process and of noise sources on the MPPT performances.

If the noise sources, which are distributed in different places of the PV system, are expressed by means of an uncertainty value (**u**) with respect to the corresponding true value of the variable, the global uncertainty on a variable that is not directly measured can be estimated by applying the *law of propagation of uncertainty* [35].

Indeed, if a *measurand* y is not directly measured (e.g., the power in the MPPT algorithm), but rather is determined from N other quantities x_1, x_2, …, x_N through a functional relation f,

$$y = f(x_1, x_2, x_3, \ldots, x_N) \tag{3.79}$$

then the combined standard uncertainty of the measurement y, designated by $\mathbf{u}_y$, is given by

$$\mathbf{u}_y^2 = \sum_{i=1}^{N} \left(\frac{\partial f}{\partial x_i} \right)^2 \cdot \mathbf{u}_{x_i}^2 \tag{3.80}$$

where it has been assumed that all the input quantities x_i are independent.

In Figure 3.31 a typical MPPT converter is shown where several sources of uncertainty have been highlighted in grey.

It is worth noting that only the most significant noise sources have been reported; they have been indicated with the following uncertainty components: the switching ripple noise ($\mathbf{u}_{Switching}$), the measurement errors ($\mathbf{u}_v$, $\mathbf{u}_i$), the errors in numerical elaboration ($\mathbf{u}_E$), and the output voltage noise $(\mathbf{u}_{v_o})$. Depending on the type of uncertainty (errors in the measurement, noise coming from the switching converter), different actions must be performed on the system in order to mitigate their effect.

The method can be easily generalized for estimating the effect of other types of uncertainty and for other MPPT algorithms.

3.4.1 Noise Attenuation by Using Low-Pass Filters

Processing the measured signals by means of a low-pass filter (LPF) is the first method for attenuating the effects of the noise. This method, however, should be applied carefully in order to avoid suppressing useful information, destabilizing the MPPT control loop, or sacrificing its promptness. Indeed,

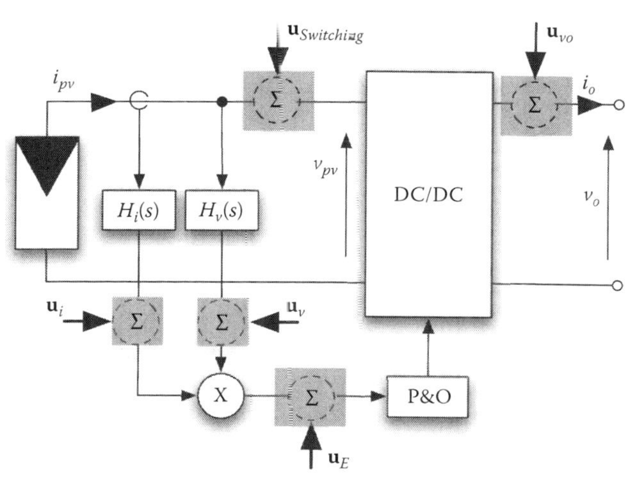

FIGURE 3.31
Uncertainty distribution in a P&O MPPT switching converter.

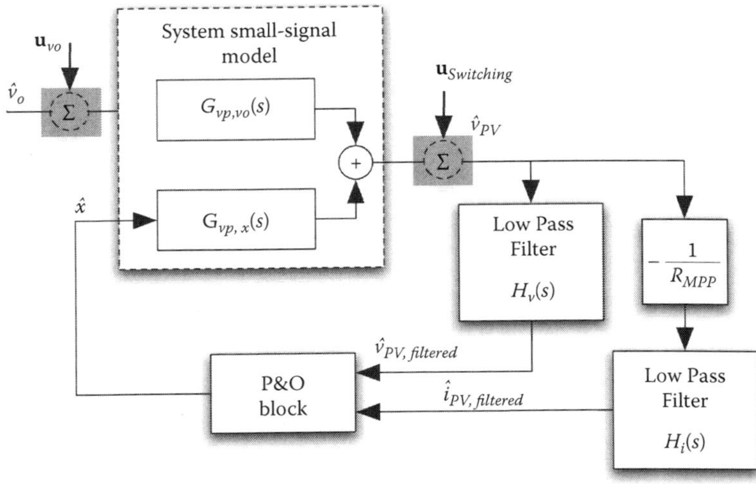

FIGURE 3.32
Effect of the LPF on the MPPT dynamics.

although the LPF can be designed to attenuate the amplitude of any noise coming from the switching converter, its intrinsic phase delay may degrade the tracking capability of the MPPT algorithm.

The scheme shown in Figure 3.32 is an extension of the scheme shown in Figure 3.4 in which the effect of the output voltage noise u_{vo} and the switching noise $u_{switching}$ have been highlighted. The LPF filters have also been included. This new scheme is useful for analyzing the MPPT dynamic behavior and for evaluating the right time interval T_p between two consecutive perturbations in a perturbative approach in more realistic operating conditions of the PV system. If we suppose that both LPFs processing the signals related to the PV voltage and the PV current have the same dynamic characteristics, then their effects on the P&O dynamics can be accounted for by considering the LPF on the PV voltage only. The equation (2.22), used to design the value of T_p in the ideal case, must be suitably corrected by accounting for the effect of the whole transfer function $G_{vp,x}(s) \cdot H_v(s)$, thus including the LPF transfer function $H_v(s)$. It is worth noting that the $G_{vp,x}(s)$ transfer function includes the whole dynamic effect between the perturbed variable x and the PV voltage.

The minimum T_p value depends on the system dynamic and also on the used LPF. A too low cutoff frequency of $H_v(s)$ might be unacceptable if a fast MPPT is required. Moreover, if the LPF is digitally implemented, the sampling time and the elaboration time are additional time delays and must be taken into account in the evaluation of T_p.

The LPF is effective for removing high-frequency noise because if its dynamic is faster than the converter dynamics it does not affect the choice of T_p.

The uncertainty ($\mathbf{u}_{v_{PV}}$) on the PV voltage, which accounts for all the possible noises affecting this variable, is related to $\mathbf{u}_{Switching}$ as follows:

$$\mathbf{u}_{v_{PV}} = |H_v(\omega_s)| \cdot \mathbf{u}_{Switching} \simeq 0 \tag{3.81}$$

where $H_v(\omega_s)$ is the gain of the PV voltage sensor evaluated at the switching frequency ω_s.

A well-designed LPF is obtained if $|H_v(\omega_s)| \simeq 0$. On the basis of the assumption given before, the PV current is also not affected by the switching noise.

The noise $\mathbf{u}_{v_o}$ is in the low-frequency range, so that the LPF does not attenuate it and its effect in the decision process of the MPPT algorithm must be taken into account. It can be accounted for as a further contribution to the uncertainty affecting the PV voltage $\mathbf{u}_{v_{PV}}$; thus

$$\mathbf{u}_{v_{PV}} = |G_{vp,vo}(\omega)| \cdot |H_v(\omega)| \cdot \mathbf{u}_{v_o} \simeq |G_{vp,vo}(\omega)| \cdot \mathbf{u}_{v_o} \tag{3.82}$$

where $|G_{vp,vo}(\omega)|$ is the magnitude of the output-to-input transfer function evaluated at the frequency of the noise and $H_v(\omega)$ is the gain of the PV voltage sensor evaluated at the same frequency. This value will be accounted for in whichever expression involving the PV voltage variable. For example, in the estimation of the PV voltage variation between two consecutive P&O perturbations, it is

$$v_{PV}^{k} = v_{PV,1} \pm \mathbf{u}_{v_{PV}} \qquad \text{and} \qquad v_{PV}^{k+1} = v_{PV,2} \pm \mathbf{u}_{v_{PV}} \tag{3.83}$$

From the propagation of uncertainty it results that

$$\Delta V_{PV,\mathbf{u}_{v_{PV}}} = v_{PV}^{k+1} - v_{PV}^{k} = \Delta V_{PV} \pm \mathbf{u}_{\Delta v_{PV}} = \Delta V_{PV} \pm 2 \cdot |G_{vp,vo}(\omega)| \cdot \mathbf{u}_{v_o} \tag{3.84}$$

where ΔV_{PV} is the PV voltage variation in an error-free environment, while $\Delta V_{PV,\mathbf{u}_{v_{PV}}}$ is the PV voltage variation corrected by an uncertainty due to the voltage noise $\mathbf{u}_{v_o}$. The difference between (3.84) and (3.16) is that the peak-to-peak ΔV_o output oscillation is now correctly represented by the double of the uncertainty $\mathbf{u}_{v_o}$. The new formalization is more suitable for including this effect in the cumulative uncertainty of ΔP_{PV} in which other noise sources can be accounted for.

The total uncertainty on the measured values of v_{PV} and i_{PV} is useful for estimating the optimal P&O step perturbation.

3.4.2 Error Compensation by Increasing the Step Perturbation

As explained in the previous section, not all of the uncertainty sources affecting the electrical variables on which the MPPT strategy is based can be removed or attenuated, so that they can affect the MPPT operation. In order to avoid any wrong decision in tracking the MPP, or to reduce at least the probability of their occurrence, the presence of uncertainty can be compensated by increasing the step amplitude of the P&O perturbation. This approach, in analogy with what is done in telecommunication systems, corresponds to an increase of the signal-to-noise ratio (SNR). Such a solution must be carefully used only if any other possible method has been used in order to improve the results, because, as already remarked, an increase of the step amplitude reduces the MPPT efficiency, so that the error compensation is paid with a loss of energy. In this section the relations among the various uncertainty sources and the MPPT parameters have been discussed in order to identify designing rules for selecting the values of the other parameters of the PV system that assure, whenever possible, the same P&O step amplitude of the system without uncertainties.

If we refer to Figure 3.31 and suppose that the errors of the measurement stage are included in the uncertainty values $\mathbf{u}_i$ and $\mathbf{u}_v$ of the measured PV current and PV voltage, then it will be

$$\begin{cases} i_m = H_i \cdot i_{PV} \pm \mathbf{u_i} \\ v_m = H_v \cdot v_{PV} \pm \mathbf{u_v} \end{cases} \tag{3.85}$$

where i_m and v_m are the measurement of the current and voltage while i_{pv} and v_{pv} are the true values of the current and voltage of the PV field. The scaling factors of the current and voltage sensors, H_i and H_v, are also taken into account. It is worth noting that the scaling factors are the static gains of the LPF transfer functions $H_i(s)$ and $H_v(s)$.

Inverting the previous equations with respect to i_{PV}, v_{PV} and considering the error-free measurements provide

$$i_{PV} = \frac{i_m}{H_i}, \qquad v_{PV} = \frac{v_m}{H_v} \qquad P_{PV} = \frac{v_m \cdot i_m}{H_v H_i} \tag{3.86}$$

The *uncertainty propagation law* (3.80) applied to the P_{PV} expression yields

$$\mathbf{u}_{P_{PV},m} = \frac{1}{H_i \cdot H_v} \cdot \sqrt{v_m^2 \cdot \mathbf{u}_i^2 + i_m^2 \cdot \mathbf{u}_v^2} \tag{3.87}$$

Accounting for additional uncertainty $\mathbf{u}_E$ due to the elaboration process, in terms of PV voltage and current, provides

$$\mathbf{u}_{P_{PV}} = \mathbf{u}_{P_{PV},m} + \mathbf{u}_E = \sqrt{v_{PV}^2 \cdot \left(\frac{\mathbf{u}_i}{H_i}\right)^2 + i_{PV}^2 \cdot \left(\frac{\mathbf{u}_v}{H_v}\right)^2} + \mathbf{u}_E \tag{3.88}$$

so that the PV power can be expressed as

$$P_{PV,\mathbf{up}} = P_{PV} \pm \mathbf{u}_{P_{PV}} \tag{3.89}$$

Equation (3.88) is used to find the best compromise among the maximum value of the uncertainties $\mathbf{u}_i$, $\mathbf{u}_v$ affecting the measured signals, the voltage and current gains H_v, H_i, and the numerical resolution $\mathbf{u}_E$ of the elaboration process that minimize the uncertainty $\mathbf{u}_{P_{PV}}$ on the PV power variation.

The P&O algorithm is based on the measurement of the power variations due to two consecutive perturbations, so that applying the *uncertainty propagation law* yields

$$\Delta P_{PV,\mathbf{u}\Delta P} = \left(P_{PV}^{k+1} \pm \mathbf{u}_{P_{PV}}\right) - \left(P_{PV}^{k} \pm \mathbf{u}_{P_{PV}}\right) = \Delta P_{PV} \pm 2\mathbf{u}_{P_{PV}} \tag{3.90}$$

where $\Delta P_{PV,U\Delta P}$ is the power variation in presence of uncertainty, P_{PV}^{k+1} and P_{PV}^{k} are the power in two consecutive steps of the MPPT perturbations in presence of uncertainty, while ΔP_{PV} is the power variation in the ideal noiseless conditions. As a consequence the uncertainty on ΔP_{PV} will be $u_{\Delta_P} = u_{P_{PV}}$. In the P&O decision process $\Delta P_{PV,\mathbf{u}_{\Delta P}}$ allows us to establish the sign of the next perturbation; thus in order to avoid wrong behavior due to the uncertainty, the sign of $\Delta P_{PV,\mathbf{u}_{\Delta P}}$ must be the same as that of ΔP_{PV} which is the real power variation of the PV field. This condition is fulfilled if

$$\left|\Delta \mathbf{P}_{PV}\right| \geq \left|\mathbf{u}_{\Delta p}\right| \tag{3.91}$$

On the other side, ΔP_{PV} itself is a combination of the PV power variations due to the P&O step perturbation, irradiance variation, and other uncertainty sources (e.g., output voltage oscillation). By means of the same procedure used for obtaining (2.32), taking into account the uncertainty, it results that

$$\Delta P_{PV} \simeq -\left(H \cdot V_{MPP} + \frac{1}{R_{MPP}}\right) \cdot \left(\Delta V_{PV,\mathbf{u}_{vPV}}\right)^2 + V_{MPP} \cdot K_{ph} \Delta G \tag{3.92}$$

where $\Delta V_{PV,UVPV}$ is the voltage variation in presence of uncertainty and it is related to the other parameters by means of (3.84).

Thus in order to estimate the right perturbation amplitude Δx, (2.23) must be modified as follows:

$$\left|\Delta \mathbf{P}_x\right| \geq \left|\Delta \mathbf{P}_G\right| + \left|\mathbf{u}_{\Delta p}\right| \tag{3.93}$$

where ΔP_x and ΔP_G have the same meaning of the parameters appearing in (2.23).

Finally, combining (2.32)–(2.35) with (3.84), (3.88), and (3.93), evaluated in the MPP, provides

$$\Delta x > \frac{1}{|G_0|} \sqrt{\frac{V_{MPP} \cdot K_{ph} \cdot |\dot{G}| \cdot T_p + 2 \cdot \sqrt{v_{PV}^2 \left(\frac{\mathbf{u}_i}{H_i}\right)^2 + i_{PV}^2 \left(\frac{\mathbf{u}_v}{H_v}\right)^2} + 2\mathbf{u}_E}{\left(H \cdot V_{MPP} + \frac{1}{R_{MPP}}\right)}} + 2 \cdot \frac{|G_{vp,vo}(\omega)|}{|G_0|} \cdot \mathbf{u}_{v_o} \tag{3.94}$$

The lower $\mathbf{u}_{v_o}$, $\mathbf{u}_i$, $\mathbf{u}_v$, and $\mathbf{u}_E$, the lower the power uncertainty, so that a right choice of the sensors, LPF parameters, CPU, and switching converter transfer functions allows minimization of the error in the estimation of the PV power and, consequently, reduction of the optimal value of Δx in the P&O algorithm.

In practice, (3.88) and the dynamic information taken from the $H_v(s)$, $H_i(s)$ transfer functions give the design criteria for designing the measurement system in a P&O-based MPPT controller in the best way. $\mathbf{u}_E$, instead, gives an idea of the CPU performances required for realizing a well-designed P&O MPPT.

3.4.3 ADC Quantization Error in the P&O Algorithm: Numerical Example

In order to show the way in which (3.94) is used, the contribution of the measurement errors $\mathbf{u}_v$, $\mathbf{u}_i$ in the increase of the step amplitude of a digitally implemented P&O MPPT algorithm is described. Table 3.2 gives the parameters of the PV field and of the boost converter used for evaluating the P&O step perturbation.

For an ideal noise-free system ($\mathbf{u}_{v_o} = \mathbf{u}_i = \mathbf{u}_v = \mathbf{u}_E = 0$), the use of (3.94) gives

$$\Delta x \geq 0.0059 \tag{3.95}$$

The uncertainty components are added one at the time so that the contribution of each one of them to the increase of Δx is shown. In the measurement stage the uncertainty due to the quantization errors introduced by the ADCs is accounted for. According to [35], some additional error sources should be accounted for, but the modeling of such terms is out of the scope

TABLE 3.2

Parameters of the PV Field and of the Boost Converter

PV Array	**Values at 300 W/m² and T = 35°C**
MPP voltage V_{MPP}	356 V
MPP current I_{MPP}	2.22 A
MPP resistance R_{MPP}	161 Ω
H_{MPP}	$1.157 \cdot 10^{-4}$ A/V²
Coefficient of the photo-induced current K_{ph}	83 mA·m²/W
PV Measurable Parameters	**Maximum Values**
PV field current $I_{PV,max}$	10 A
PV field voltage $V_{PV,max}$	600 V
MPPT Design Parameters	**Nominal Values**
Irradiance variation	100 W/m²/s
P&O time interval $T_p\dot{G}$	2 ms
Boost output voltage $V_o = G_o$	600 V
Output voltage oscillation $\Delta V_o = \mathbf{u}_{v_o}$	0 V

of this book. It is assumed that the uncertainty is directly related to the last significant bit (LSB) and to ADC voltage resolution, so that

$$\mathbf{u}_i = \mathbf{u}_v = \frac{1}{2} LSB = \frac{1}{2} \frac{V_{FS}}{2^N} \tag{3.96}$$

where V_{FS} is the full-scale input voltage and N is the number of bits of the ADCs. Moreover, in order to map the PV voltage and current in the V_{FS} range, the H_v and H_i scaling factors must be chosen according to the following equations:

$$H_v = \frac{V_{FS}}{V_{PV,max}} \qquad H_i = \frac{V_{FS}}{I_{PV,max}} \tag{3.97}$$

If $N = 10$ bits and $V_{FS} = 5$ V are adopted, which are typical values for the ADC parameters from Table 3.2, it results that

$$\begin{gathered} \mathbf{u}_\mathrm{i} = \mathbf{u}_\mathrm{v} = \frac{1}{2} \frac{V_{FS}}{2^N} \cong 2.4mV \\ H_v = \frac{5}{600} \cong 0.0083 \qquad H_i = \frac{5}{10} = 0.5 \frac{V}{A} \end{gathered} \tag{3.98}$$

Such values are put in (3.94), so that the new value for Δx must fulfill the following constraint:

$$\Delta x \geq 0.0157 \tag{3.99}$$

which is almost three times greater than the value in noiseless conditions given in (3.95).

Of course this is the condition in which the quantization errors, introduced in the measurement process, produce a wrong decision in the MPPT algorithm with a probability almost zero. If a lower level of confidence in the P&O decision is accepted, Δx can be reduced progressively.

Alternatively, in order to improve the MPPT performances as much as possible, the optimal Δx can be reduced by using different ADCs. Table 3.3 shows the results for three ADCs characterized by different numbers of bits but with the same V_{FS}. Based on (3.88), it is also possible to evaluate separately the effect of $\mathbf{u}_v$ and $\mathbf{u}_i$ on the PV power estimation; this can be useful for identifying the channel that is more sensible to quantization error.

Table 3.3 shows clearly which is the minimum number of bits in the ADC that makes the quantization error negligible. It is worth noting that the uncertainty coming from the two measurement channels can be significantly different. Indeed, in Table 3.3, such a difference has been highlighted by showing the values of the power uncertainty due to the two channels

TABLE 3.3

ADC Resolution on the MPPT Performances

	ADC 10 Bit	ADC 16 Bit	ADC 20 Bit
$\mathbf{u}_i, \mathbf{u}_v$	2.4 mV	38 μV	2.4 μV
$\Delta P_{\mathbf{u}_i}$	≃3.5 W	≃56 mW	≃3.4 mW
$\Delta P_{\mathbf{u}_v}$	≃1 W	≃17 mW	≃1 mW
Δx	0.0157	0.0062	0.0059

separately. In particular, ΔP_{Ui} is the increase in the power variation due to uncertainty on the measurement of the PV current (u_i), and ΔP_{Uv} is the increase in the power variation due to uncertainty on the measurement of the PV voltage (u_v). The sensibility of the MPPT performances with respect to quantization error is such a critical aspect that some manufacturers produce ADC with high resolution specifically designed for power monitoring in PV application [36]. Alternatively, digital processing combined with multiple sampling of PV variables can be used for increasing fictitiously the equivalent number of the ADC bit [37]. In conclusion, it is possible to state that the action concerning the minimization of uncertainty must be focused on the noise sources that influence more heavily the decision process of the MPPT. Equation (3.94) shows clearly this dependency for the P&O algorithm.

References

1. S.B. Kjær. Design and control of an inverter for photovoltaic applications. PhD thesis, Aalborg University, Denmark Institute of Energy Technology, January 2005.
2. N. Femia, G. Petrone, G. Spagnuolo, and M. Vitelli. A technique for improving P&O MPPT performances of double stage grid-connected photovoltaic systems. *IEEE Transactions on Industrial Electronics*, 56(11):4473–4482, 2009.
3. R.-J. Wai and C.-Y. Lin. Active low-frequency ripple control for clean-energy power-conditioning mechanism. *IEEE Transactions on Industrial Electronics*, 57:3780–3792, 2010.
4. R.S. Gemmen. Analysis for the effect of inverter ripple current on fuel cell operating condition. *Journal of Fluids Engineering*, 125(3):576–585, 2003.
5. C. Liu and J.-S. Lai. Low frequency current ripple reduction technique with active control in a fuel cell power system with inverter load. *IEEE Transactions on Power Electronics*, 22(4):1429–1436, 2007.
6. S.K. Mazumder, R.K. Burra, and K. Acharya. A ripple-mitigating and energy-efficient fuel cell power-conditioning system. *IEEE Transactions on Power Electronics*, 22(4):1437–1452, 2007.

7. K. Fukushima, I. Norigoe, M. Shoyama, T. Ninomiya, Y. Harada, and K. Tsukakoshi. Input current-ripple consideration for the pulse-link DC-AC converter for fuel cells by small series LC circuit. In *Twenty-Fourth Annual IEEE Applied Power Electronics Conference and Exposition (APEC 2009)*, February 2009, pp. 447–451.
8. J.-I. Itoh and F. Hayashi. Ripple current reduction of a fuel cell for a single-phase isolated converter using a dc active filter with a center tap. In *Twenty-Fourth Annual IEEE Applied Power Electronics Conference and Exposition (APEC 2009)*, February 2009, pp. 1813–1818.
9. J.-M. Kwon, E.-H. Kim, B.-H. Kwon, and K.-H. Nam. High-efficiency fuel cell power conditioning system with input current ripple reduction. *IEEE Transactions on Industrial Electronics*, 56(3):826–834, 2009.
10. E. Mamarelis, C.A. Ramos-Paja, G. Petrone, G. Spagnuolo, M. Vitelli, and R. Giral. FPGA-based controller for mitigation of the 100 Hz oscillation in grid connected PV systems. In *2010 IEEE International Conference on Industrial Technology (ICIT)*, March 2010, pp. 925–930.
11. T. Brekken, N. Bhiwapurkar, M. Rathi, N. Mohan, C. Henze, and L.R. Moumneh. Utility-connected power converter for maximizing power transfer from a photovoltaic source while drawing ripple-free current. In *2002 IEEE 33rd Annual Power Electronics Specialists Conference, 2002*, vol. 3, pp. 1518–1522.
12. R.W. Erikson and D. Maksimovic. *Fundamentals of power electronics*. 2nd ed. Kluwer Academic Publishers, Norwell, MA, 2000.
13. L. Castanier and S. Silvestre. *Modelling photovoltaic systems using pspice*. Vol. 1. John Wiley & Sons, West Sussex, UK, 2002.
14. V. Salas, E. Olìas, A. Barrado, and A. Làzaro. Review of the maximum power point tracking algorithms for stand-alone photovoltaic systems. *Solar Energy Materials and Solar Cells*, 90:1555–1576, 2006.
15. J.P. Gaubert, R. Kadri, and G. Champenois. An improved maximum power point tracking for photovoltaic grid-connected inverter based on voltage-oriented control. *IEEE Transactions on Industrial Electronics*, 58(1):66–75, 2011.
16. O. Lopez-Lapena, M.T. Penella, and M. Gasulla. A new MPPT method for low-power solar energy harvesting. *IEEE Transactions on Industrial Electronics*, 57(9):3129–3138, 2010.
17. O. Lopez-Lapena, M. Penella, and M. Gasulla. A close-loop maximum power point tracker for sub-watt photovoltaic panels. *IEEE Transactions on Industrial Electronics*, PP(99):1, 2011.
18. Y. Xue, L. Chang, S.B. Kjær, J. Bordonau, and T. Shimizu. Topologies of single-phase inverters for small distributed power generators: An overview. *IEEE Transactions on Power Electronics*, 19(5):1305–1314, 2004.
19. H. Sira-Ramirez. Sliding motions in bilinear switched networks. *IEEE Transactions on Circuits and Systems*, 34(8):919–933, 1987.
20. L. Rossetto, G. Spiazzi, P. Tenti, B. Fabiano, and C. Licitra. Fast-response high-quality rectifier with sliding mode control. *IEEE Transactions on Power Electronics*, 9(2):146–152, 1994.
21. S.-C. Tan, Y.M. Lai, and C.K. Tse. General design issues of sliding-mode controllers in dc dc converters. *IEEE Transactions on Industrial Electronics*, 55(3):1160–1174, 2008.

22. M.A. Sofla and G.B. Gharehpetian. Dynamic performance enhancement of microgrids by advanced sliding mode controller. *International Journal of Electrical Power and Energy Systems*, 33(1):1–7, 2011.
23. J. Fernandez-Vargas and G. Ledwich. Variable structure control for power systems stabilization. *International Journal of Electrical Power and Energy Systems*, 32(2):101–107, 2010.
24. R.-J. Wai, W.-H. Wang, and C.-Y. Lin. High-performance stand-alone photovoltaic generation system. *IEEE Transactions on Industrial Electronics*, 55(1):240–250, 2008.
25. R.-J. Wai and W.-H. Wang. Grid-connected photovoltaic generation system. *IEEE Transactions on Circuits and Systems I: Regular Papers*, 55(3):953–964, 2008.
26. Y.-K. Lo, S.-C. Yen, and C.-Y. Lin. A high-efficiency AC-to-DC adaptor with a low standby power consumption. *IEEE Transactions on Industrial Electronics*, 55(2):963–965, 2008.
27. Linear Company. Burst mode in Linear products. LTC3458. http://cds.linear.com/docs/datasheet/3442fa.pdf.
28. Powersim. Power electronics simulation software, 2008. http://www.powersys.fr/psimpresent.php.
29. G. Spagnuolo, G. Petrone, M. Vitelli, J. Calvente, C. Ramos-Paja, R. Giral, E. Mamarelis, and E. Bianconi. A fast current-based MPPT technique employing sliding mode control. *IEEE Transactions on Industrial Electronics*, PP(99):1, 2012.
30. E. Bianconi, J. Calvente, R. Giral, G. Petrone, C.A. Ramos-Paja, G. Spagnuolo, and M. Vitelli. A fast current-based MPPT technique based on sliding mode control. In *2011 IEEE International Symposium on Industrial Electronics (ISIE)*, June 2011, pp. 59–64.
31. E. Bianconi, J. Calvente, R. Giral, G. Petrone, C.A. Ramos-Paja, G. Spagnuolo, and M. Vitelli. Improving the perturb and observe maximum power point tracking by using sliding mode control. In *2011 IEEE International Symposium on Industrial Electronics (ISIE)*, June 2011, pp. 310–315.
32. N. Femia, G. Petrone, G. Spagnuolo, and M. Vitelli. Optimization of perturb and observe maximum power point tracking method. *IEEE Transactions on Power Electronics*, 20(4):963–973, 2005.
33. D. Arnold and H.-G. Beyer. A comparison of evolution strategies with other direct search methods in the presence of noise. *Computational Optimization and Applications*, 24:135–159, 2003. 10.1023/A:1021810301763.
34. H. Al-Atrash, I. Batarseh, and K. Rustom. Effect of measurement noise and bias on hill-climbing MPPT algorithms. *IEEE Transactions on Aerospace and Electronic Systems*, 46(2):745–760, 2010.
35. JCGM/WG 1. Evaluation of measurement data—Guide to the expression of uncertainty in measurement. Working Group 1 of the Joint Committee for Guides in Metrology, September 2008.
36. Texas Instruments. High-resolution analog-to-digital converter—ads1282. http://www.ti.com/lit/ds/symlink/ads1282.pdf, 2012.
37. E. Balestrieri, P. Daponte, and S. Rapuano. A state of the art on ADC error compensation methods. *IEEE Transactions on Instrumentation and Measurement*, 54(4):1388–1394, 2005.

4

Distributed Maximum Power Point Tracking of Photovoltaic Arrays

4.1 Limitations of Standard MPPT

Photovoltaic (PV) systems usually adopt field maximum power point tracking (FMPPT), that is, the tracking of the maximum power point (MPP) of the power vs. voltage (P-V) characteristic of the whole field composed by paralleled PV strings (Figure 4.1).

In case of mismatch (due to clouds, shadows, dirtiness, manufacturing tolerances, aging, different orientation of parts of the PV field, etc.), the P-V characteristic of the PV field may exhibit more than one peak, due to the adoption of bypass diodes, and MPPT algorithms can fail, causing a severe decrease in the overall system efficiency [1–34], unless the whole P-V characteristic is periodically swept. Even when FMPPT is able to catch the absolute maximum power of the mismatched PV field, such a power is lower than the sum of the available maximum powers that the mismatched modules are able to provide.

4.2 A New Approach: Distributed MPPT

The solution adopted to overcome the drawbacks associated to mismatching phenomena in PV applications is called distributed maximum power point tracking (DMPPT). Two different DMPPT approaches can be used. The first one is based on the adoption of module-dedicated DC/AC converters, called microinverters, and realizing the MPPT for each PV module [35–37]. The second approach is instead based on the use of module-dedicated DC/DC converters, realizing the MPPT for each module and centralized inverters [45–64].

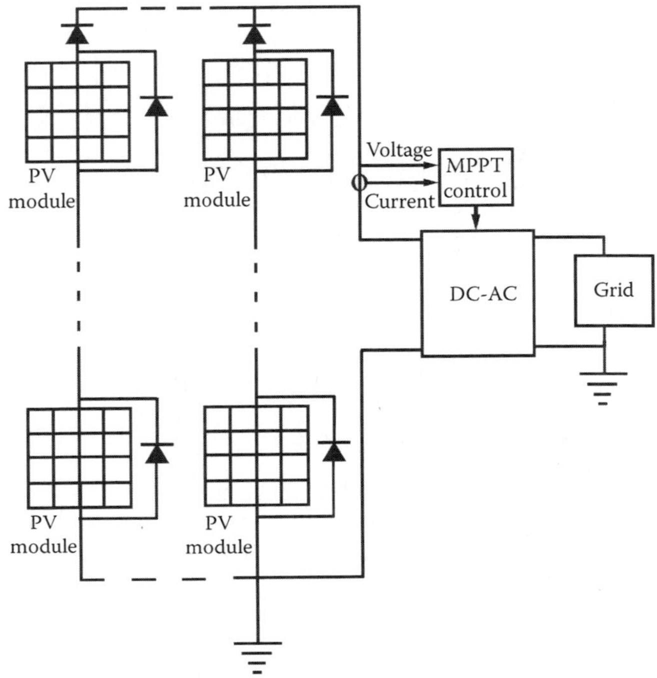

FIGURE 4.1
Grid-connected PV system with FMPPT. Strings of PV modules are put in parallel and connected to a single DC/AC inverter.

4.2.1 DMPPT by Means of Microinverters

A microinverter is a DC/AC grid-connected device that converts the DC power from a single PV module to grid-compliant AC power and carries out the MPPT on such a module (Figure 4.2) [35–37].

The advantages provided by the adoption of microinverters are the following:

- Almost no waste of available PV energy in case of mismatching.
- Modularity, because PV systems can be easily expanded by simply paralleling microinverters at the AC side.
- Each module works independently. If one module fails, the other modules will continue to deliver power to the grid.
- Low minimum system size. That is, the threshold for people to start their own PV plant is lowered.
- Use of standard AC installation material (no DC cabling), which reduces costs of installation material and system design.
- No need for string diodes, and therefore no additional conduction losses due to the presence of blocking diodes.

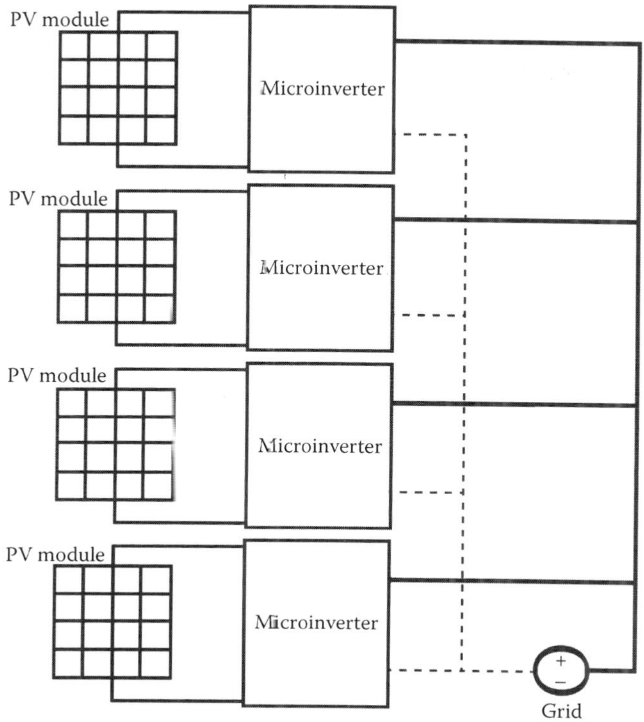

FIGURE 4.2
Grid-connected PV system with DMPPT. Approach based on the adoption of microinverters with the output ports in parallel to the grid.

Microinverters also have some disadvantages. First, the small level of power (150 W ÷ 250 W), which allows low purchase investment, also strongly influences the cost since the lower the power rating, the higher the cost per produced kWh. In order to reduce costs, mass production is a mandatory condition and may only be achieved by means of flexible solutions capable of operating with most of the available panels in the market. This leads to the necessity of high-voltage-gain capability since PV panels usually have output voltages in the range (20, 50) V. One of the difficulties related to the design of a microinverter is represented by the need of the voltage amplification from the low DC level of the PV module terminals to the AC level typical of European grid 230 Vrms (worst case since the U.S. grid AC level is lower). Moreover, AC grid connection details such as filtering, protection, and control, including anti-islanding, must all be distributed to the per-panel converters, and this contributes to keeping high the economic cost per kWp of PV systems adopting microinverters. Another disadvantage is represented by the mandatory design requirement of highly robust power electronics with long lifetimes (hopefully the expected lifetime of the PV module, generally considered 20 years). Lifetime is closely related to reliability and expected degree of failure of components. Moreover, harsh ambient conditions are expected with a wide range of temperature variations during a single day, and consequently also mechanical stresses on components and insulation materials. Among all passive or active elements, electrolytic capacitors are the ones with the shortest lifetime and with the further drawback that aging increases the values of the equivalent series resistance (ESR) and consequently also the losses. The energy storage capability needed for single-phase AC connection in order to decouple the DC side (characterized by a constant power flow) from the AC side (characterized by the presence of 100 Hz power fluctuations) requires the presence of energy storage capacitors—in most cases electrolytic capacitors. On the basis of the above considerations, inverter topologies that do not require high capacitance values in order to allow the use of the film technology (longer lifetime, good thermal and electrical stability, and lower ESR) are more suitable for microinverter applications.

An AC module is the combination of a single PV module and a single DC/AC converter (a microinverter) that converts light into AC power when it is connected in parallel to the network. In the late 1980s research on AC modules was started by Professor Kleinkauf at the Institut für Solar Energieversorgungstechnik (ISET); he was one of the co-founders of SMA, the worldwide market leader for solar inverters.

Mass production of AC modules and microinverters began in the late 1990s. Some examples: NKF Kabel began shipping the OK4 100 W microinverter in 1996, Evergreen Solar 112 W AC module was on the California Energy Commission (CEC) list of certified modules in 2000, Ascension Technology shipped the first 300 W AC modules in 1997. AC modules and microinverters almost completely disappeared from the market in 2003, mainly due to the relatively high costs and the failing of inverters. Recently, however, new commercial developments in the field of AC modules and microinverters have taken place: Enphase 230 W Micro-Inverter (2008, United States) [38], Greenray Solar 200 W SunSine AC module (2009, United States) [39], Petra Solar SunWave 200 W AC module (2009, United States) [40], Enecsys 280 W microinverter (2010, UK) [41], Exeltech 240 W AC module (2010, United States) [42], Solarbridge 235 W microinverter (2010, United States) [43], and INVOLAR 250 W microinverter (2010, China) [44].

In theory, series connection of the output ports of microinverters to achieve direct (transformer-less) 230 V AC grid connection is also possible (Figure 4.3). Such an architecture is promising and deserves further consideration and research. No commercial devices of this kind are currently available on the market.

4.2.2 DMPPT by Means of DC/DC Converters

As already written in the beginning of Section 4.2, the second approach adopted to face the drawbacks associated with mismatching phenomena is based on the use of MPPT module-dedicated DC/DC converters (Figure 4.4) and centralized inverters [45–64].

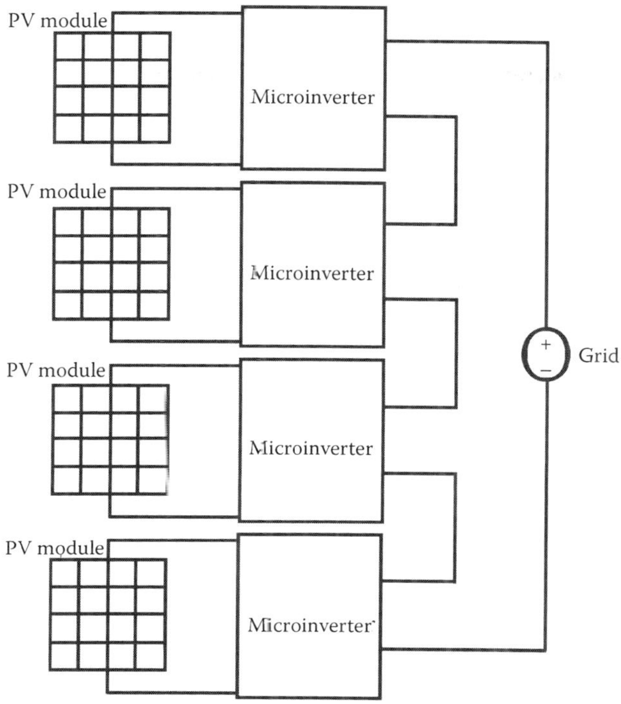

FIGURE 4.3
Grid-connected PV system with DMPPT. Approach based on the adoption of microinverters with the output ports connected in series.

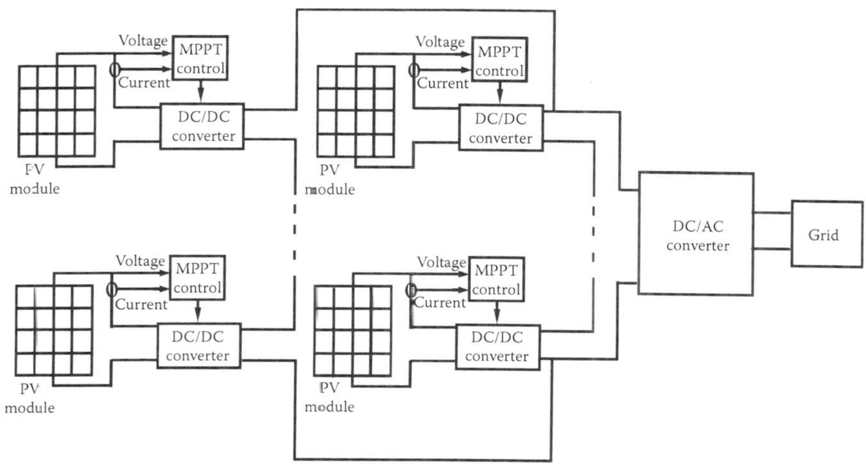

FIGURE 4.4
Grid-connected PV system with DMPPT. Approach based on the adoption of MPPT DC/DC converters with the output ports connected in series.

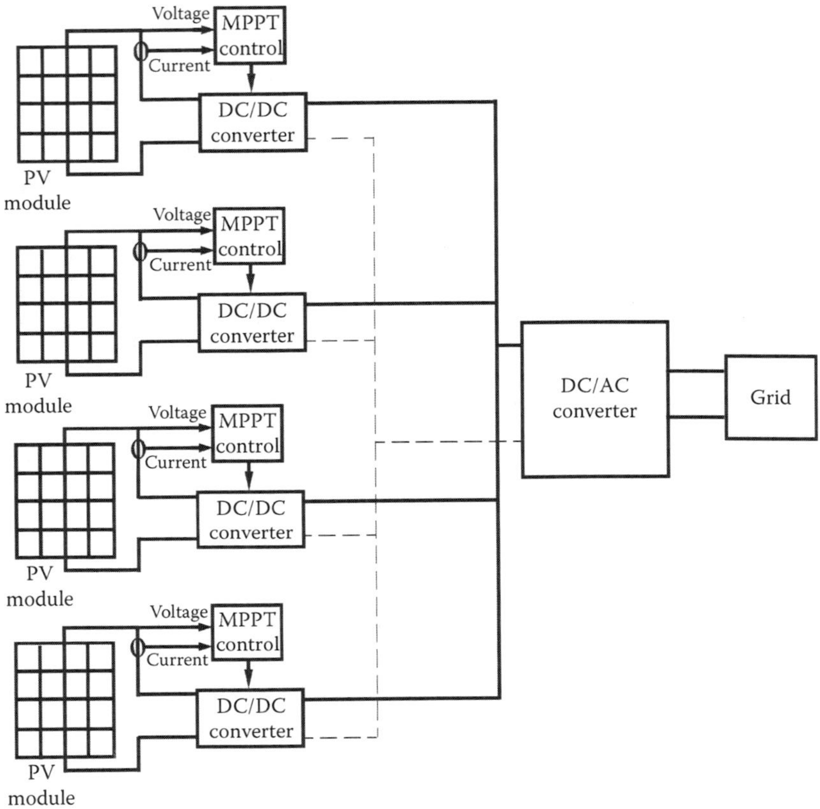

FIGURE 4.5
Grid-connected PV system with DMPPT. Approach based on the adoption of MPPT DC/DC converters with the output ports connected in parallel.

Examples of commercial devices developed with reference to the architecture shown in Figure 4.4 are SolarMagic Power Optimizers by National Semiconductors [65], SolarEdge Power Box [66], Tigo Energy Module Maximizers [67], and Xandex SunMizers [68].

In theory the architecture shown in Figure 4.5 could also be adopted, but the price to pay is represented by the need of using DC/DC converters characterized by two contrasting requirements: a high-voltage conversion ratio (e.g., from 30 V to 350 V) and, at the same time, a high conversion efficiency. To the best of the authors' knowledge, only one company, eIQ Energy, has launched on the market a device, called Parallux Vboost, that has been developed for such an architecture [69].

4.3 DC Analysis of a PV Array with DMPPT

In the following, PV systems adopting the architecture shown in Figure 4.4 will be discussed and analyzed in detail because the connection in series of the output ports of the DC/DC converters imposes some nontrivial constraints that must be carefully taken into account if the efficient working of the PV system is desired. Hereafter, a system composed by a PV module with a dedicated DC/DC converter will be referred to as self-controlled PV module (SCPVM). There are essentially two main drawbacks of DMPPT applications. The first is represented by the fact that DMPPT is able to ensure higher energy efficiency than FMPPT, in the presence of mismatching phenomena, only if the efficiency of the power stage of MPPT DC/DC converters is high enough [70]. The second drawback is represented by the fact that conditions exist in which the DMPPT approach also does not allow the working of each PV module of the field in its MPP. In other words, by using the DMPPT approach, the total power extracted by the PV field can be lower than the sum of the maximum available powers of the single modules, and in addition, it can be also lower than the power extracted by using the FMPPT approach. This aspect is not only related to constraints associated to the more or less limited voltage conversion ratio of the particular DC/DC converters adopted. Indeed, limitations of DMPPT performances may occur not only when using a topology (e.g., the boost one) that is able only to step up the output voltage with respect to the input voltage, but also when using DC/DC topologies, such as the buckboost and Cuk topologies, which are able to step up or step down the output voltage with respect to the input voltage, and that therefore, at first sight, could be considered able to allow the desired performances of DMPPT applications in whichever operating condition. In fact, lower than expected DMPPT performances can also take place due to the finite ratings of devices used in the power stage of SCPVMs or due to a nonoptimal value of the string voltage [46, 47, 71]. This aspect will be discussed in greater detail in the following sections.

4.3.1 Feasible Operating Regions

In the following, without loss of generality we will refer only to a step-up topology (boost) and a step-up/down topology (buckboost). Let's consider a string of N SCPVMs with the output ports connected in series (Figure 4.4). It is worth noting that, depending on the adopted converter topology, the ratings of the adopted devices, the irradiance G_k, and the temperature T_k characterizing the k-th module (a rough estimate of T_k can be obtained by means of the following formula $T_k = T_a + (\text{NOCT} - 20°)*G_k/800$, where T_a is the ambient temperature and NOCT is the so-called nominal operating cell temperature

[1]), not all the possible sets of PV operating voltages [V_{pan1}, …, V_{panN}] allow us to fulfill the following set of conditions:

$$0 < V_{pank} < V_{ock} \tag{4.1a}$$

$$\sum_{k=1}^{N} P_{pank}(V_{pank}) = P_{tot} \tag{4.1b}$$

$$I_{out} = \frac{P_{tot}}{V_{bulk}} \tag{4.1c}$$

$$V_{outk} = \frac{P_{pank}}{I_{out}} \tag{4.1d}$$

$$I_{outk} = I_{out} \tag{4.1e}$$

$$M_k = \frac{V_{outk}}{V_{pank}} \tag{4.1f}$$

$$M_{min} < M_k < M_{max} \tag{4.1g}$$

$$V_{offk} < \mathrm{V}_{\mathrm{ds\ max}} \tag{4.1h}$$

$$I_{onk} < I_{ds\ max} \tag{4.1i}$$

where V_{pank} is the voltage of the *k*-th PV module, $P_{pank}(V_{pank})$ is the corresponding extracted power, V_{ock} is the open-circuit voltage of the *k*-th module, V_{outk} and I_{outk}, respectively, are the output voltage and current of the *k*-th SCPVM, V_{bulk} is the inverter DC input voltage, M_k is the voltage conversion ratio of the *k*-th converter, and M_{min} and M_{max} are the corresponding minimum and maximum values ($M_{min} = 1$ and $M_{max} \rightarrow \infty$ for the ideal boost converter, $M_{min} = 0$ and $M_{max} \rightarrow \infty$ for the ideal buckboost converter). V_{offk} represents the value of the voltage across the switches of the converter when they are in the OFF state, and $V_{ds\,max}$ is the corresponding rated maximum allowed value, which depends on the voltage rating of the adopted devices. I_{onk} represents the peak value of the current in the switches of the converter when they are in the ON state, and $I_{ds\,max}$ is the corresponding maximum allowed value, which also depends on the ratings of the adopted devices. It is worth noting that, whereas in the boost topology the voltage across the active switch and the output capacitor, during its OFF subinterval, is equal to the output voltage V_{outk} [72], in the case of the buckboost topology, the voltage across the active switch, during its OFF subinterval, is equal to the sum of the input and output voltages $V_{pank} - V_{outk}$ (the buckboost is an inverting topology; therefore $V_{outk} < 0$) [72]. Moreover, it is worth noting that constraint (4.1i) plays a strong effect only in the case of the buckboost converter. In fact, when using the boost converters, it is enough to ensure that their switches are able to conduct the short-circuit currents of the PV modules connected at their input ports (the peak value of the currents in the switches is equal to the PV current). Instead, in the case of the buckboost converters, when the value of the duty cycle D decreases, the peak value of the switches currents increase (the peak value of the currents in the switches is equal to the PV current divided by D [72]). Thus, especially when stepping down the output voltage with respect to the input voltage ($D < 0.5$), the currents in the switches may exceed the safety threshold $I_{ds\,max}$. In the following, a set of PV operating voltages (V_{pan1}, …, V_{panN}) will be called feasible if it fulfills all constraints 4.1a – 4.1i. Moreover, the term *DMPPT efficiency* will refer to the ratio η_{DMPPT} between the PV power that can be extracted by adopting DMPPT and the maximum available power, while the term *FMPPT efficiency* will refer to the ratio η_{FMPPT} between the PV power that can be extracted by adopting FMPPT and the maximum available power.

4.3.2 Examples of Feasible Operating Regions

In order to show that, depending on the environmental conditions and system parameters, the equality $\eta_{DMPPT} = 1$ is not always achievable, some useful maps are reported in Figures 4.6–4.11. They refer for simplicity to a string of 2 SCPVMs adopting Sunmodule SW225 PV modules ($V_{oc} = 36.8$ V, $I_{sc} = 8.17$ A, $V_{MPP} = 29.5$ V, $I_{MPP} = 7.63$ A, NOCT = 46°C). In the maps of Figures 4.6–4.11 the horizontal and vertical axes represent the operating voltages V_{pan1} and V_{pan2} of the two PV modules. G_1 (G_2) is the irradiance level of PV module 1 (2). The colored region is the locus of all the feasible operating voltages; the colors are used to identify the levels of the total PV power associated to each feasible operating point, as indicated by the right vertical bar. The remaining part of the plane $V_{pan1} - V_{pan2}$ is of course composed by nonfeasible operating points. Different symbols are adopted to identify, for each nonfeasible operating point, the not-fulfilled constraint as indicated by "Legend 1" for the boost converter and "Legend 2" for the buckboost converter. The absolute MPP

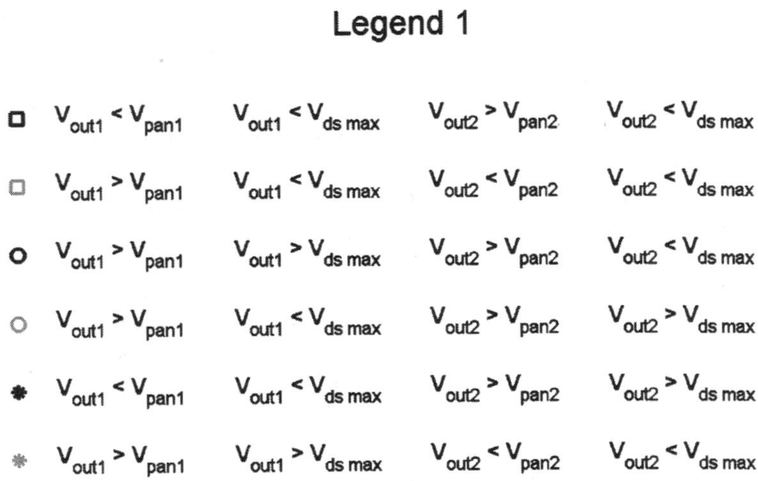

is the point, marked with an asterisk and with coordinates (V_{MPP1}, V_{MPP2}); instead, the feasible MPP is the point belonging to the feasibility region and characterized by the maximum value of the PV power. In the case shown in Figure 4.6 (boost converter, $G_1 = G_2 = 1000$ W/m^2, $V_{ds\ max} = 50$ V, $V_{bulk} = 80$ V) it is $\eta_{DMPPT} = \eta_{FMPPT} = 1$ since the absolute MPP and the feasible MPP coincide. Instead, in the case shown in Figure 4.7 (boost converter, $G_1 = 1000$ W/m^2, $G_2 = 500$ W/m^2, $V_{ds\ max} = 50$ V, $V_{bulk} = 80$ V), it is $\eta_{DMPPT} = 0.9$, $\eta_{FMPPT} = 0.72$, since the absolute MPP does not belong to the feasibility region. This is a simple example of a situation in which not only η_{FMPPT} but also η_{DMPPT} is lower than 1. By adopting the buckboost converter and the same device voltage rating, not only in the mismatched case of Figure 4.9 ($G_1 = 1000$ W/m^2, $G_2 = 500$ W/m^2, $V_{ds\ max} = 50$ V, $V_{bulk} = 80$ V, $\eta_{DMPPT} = 0.31$, $\eta_{FMPPT} = 0.72$), but also in the uniform case of Figure 4.8 ($G_1 = G_2 = 1000$ W/m^2, $V_{ds\ max} = 50$ V, $V_{bulk} = 80$ V,

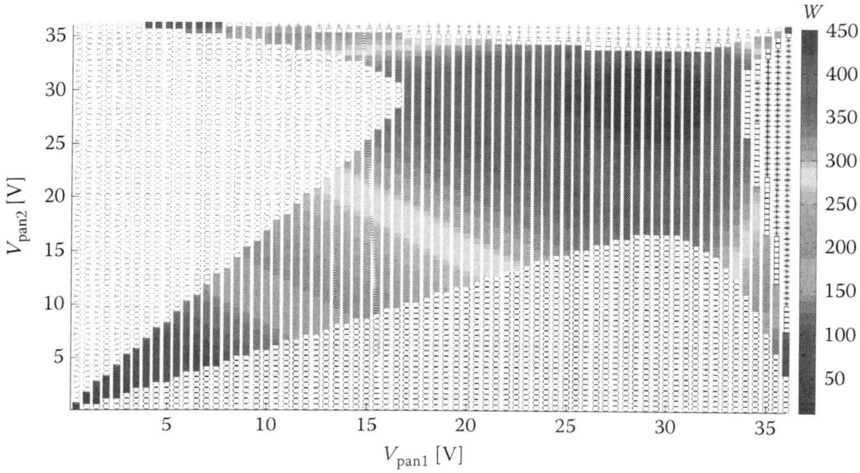

FIGURE 4.6 (See color insert)
(Boost) $G_1 = G_2 = 1000$ W/m², $V_{bulk} = 80$ V, $V_{ds\ max} = 50$ V.

$\eta_{DMPPT} = 0.36$, $\eta_{FMPPT} = 1$), we get that DMPPT is not able to allow the working of each module in its MPP, due to the limited voltage rating (50 V) that has been considered. In order to get larger feasibility regions, the buckboost converter requires the adoption of devices with higher voltage ratings, as shown in Figure 4.10 ($V_{ds\ max} = 80$ V, $\eta_{DMPPT} = 1$, $\eta_{FMPPT} = 1$). Figures 4.6–4.10 have been obtained with $I_{ds\ max} \rightarrow \infty$. The effect of constraint 4.1i can be easily understood by comparing Figures 4.10 and 4.11, both referring to the

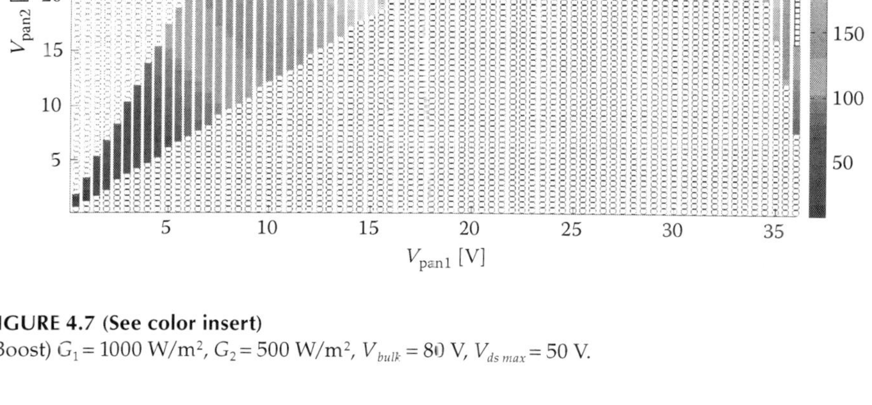

FIGURE 4.7 (See color insert)
(Boost) $G_1 = 1000$ W/m², $G_2 = 500$ W/m², $V_{bulk} = 80$ V, $V_{ds\ max} = 50$ V.

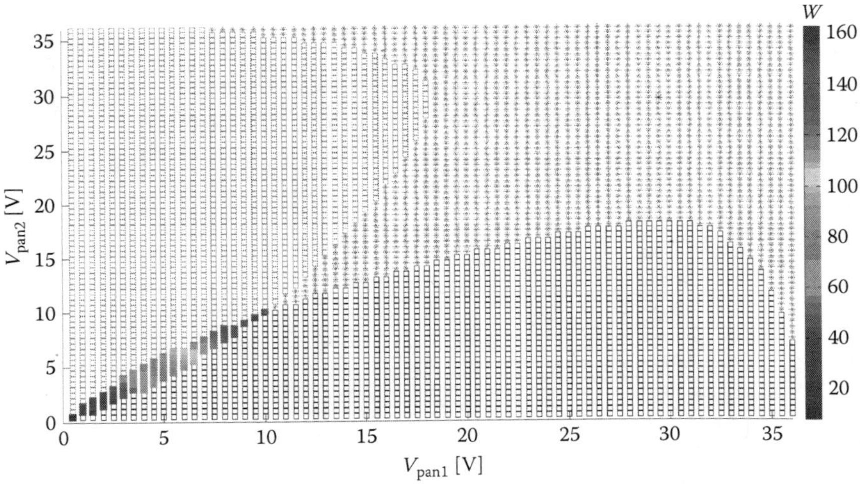

FIGURE 4.8 (See color insert)
(Buckboost) $G_1 = G_2 = 1000$ W/m^2, $V_{bulk} = 80$ V, $V_{ds\,max} = 50$ V, $I_{ds\,max} \rightarrow \infty$.

buckboost converter, under the same operating conditions. Figure 4.10 has been obtained by considering $I_{ds\,max} \rightarrow \infty$, while Figure 4.11 has been obtained by considering $I_{ds\,max} = 16$ A.

The above results are useful in order to stress two fundamental aspects. First, in DMPPT PV applications, it is necessary to adopt suitable additional protection circuitries able to avoid voltages or current constraints, related to the finite ratings of switching devices adopted in SCPVMs, being

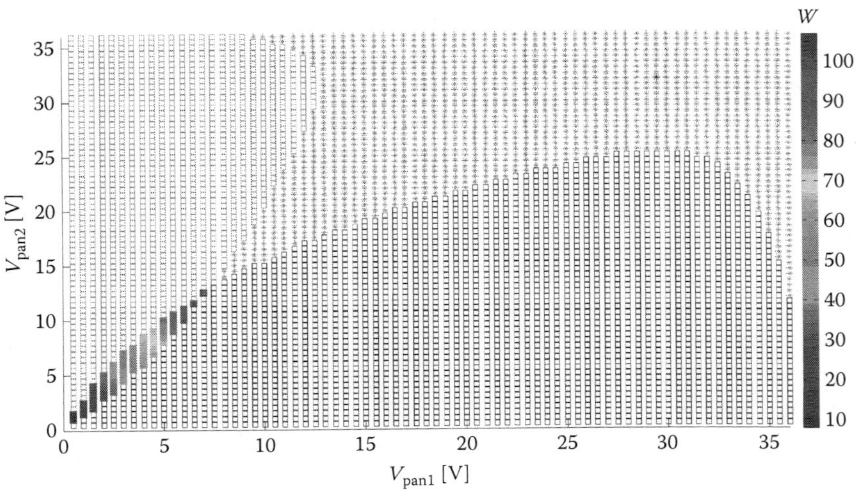

FIGURE 4.9 (See color insert)
(Buckboost) $G_1 = 1000$ W/m^2, $G_2 = 500$ W/m^2, $V_{bulk} = 80$ V, $V_{ds\,max} = 50$ V, $I_{ds\,max} \rightarrow \infty$.

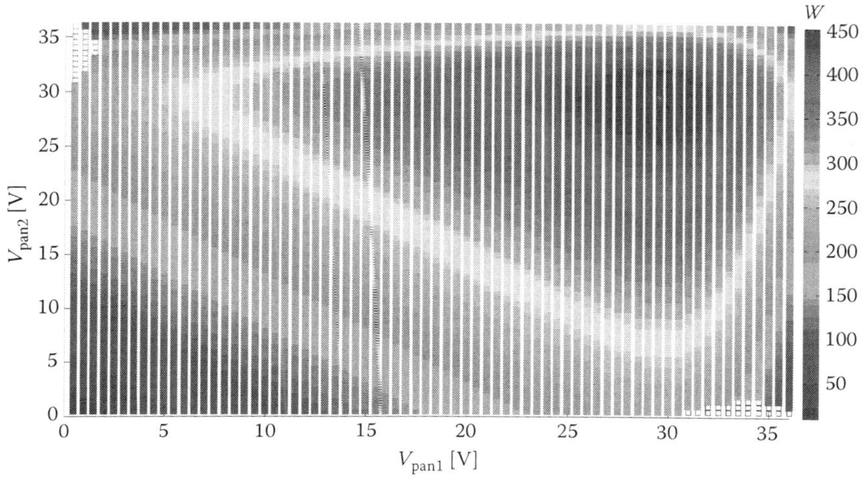

FIGURE 4.10 (See color insert)
(Buckboost) $G_1 = G_2 = 1000$ W/m^2, $V_{bulk} = 50$ V, $V_{ds\,max} = 80$ V, $I_{ds\,max} \to \infty$.

violated. In the absence of such protection circuitries, when the absolute MPP does not belong to the feasibility region, the MPPT circuitries would force an operating point nearly coincident with the absolute MPP rather than with the feasible MPP, thus leading to the possible damage of one or more SCPVMs.

Moreover, many inverter manufacturers state that DMPPT is beneficial since it allows us to have a fixed voltage at the inverter input because the

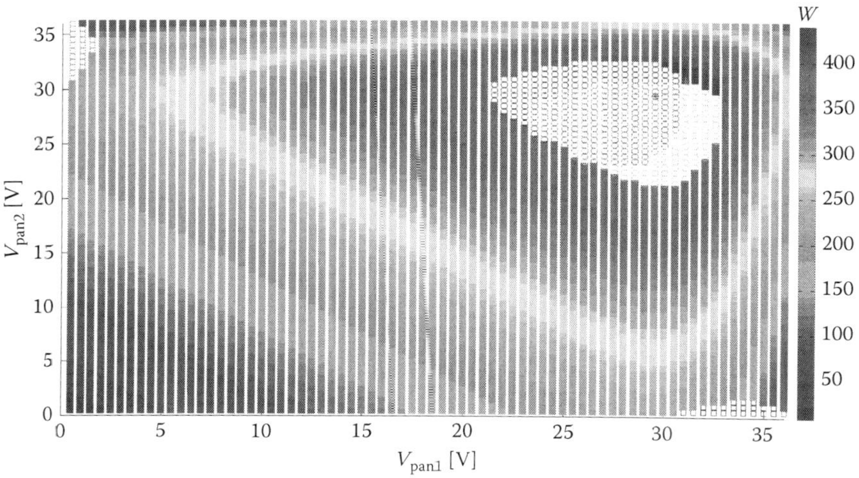

FIGURE 4.11 (See color insert)
(Buckboost) $G_1 = G_2 = 1000$ W/m^2, $V_{bulk} = 50$ V, $V_{ds\,max} = 80$ V, $I_{ds\,max} = 16$ A.

tracking of the MPP is performed on each PV module, independently from the other ones, so that the inverter is allowed to operate at the input voltage value where the inverter itself exhibits its peak efficiency. In order to be really effective instead, the MPPT function must be performed not only by each SCPVM, but also by the inverter on its own input voltage. This aspect will be clarified in the following.

4.3.3 I-V and P-V Characteristics of Boost-Based SCPVMs

The first step is to show which is the shape of the current vs. voltage characteristic at the output of an ideal boost MPPT converter fed by a PV module, that is, at the output of a boost-based SCPVM. In other words, we are interested in understanding how the current vs. voltage characteristic of a PV module (input characteristic of the boost MPPT converter) is modified when adopting a boost-based SCPVM. If we refer to an ideal lossless boost converter made with devices characterized by unlimited voltage and current ratings, it is easy to understand that if the boost output voltage is greater than (or at most equal to) V_{MPP}, then the current vs. voltage characteristic of the SCPVM is a hyperbole of equation $V^*I = P_{MPP}$, where V is the output voltage of the SCPVM, I is the output current of the SCPVM, and P_{MPP} is the maximum power that can be provided by the adopted PV module in the considered atmospheric conditions. In fact, the boost MPPT converter can keep the PV module in its MPP and, at the same time, is able to provide an output voltage greater than (or at most equal to) its input voltage, that is, greater than (or at most equal to) V_{MPP}. Instead, if the output voltage V of the SCPVM is lower than V_{MPP}, then the current vs. voltage characteristic of the SCPVM is coincident with that of the adopted PV module in the considered atmospheric conditions. In fact, in such a case ($V < V_{MPP}$), of course, the operating voltage of the PV module must also be located at the left of the MPP, where the power vs. voltage characteristic of the PV module is monotonically increasing with the voltage. Therefore the PV operating point characterized by the highest available power is also characterized by the highest possible PV voltage, which in turn cannot exceed V. This means that it must be $V_{pan} = V$ and hence also $I_{pan} = I$; therefore the portion for $V < V_{MPP}$ of the current vs. voltage output characteristic of the SCPVM must be coincident with the portion for $V_{pan} < V_{MPP}$ of the current vs. voltage characteristic of the PV module. In particular, in Figure 4.12 the I-V characteristic of an SW225 PV module (G = 1000 W/m^2, ambient temperature T_a = 25°C) is reported together with the corresponding I-V output characteristic of the associated boost-based SCPVM. In the sequel, the PV module temperature T will be always evaluated by means of the following approximated formula: $T = T_a$ + (NOCT – 20°)*G/800, where T_a is the ambient temperature and NOCT is the nominal operating cell temperature [1].

The next step is that of obtaining the equivalent current vs. voltage characteristic of the series connection of two boost-based SCPVMs. As an example,

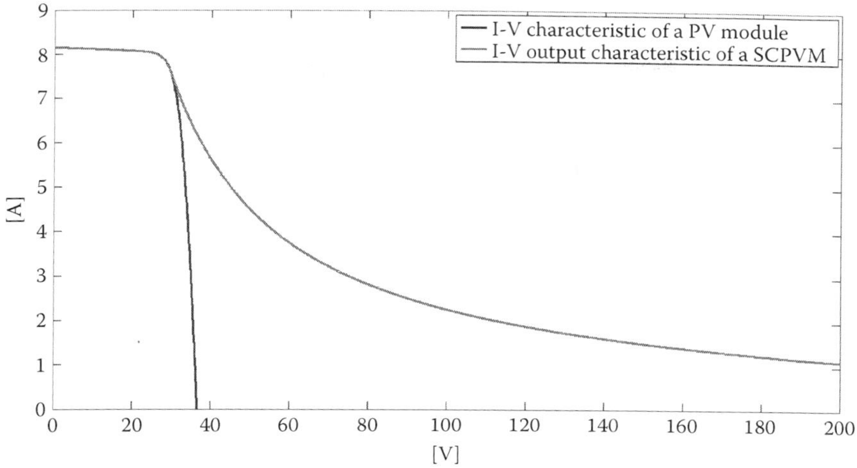

FIGURE 4.12
$G = 1000$ W/m², $T_a = 25°$C; boost-based SCPVM ($V_{ds\,max} \rightarrow \infty$).

in Figure 4.13, the I-V characteristics of two SW225 PV modules respectively operating at $G_1 = 1000$ W/m², $T_a = 25°$C, and at $G_2 = 800$ W/m², $T_a = 25°$C are reported together with the corresponding I-V output characteristics of the two associated boost-based SCPVMs.

It is easy to obtain the I-V equivalent characteristic of two electrical devices that are connected in series, once the I-V characteristic of each of the two devices is known. Basically, for each value of the current (which is common

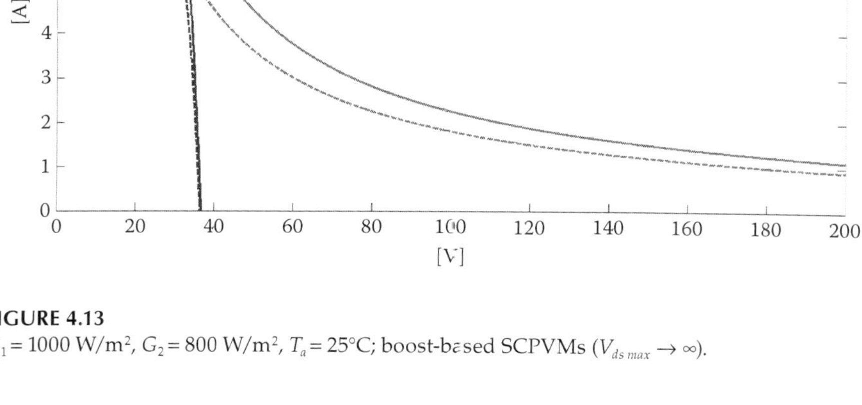

FIGURE 4.13
$G_1 = 1000$ W/m², $G_2 = 800$ W/m², $T_a = 25°$C; boost-based SCPVMs ($V_{ds\,max} \rightarrow \infty$).

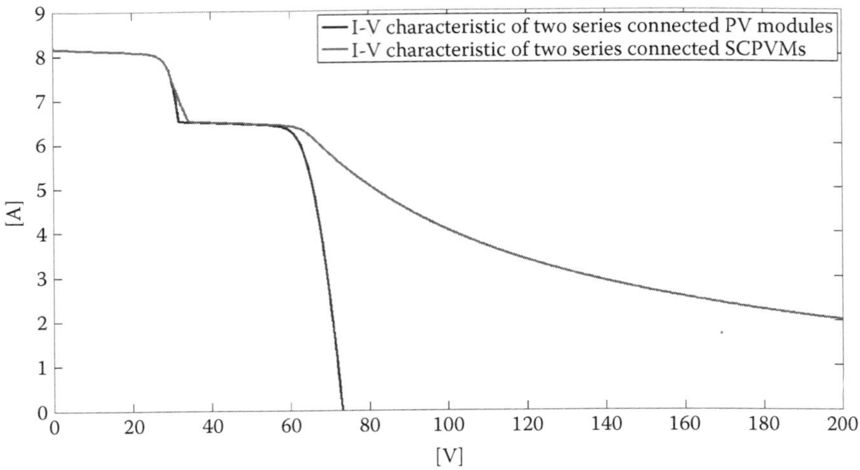

FIGURE 4.14
$G_1 = 1000$ W/m², $G_2 = 800$ W/m², $T_a = 25$°C; boost-based SCPVMs ($V_{ds\,max} \rightarrow \infty$).

to the two devices since they are connected in series), the corresponding value of the voltage is given by the sum of the two voltages, which can be obtained by the two characteristics of the two devices in correspondence of the considered value of the current. By adopting the above technique it is possible to obtain, as shown in Figure 4.14, the I-V characteristic of the series connection of two PV modules (G_1 = 1000 W/m², G_2 = 800 W/m², T_a = 25°C)

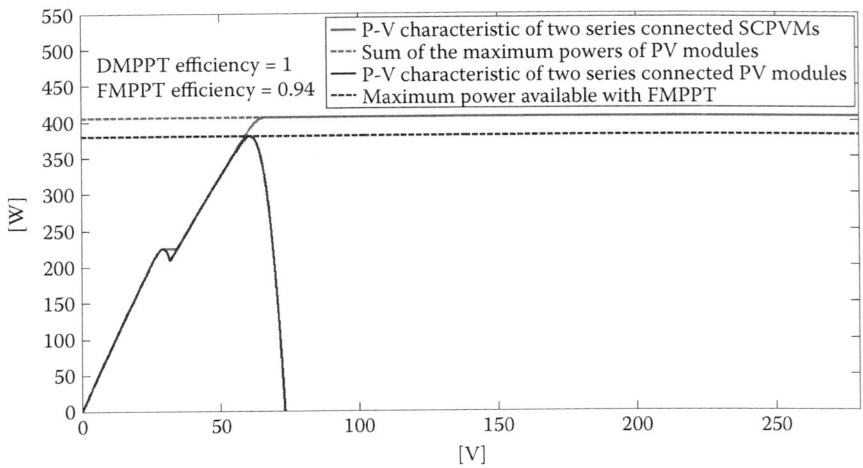

FIGURE 4.15
$G_1 = 1000$ W/m², $G_2 = 800$ W/m², $T_a = 25$°C; boost-based SCPVMs ($V_{ds\,max} \rightarrow \infty$).

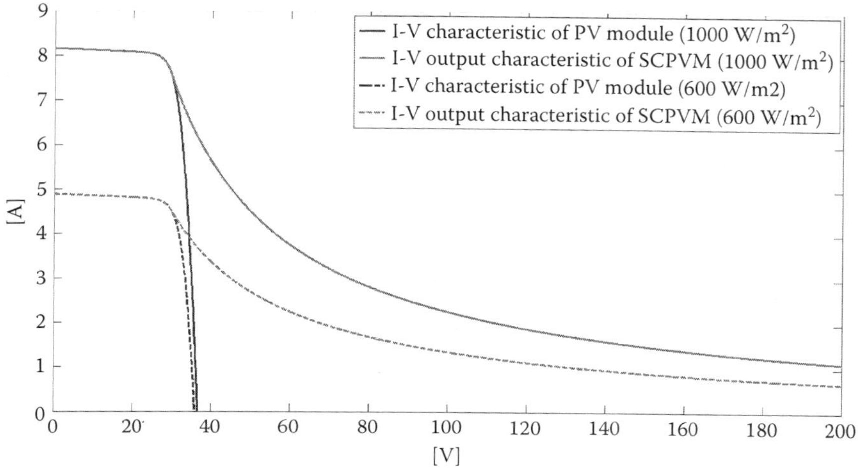

FIGURE 4.16
$G_1 = 1000$ W/m^2, $G_2 = 600$ W/m^2, $T_a = 25$°C; boost-based SCPVMs ($V_{ds\,max} \rightarrow \infty$).

and the I-V characteristic of the series connection of the associated boost-based SCPVMs.

In Figure 4.15 the power vs. voltage (P-V) characteristic of the series connection of the two PV modules ($G_1 = 1000$ W/m^2, $G_2 = 800$ W/m^2, $T_a = 25$°C) and the P-V characteristic of the series connection of the associated boost-based SCPVMs is reported. In the case shown in Figure 4.15 it is $\eta_{FMPPT} = 0.94$; instead, it is $\eta_{DMPPT} = 1$, but only if the voltage V_{bulk} exceeds a threshold of

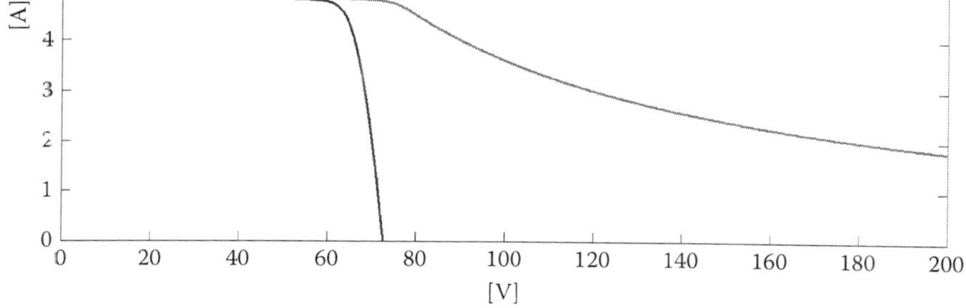

FIGURE 4.17
$G_1 = 1000$ W/m^2, $G_2 = 600$ W/m^2, $T_a = 25$°C; boost-based SCPVMs ($V_{ds\,max} \rightarrow \infty$).

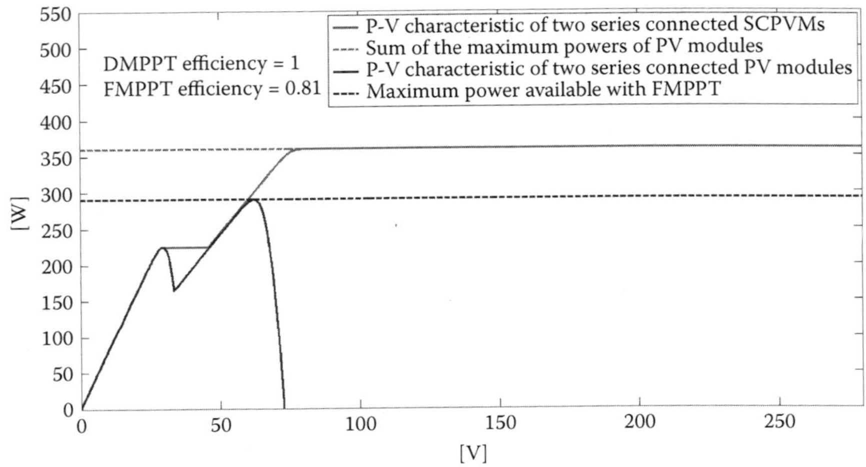

FIGURE 4.18
$G_1 = 1000$ W/m², $G_2 = 600$ W/m², $T_a = 25$°C; boost-based SCPVMs ($V_{ds\ max} \rightarrow \infty$).

about 67 V. Therefore, in such a case, DMPPT is beneficial only if the efficiency of the power stage of the two MPPT converters is higher than 0.94 and the operating voltage of the series connection is higher than about 67 V.

Similar considerations also hold for Figures 4.16–4.18, which refer to the case $G_1 = 1000$ W/m², $G_2 = 600$ W/m², and $T_a = 25$°C. In such a case it

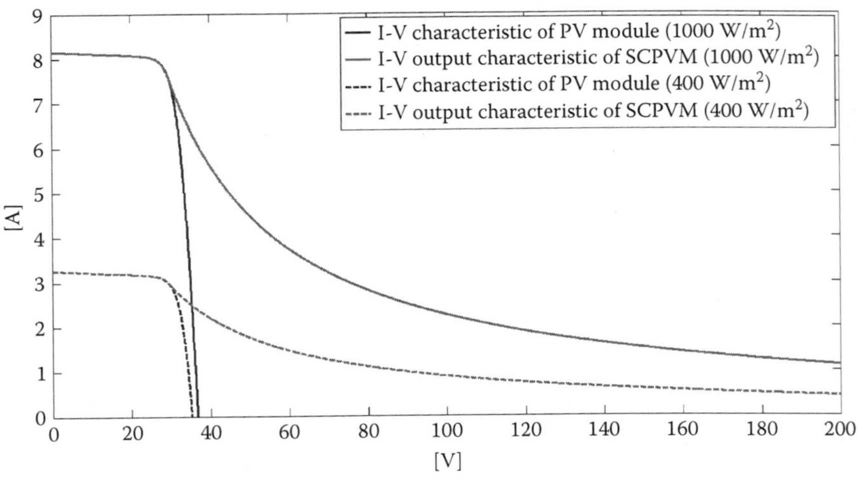

FIGURE 4.19
$G_1 = 1000$ W/m², $G_2 = 400$ W/m², $T_a = 25$°C; boost-based SCPVMs ($V_{ds\ max} \rightarrow \infty$).

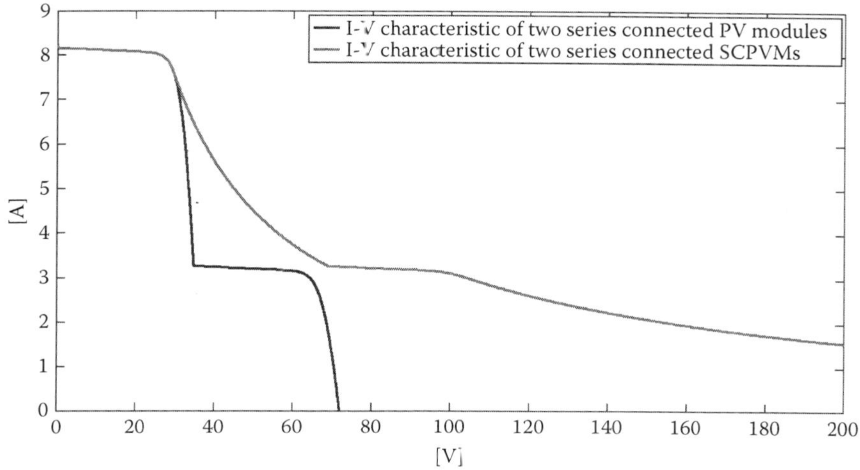

FIGURE 4.20
$G_1 = 1000$ W/m², $G_2 = 400$ W/m², $T_a = 25$°C; boost-based SCPVMs ($V_{ds\,max} \to \infty$).

is $\eta_{FMPPT} = 0.81$; instead, $\eta_{DMPPT} = 1$, but only if the voltage V_{bulk} exceeds a threshold of about 80 V.

Figures 4.19–4.21 refer instead to the case $G_1 = 1000$ W/m², $G_2 = 400$ W/m², and $T_a = 25$°C. In such a case it is $\eta_{FMPPT} = 0.72$; instead, $\eta_{DMPPT} = 1$, but only if $V_{bulk} > 104$ V.

Eventually Figures 4.22–4.24 refer to the case $G_1 = 1000$ W/m², $G_2 = 200$ W/m², and $T_a = 25$°C. In such a case it is $\eta_{FMPPT} = 0.84$; instead, $\eta_{DMPPT} = 1$ if

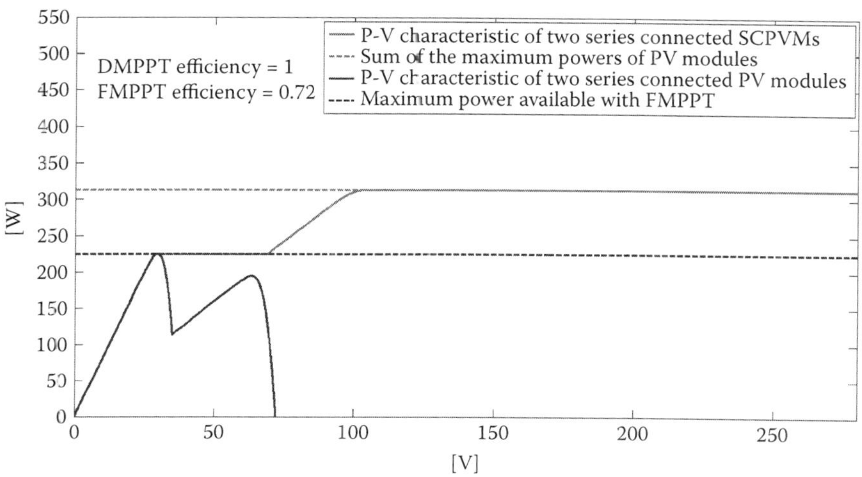

FIGURE 4.21
$G_1 = 1000$ W/m², $G_2 = 400$ W/m², $T_a = 25$°C; boost-based SCPVMs ($V_{ds\,max} \to \infty$).

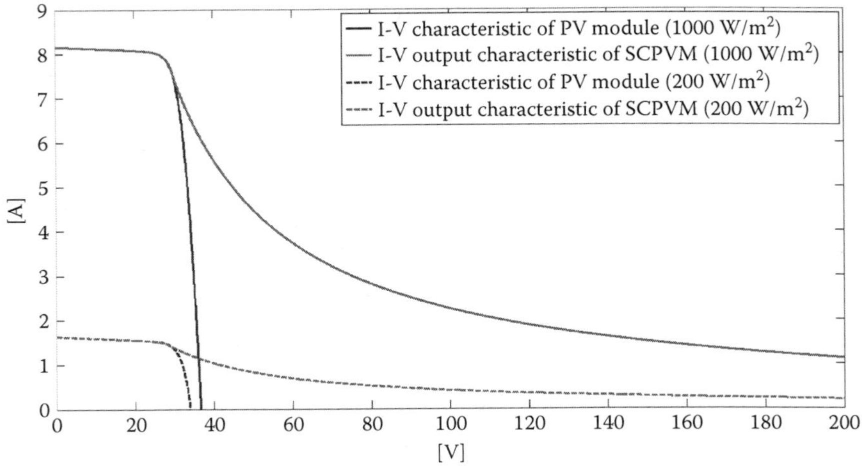

FIGURE 4.22
$G_1 = 1000$ W/m², $G_2 = 200$ W/m², $T_a = 25$°C; boost-based SCPVMs ($V_{ds\ max} \to \infty$).

$V_{bulk} > 184$ V. By analyzing Figures 4.13–4.24 it is evident that the higher the ratio G_1/G_2, the higher the threshold that V_{bulk} must exceed in order to get $\eta_{DMPPT} = 1$, and consequently, the higher the voltage stress the components of the boost converters must sustain.

In Figure 4.25 the I-V characteristic of an SW225 PV module ($G = 1000$ W/m², $T_a = 25$°C) is reported together with the corresponding I-V output characteristic of the associated boost-based SCPVM. The boost converter is lossless

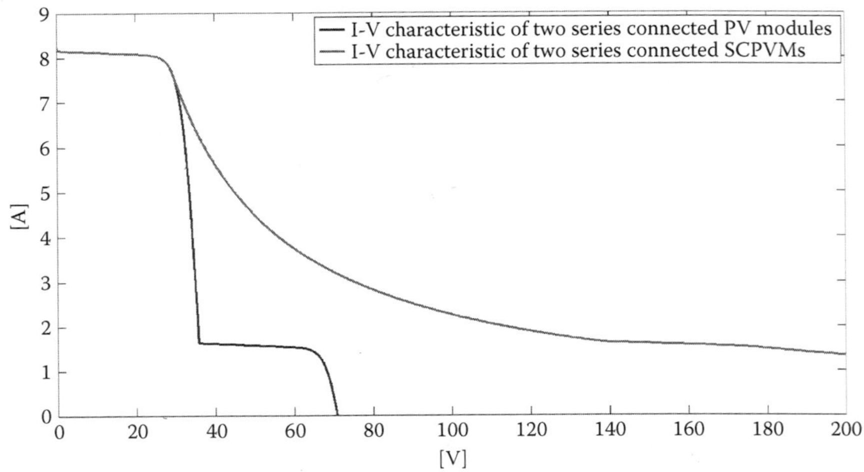

FIGURE 4.23
$G_1 = 1000$ W/m², $G_2 = 200$ W/m², $T_a = 25$°C; boost-based SCPVMs ($V_{ds\ max} \to \infty$).

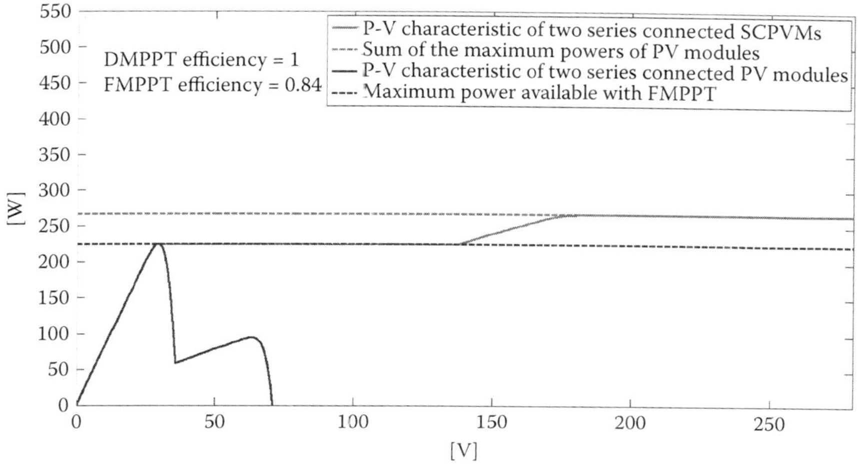

FIGURE 4.24
$G_1 = 1000$ W/m², $G_2 = 200$ W/m², $T_a = 25$°C; boost-based SCPVMs ($V_{ds\ max} \to \infty$).

but is now characterized by a finite value of $V_{ds\ max}$ ($V_{ds\ max} = 60$ V). Therefore the current vs. voltage output characteristic of the SCPVM is now truncated at V = 60 V. Figures 4.25–4.37 have been obtained by considering such a finite voltage rating ($V_{ds\ max} = 60$ V). In Figure 4.26, the I-V characteristics of two SW225 PV modules respectively operating at $G_1 = 1000$ W/m², $T_a = 25$°C and at $G_2 = 800$ W/m², $T_a = 25$°C are reported together with the corresponding I-V output characteristics of the two associated boost-based SCPVMs.

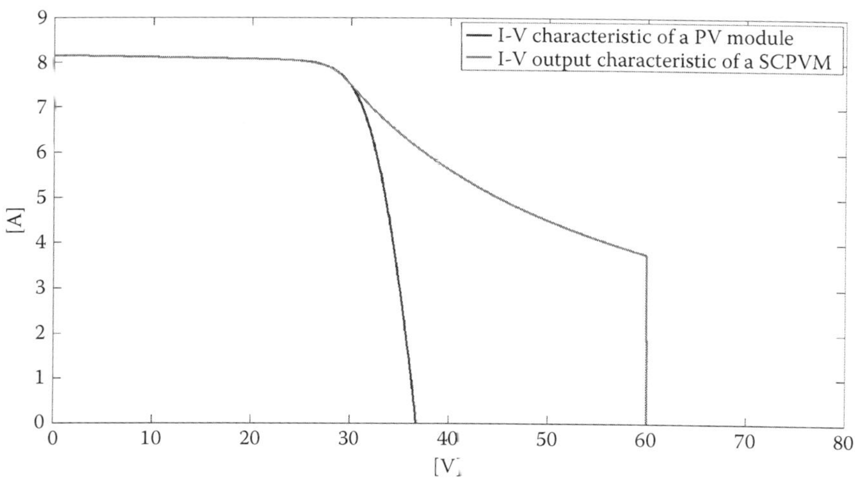

FIGURE 4.25
$G = 1000$ W/m², $T_a = 25$°C; boost-based SCPVM ($V_{ds\ max} = 60$ V).

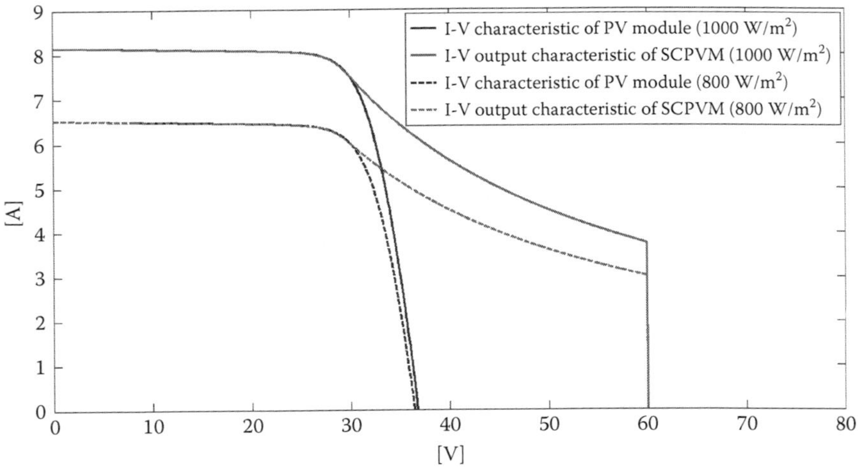

FIGURE 4.26
$G_1 = 1000$ W/m², $G_2 = 800$ W/m², $T_a = 25$°C; boost-based SCPVMs ($V_{ds\,max} = 60$ V).

In Figure 4.27, the I-V characteristic of the series connection of two PV modules ($G_1 = 1000$ W/m², $G_2 = 800$ W/m², $T_a = 25$°C) and the I-V characteristic of the series connection of the associated boost-based SCPVMs are shown.

In Figure 4.28 the P-V characteristic of the series connection of the two PV modules ($G_1 = 1000$ W/m², $G_2 = 800$ W/m², $T_a = 25$°C) and the P-V characteristic of the series connection of the associated boost-based SCPVMs are

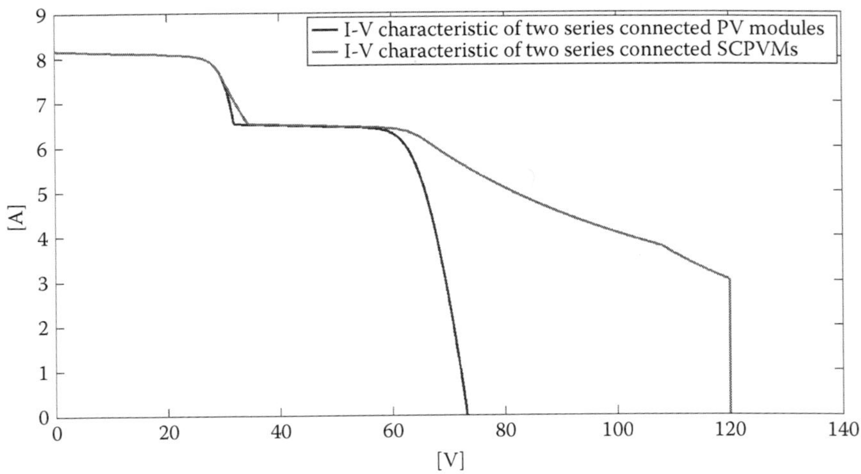

FIGURE 4.27
$G_1 = 1000$ W/m², $G_2 = 800$ W/m², $T_a = 25$°C; boost-based SCPVMs ($V_{ds\,max} = 60$ V).

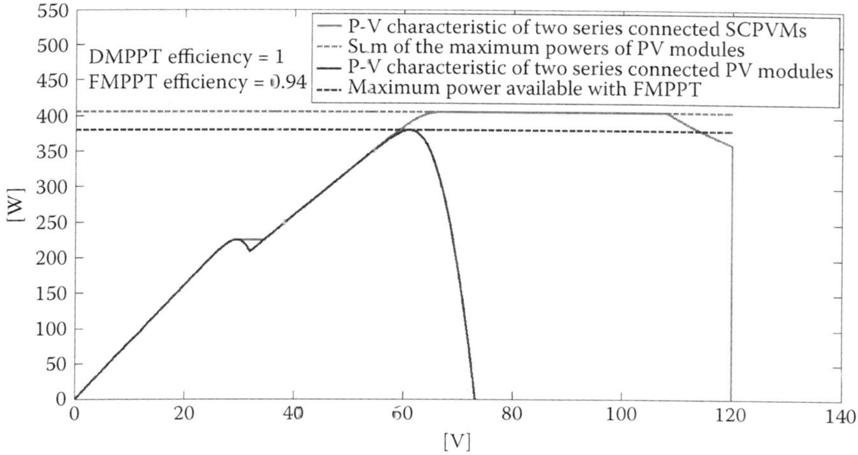

FIGURE 4.28
$G_1 = 1000$ W/m², $G_2 = 800$ W/m², $T_a = 25$°C; boost-based SCPVMs ($V_{ds\,max} = 60$ V).

reported. In the case shown in Figure 4.28 it is $\eta_{FMPPT} = 0.94$; instead, it is $\eta_{DMPPT} = 1$, but only if 67 V $< V_{bulk} <$108 V.

Similar considerations also hold for Figures 4.29–4.31, which refer to the case $G_1 = 1000$ W/m², $G_2 = 600$ W/m², and $T_a = 25$°C.

In such a case it is $\eta_{FMPPT} = 0.81$; instead, $\eta_{DMPPT} = 1$, but only if 80 V $< V_{bulk}$ <96 V.

Figures 4.32–4.34 refer instead to the case $G_1 = 1000$ W/m², $G_2 = 400$ W/m², and $T_a = 25$°C. In such a case it is $\eta_{FMPPT} = 0.72$; instead, the maximum value

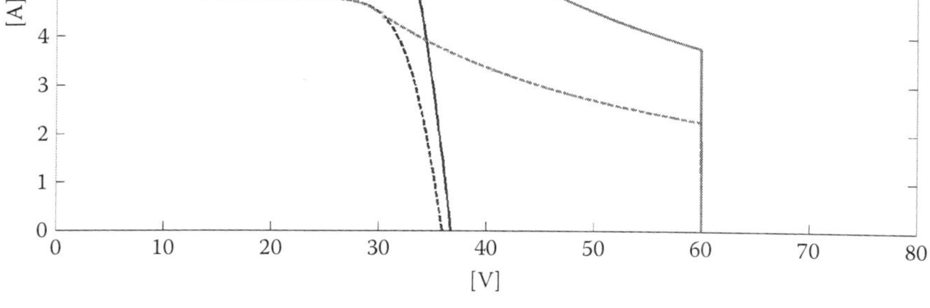

FIGURE 4.29
$G_1 = 1000$ W/m², $G_2 = 600$ W/m², $T_a = 25$°C; boost-based SCPVMs ($V_{ds\,max} = 60$ V).

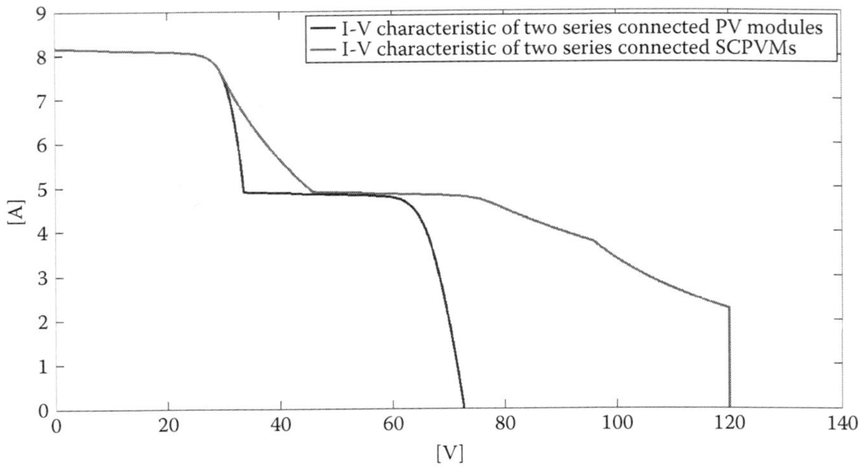

FIGURE 4.30
$G_1 = 1000$ W/m^2, $G_2 = 600$ W/m^2, $T_a = 25$°C; boost-based SCPVMs ($V_{ds\,max} = 60$ V).

of the DMPPT efficiency is $\eta_{DMPPT} = 0.87$ and it can be obtained only if V_{bulk} is nearly equal to 88 V.

Eventually Figures 4.35–4.37 refer to the case G_1 = 1000 W/m^2, G_2 = 200 W/m^2, and $T_a = 25$°C. In such a case it is $\eta_{FMPPT} = 0.84$; instead, the maximum value of the DMPPT efficiency is $\eta_{DMPPT} = 0.84$, and it can be obtained only if 30 V < V_{bulk} < 60 V.

By analyzing Figures 4.28, 4.31, 4.34, and 4.37 it is evident that not only is it not always possible to get η_{DMPPT} = 1, but also, in all four considered cases,

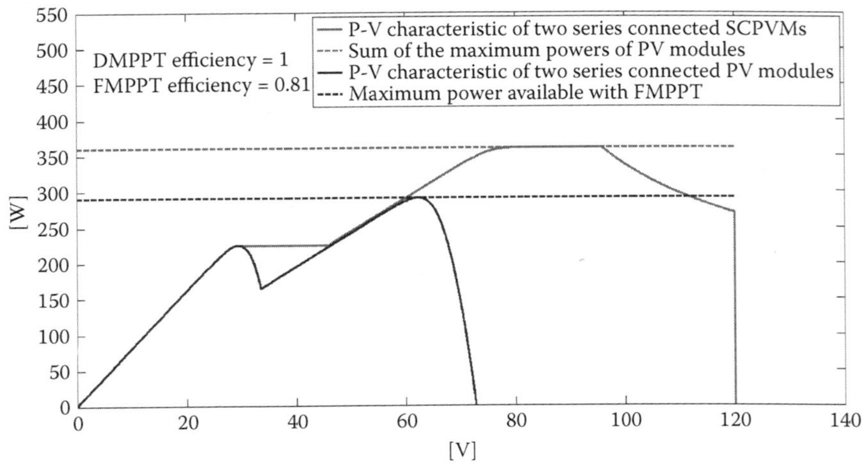

FIGURE 4.31
$G_1 = 1000$ W/m^2, $G_2 = 600$ W/m^2, $T_a = 25$°C; boost-based SCPVMs ($V_{ds\,max} = 60$ V).

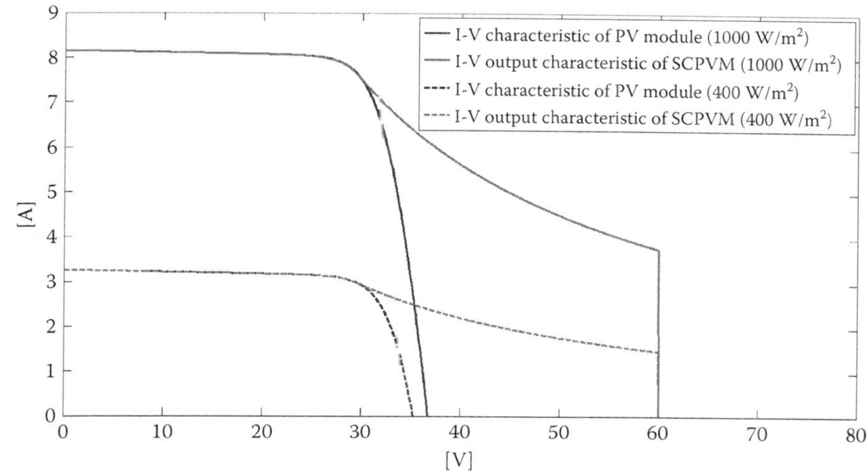

FIGURE 4.32
$G_1 = 1000$ W/m², $G_2 = 400$ W/m², $T_a = 25$°C; boost-based SCPVMs ($V_{ds\,max} = 60$ V).

voltage operating regions exist where the DMPPT efficiency is lower than the maximum obtainable FMPPT efficiency.

4.3.4 I-V and P-V Characteristics of Buckboost-Based SCPVMs

In the following a similar analysis will be carried out with reference to buckboost-based SCPVMs. Even in this case the first step is to show which

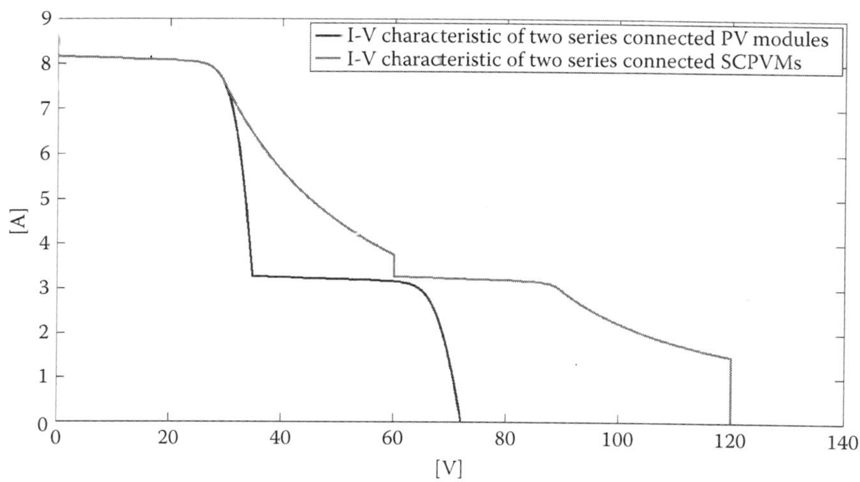

FIGURE 4.33
$G_1 = 1000$ W/m², $G_2 = 400$ W/m², $T_a = 25$°C; boost-based SCPVMs ($V_{ds\,max} = 60$ V).

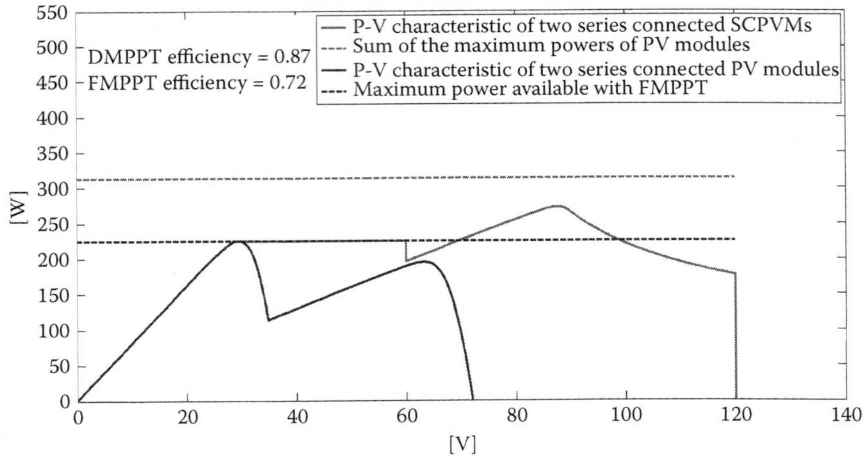

FIGURE 4.34
$G_1 = 1000$ W/m^2, $G_2 = 400$ W/m^2, $T_a = 25$°C; boost-based SCPVMs ($V_{ds\,max} = 60$ V).

is the shape of the current vs. voltage characteristic at the output of an ideal buckboost MPPT converter fed by a PV module, that is, at the output of a buckboost-based SCPVM. If we refer to an ideal lossless buckboost converter made with devices characterized by unlimited voltage and current ratings, it is easy to understand that whichever the value of V (in the following, since the buckboost converter is an inverting DC/DC converter, V will label the absolute value of the output voltage of the SCPVM), the I-V characteristic of the SCPVM is a hyperbole of equation $V^*I = P_{MPP}$,

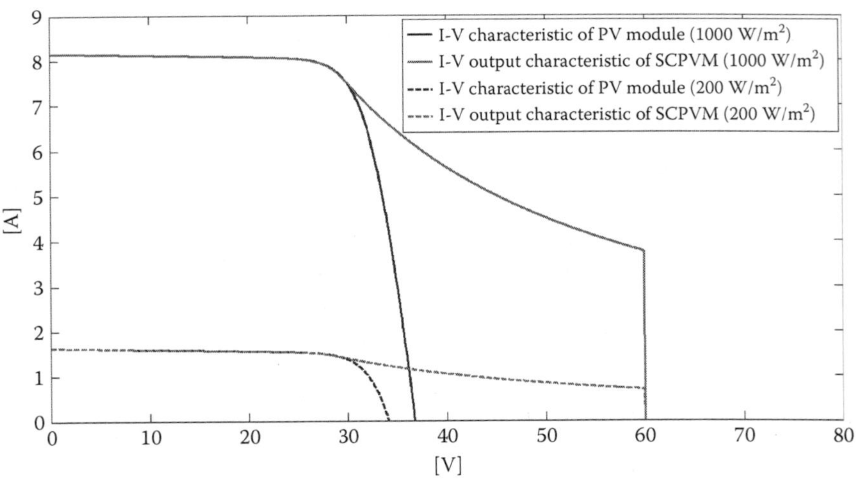

FIGURE 4.35
$G_1 = 1000$ W/m^2, $G_2 = 200$ W/m^2, $T_a = 25$°C; boost-based SCPVMs ($V_{ds\,max} = 60$ V).

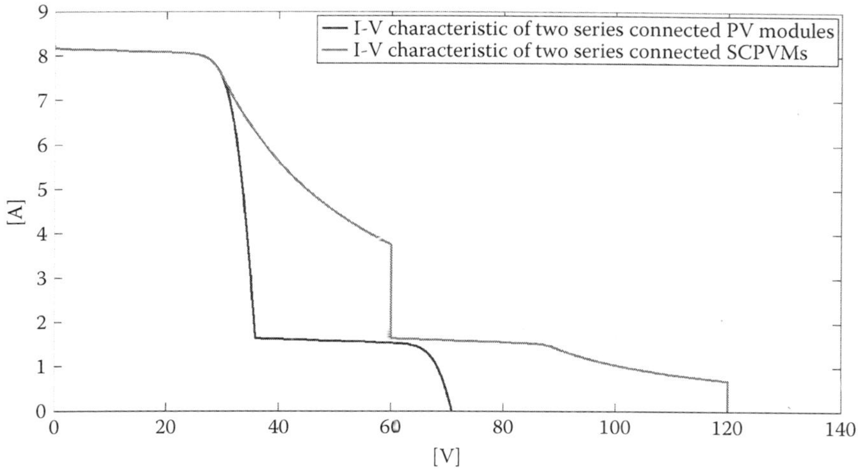

FIGURE 4.36
$G_1 = 1000$ W/m², $G_2 = 200$ W/m², $T_a = 25$°C; boost-based SCPVMs ($V_{ds\ max} = 60$ V).

where I is the output current of the SCPVM and P_{MPP} is the maximum power that can be provided by the adopted PV module in the considered atmospheric conditions.

In fact, the buckboost MPPT converter can keep the PV module in its MPP, and at the same time, it is able to provide an output voltage V greater or lower than its input voltage, that is, greater or lower than V_{MPP}. As an example, in Figure 4.38 the I-V characteristic of an SW225 PV module ($G = 1000$ W/m², $T_a = 25$°C) is reported together with the corresponding I-V output

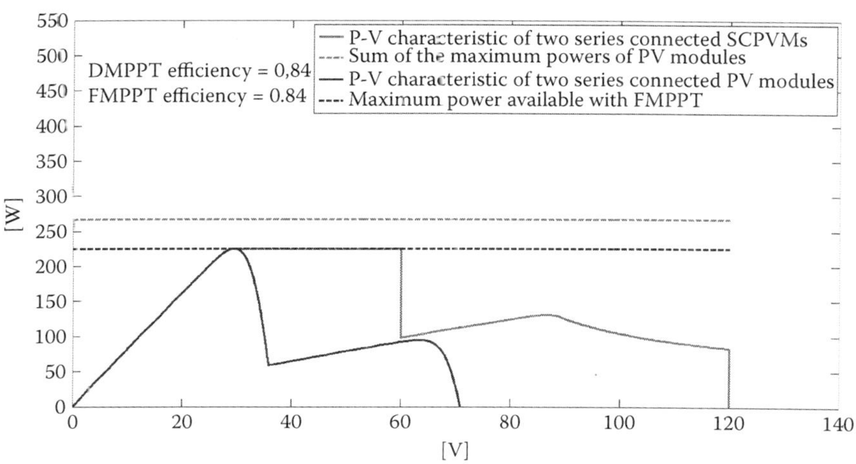

FIGURE 4.37
$G_1 = 1000$ W/m², $G_2 = 200$ W/m², $T_a = 25$°C; boost-based SCPVMs ($V_{ds\ max} = 60$ V).

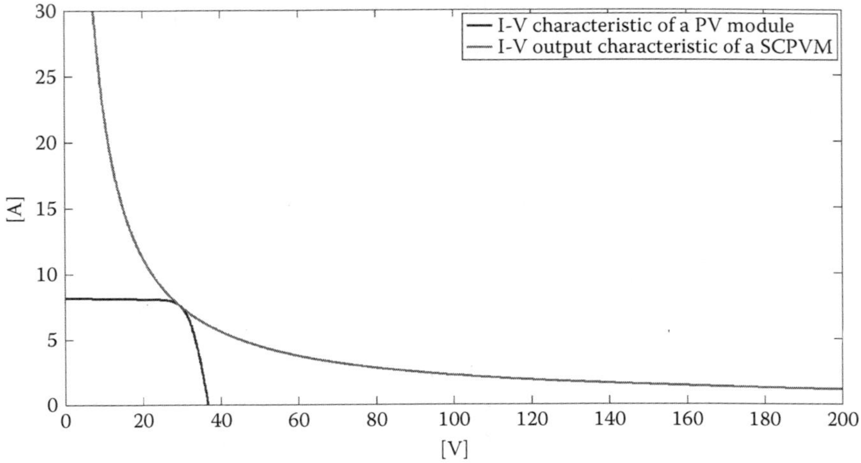

FIGURE 4.38
$G = 1000$ W/m², $T_a = 25$°C; buckboost-based SCPVM ($V_{ds\ max} \to \infty$, $I_{ds\ max} \to \infty$).

characteristic of the associated buckboost-based SCPVM. Figures 4.38–4.50 refer to the case of ideal lossless buckboost MPPT converters characterized by $V_{ds\ max} \to \infty$, $I_{ds\ max} \to \infty$.

Of course, the two characteristics of Figure 4.38 are tangent in the MPP. In Figure 4.39, the I-V characteristics of two SW225 PV modules respectively operating at $G_1 = 1000$ W/m², $T_a = 25$°C, and at $G_2 = 800$ W/m², $T_a = 25$°C are reported together with the corresponding I-V output characteristics of the two associated buckboost-based SCPVMs.

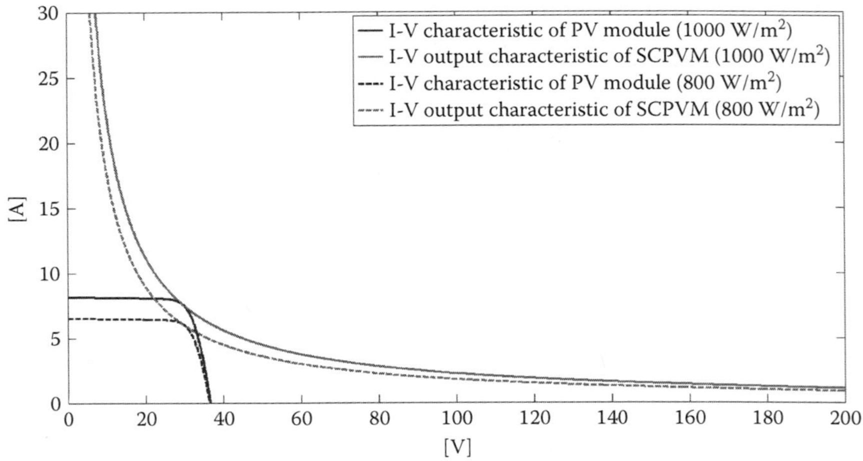

FIGURE 4.39
$G_1 = 1000$ W/m², $G_2 = 800$ W/m², $T_a = 25$°C; buckboost-based SCPVMs ($V_{ds\ max} \to \infty$, $I_{ds\ max} \to \infty$).

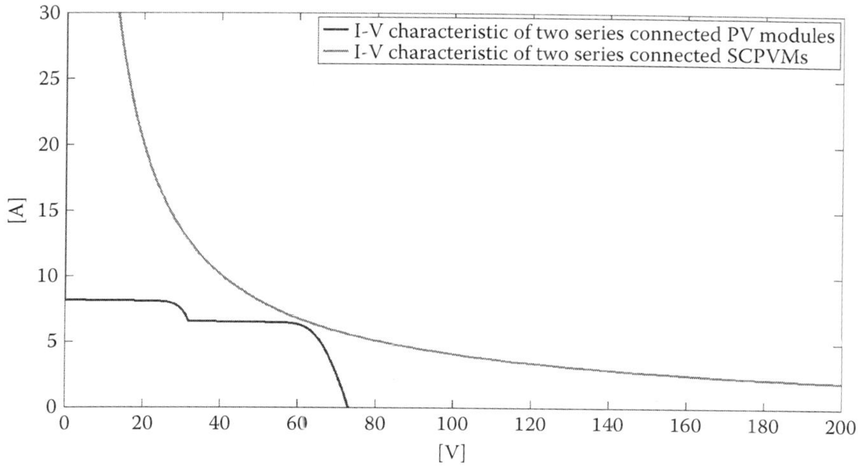

FIGURE 4.40
$G_1 = 1000$ W/m², $G_2 = 800$ W/m², $T_a = 25$°C; buckboost-based SCPVMs ($V_{ds\,max} \rightarrow \infty$, $I_{ds\,max} \rightarrow \infty$).

The next step is that of obtaining the equivalent current vs. voltage characteristic of the series connection of two buckboost-based SCPVMs. Such a characteristic can be obtained as previously discussed with reference to boost-based SCPVMs. Therefore it is possible to obtain, as shown in Figure 4.40, the I-V characteristic of the series connection of two PV modules ($G_1 = 1000$ W/m², $G_2 = 800$ W/m², $T_a = 25$°C) and the I-V characteristic of the series connection of the associated buckboost-based SCPVMs.

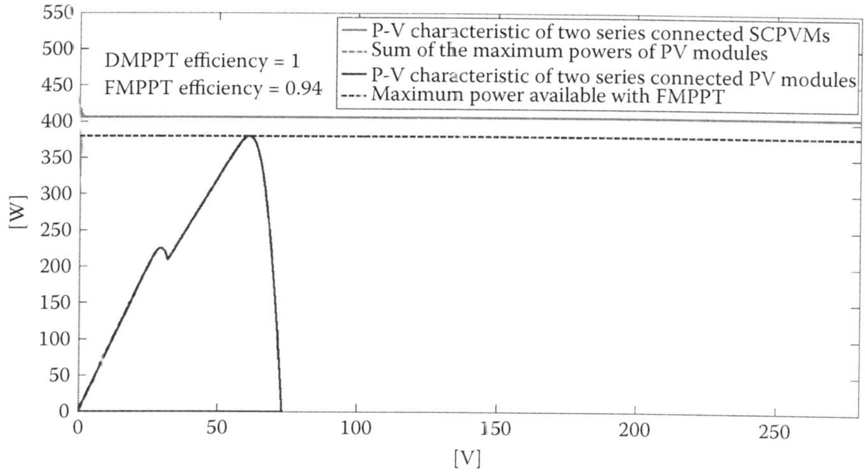

FIGURE 4.41
$G_1 = 1000$ W/m², $G_2 = 800$ W/m², $T_a = 25$°C; buckboost-based SCPVMs ($V_{ds\,max} \rightarrow \infty$, $I_{ds\,max} \rightarrow \infty$).

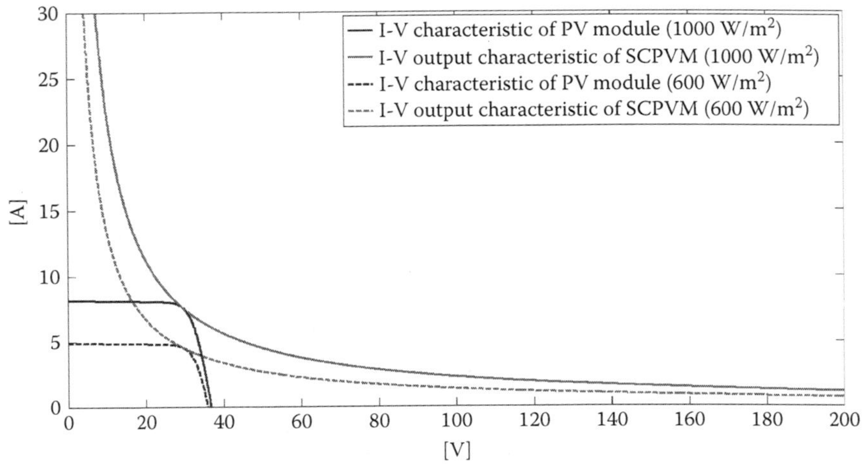

FIGURE 4.42
$G_1 = 1000$ W/m^2, $G_2 = 600$ W/m^2, $T_a = 25$°C; buckboost-based SCPVMs ($V_{ds\,max} \rightarrow \infty$, $I_{ds\,max} \rightarrow \infty$).

In Figure 4.41 the P-V characteristic of the series connection of the two PV modules ($G_1 = 1000$ W/m^2, $G_2 = 800$ W/m^2, $T_a = 25$°C) and the P-V characteristic of the series connection of the associated buckboost-based SCPVMs are reported. They are $\eta_{FMPPT} = 0.94$, while $\eta_{DMPPT} = 1$, whichever the value of V_{bulk}.

Similar considerations also hold for Figures 4.42–4.44, which refer to the case $G_1 = 1000$ W/m^2, $G_2 = 600$ W/m^2, and $T_a = 25$°C.

In such a case it is $\eta_{FMPPT} = 0.81$; instead, $\eta_{DMPPT} = 1$, whichever the value of V_{bulk}.

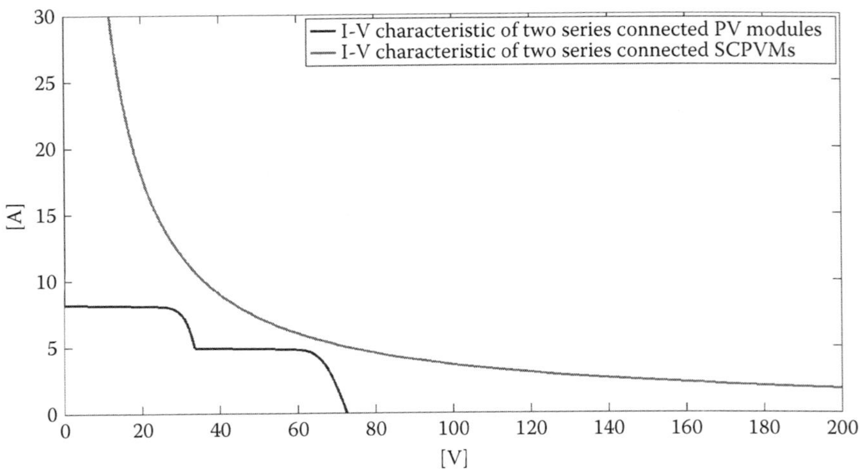

FIGURE 4.43
$G_1 = 1000$ W/m^2, $G_2 = 600$ W/m^2, $T_a = 25$°C; buckboost-based SCPVMs ($V_{ds\,max} \rightarrow \infty$, $I_{ds\,max} \rightarrow \infty$).

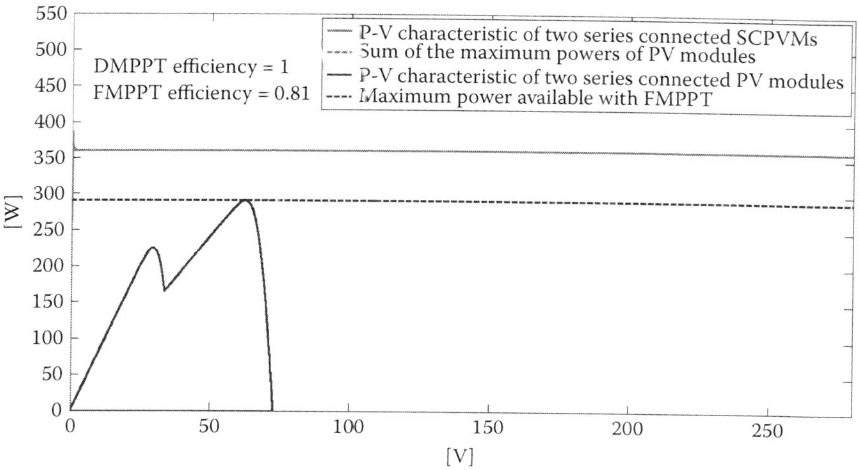

FIGURE 4.44
$G_1 = 1000$ W/m², $G_2 = 600$ W/m², $T_a = 25$°C; buckboost-based SCPVMs ($V_{ds\,max} \to \infty$, $I_{ds\,max} \to \infty$).

Figures 4.45–4.47 refer instead to the case $G_1 = 1000$ W/m², $G_2 = 400$ W/m², and $T_a = 25$°C. In such a case it is $\eta_{FMPPT} = 0.72$; once again, $\eta_{DMPPT} = 1$, whichever the value of V_{bulk}.

Eventually Figures 4.48–4.50 refer to the case $G_1 = 1000$ W/m², $G_2 = 200$ W/m², and $T_a = 25$°C. In such a case it is $\eta_{FMPPT} = 0.84$; also in this case $\eta_{DMPPT} = 1$, whichever the value of V_{bulk}.

By analyzing Figures 4.38–4.50 and comparing them with the corresponding Figures 4.12–4.24, it is evident that buckboost-based SCPVMs seem to be much

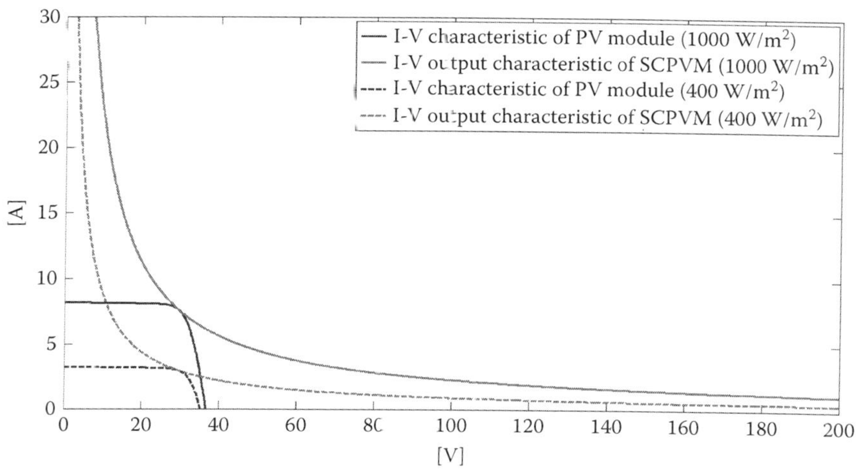

FIGURE 4.45
$G_1 = 1000$ W/m², $G_2 = 400$ W/m², $T_a = 25$°C; buckboost-based SCPVMs ($V_{ds\,max} \to \infty$, $I_{ds\,max} \to \infty$).

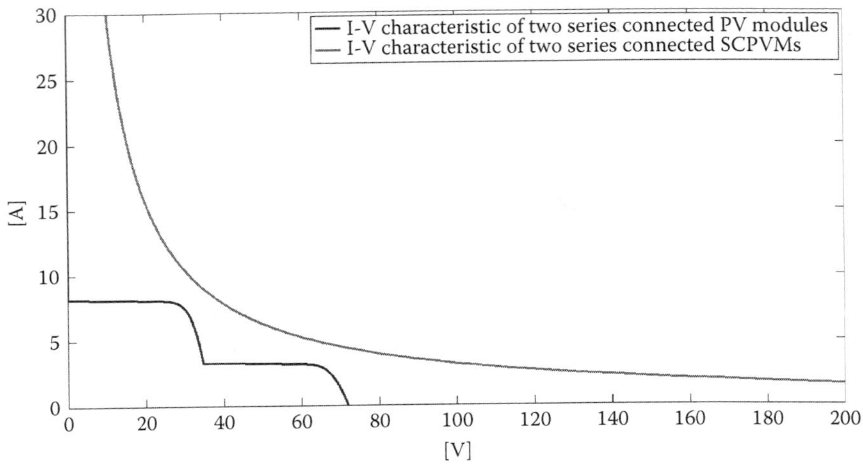

FIGURE 4.46
$G_1 = 1000$ W/m², $G_2 = 400$ W/m², $T_a = 25$°C; buckboost-based SCPVMs ($V_{ds\ max} \rightarrow \infty$, $I_{ds\ max} \rightarrow \infty$).

more flexible than boost-based SCPVMs. But this is true only if we consider switching devices characterized by unlimited voltage and current ratings.

In the following we discuss the effect of finite voltage and current ratings of the switching devices adopted in buckboost-based SCPVMs. In particular, Figures 4.51–4.63 refer to the case $V_{ds\ max} = 60$ V and $I_{ds\ max} = 16$ A; the considered buckboost converters are lossless. As usual, the first step is to show which is the shape of the I-V characteristic at the output of a lossless

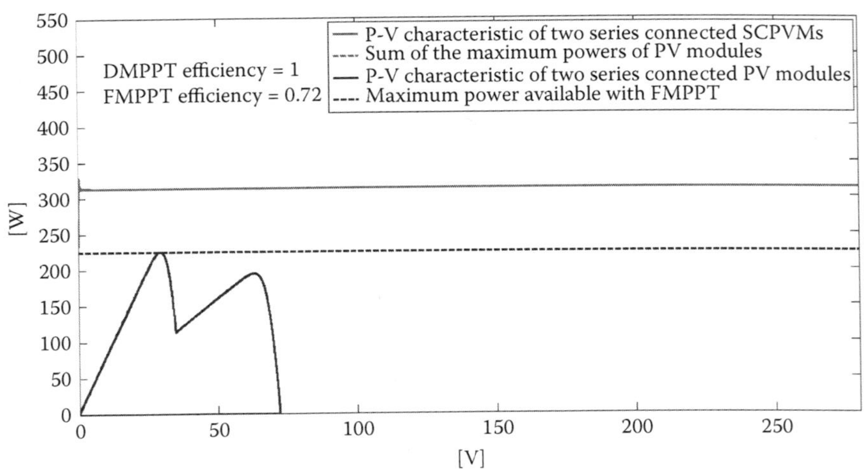

FIGURE 4.47
$G_1 = 1000$ W/m², $G_2 = 400$ W/m², $T_a = 25$°C; buckboost-based SCPVMs ($V_{ds\ max} \rightarrow \infty$, $I_{ds\ max} \rightarrow \infty$).

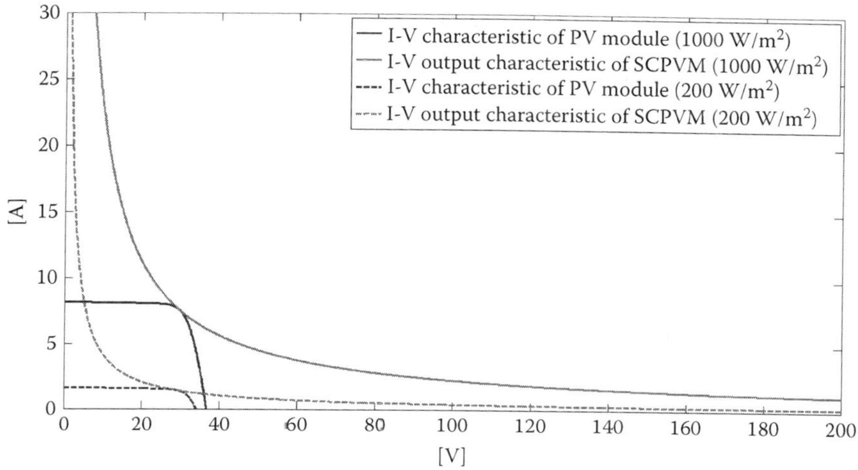

FIGURE 4.48
$G_1 = 1000$ W/m², $G_2 = 200$ W/m², $T_a = 25°C$; buckboost-based SCPVMs ($V_{ds\ max} \rightarrow \infty$, $I_{ds\ max} \rightarrow \infty$).

buckboost MPPT converter fed by a PV module, by taking into account the finite values assumed in this case by $V_{ds\ max}$ and $I_{ds\ max}$.

If $I < (I_{ds\ max} - I_{MPP})$ and $V < (V_{ds\ max} - V_{MPP})$ the current vs. voltage characteristic of the SCPVM is a hyperbole of equation $V^*I = P_{MPP}$, where V is the absolute value of the output voltage of the SCPVM, I is the output current of the SCPVM, and P_{MPP} is the maximum power, which can be provided by the adopted PV module in the considered atmospheric conditions. In fact, if the above inequalities are fulfilled, the buckboost MPPT converter can keep

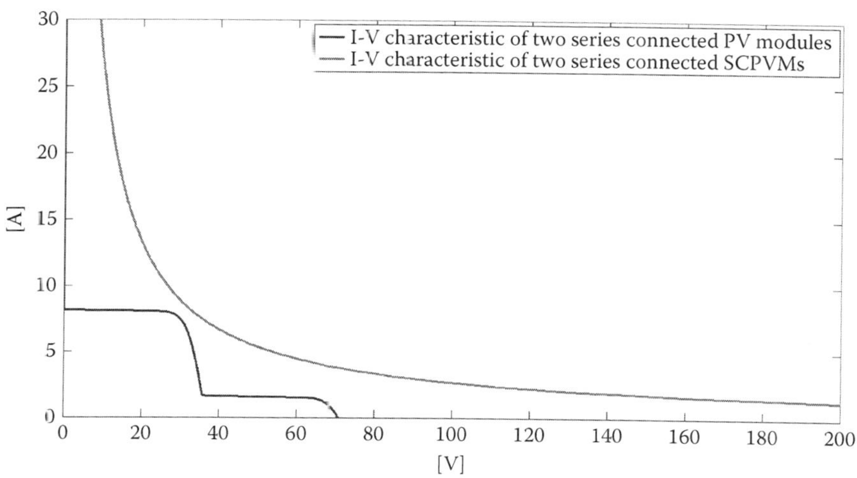

FIGURE 4.49
$G_1 = 1000$ W/m², $G_2 = 200$ W/m², $T_a = 25°C$; buckboost-based SCPVMs ($V_{ds\ max} \rightarrow \infty$, $I_{ds\ max} \rightarrow \infty$).

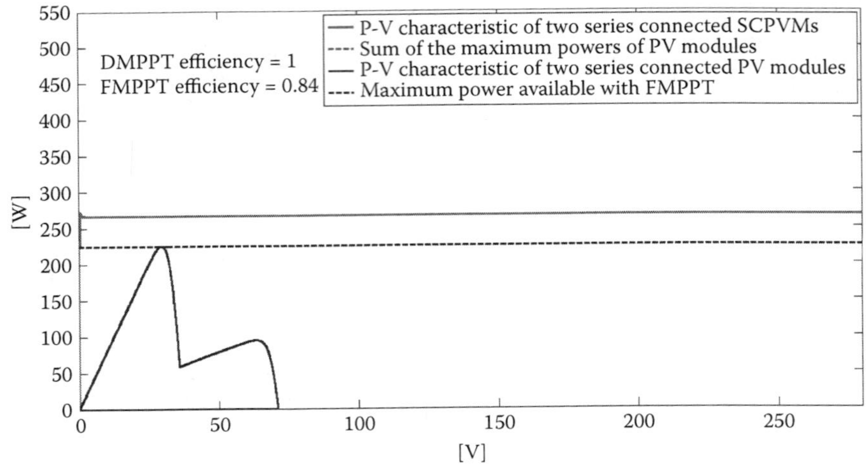

FIGURE 4.50
$G_1 = 1000$ W/m², $G_2 = 200$ W/m², $T_a = 25$°C; buckboost-based SCPVMs ($V_{ds\ max} \rightarrow \infty$, $I_{ds\ max} \rightarrow \infty$).

the PV module in its MPP, and at the same time, it is able to grant that the voltage and current ratings of its switching devices are not exceeded.

For $V > (V_{ds\ max} - V_{MPP})$ it is $V_{pan} = V_{ds\ max} - V$ and $I = P_{pan}(V_{pan})/V$. Finally, for $V < P_{MPP}/(I_{ds\ max} - I_{MPP})$ it is $I = I_{ds\ max} - I_{MPP}$.

As an example, in Figure 4.51 the I-V characteristic of an SW225 PV module (G = 1000 W/m², T_a = 25°C) is reported together with the corresponding I-V output characteristic of the associated buckboost-based SCPVM.

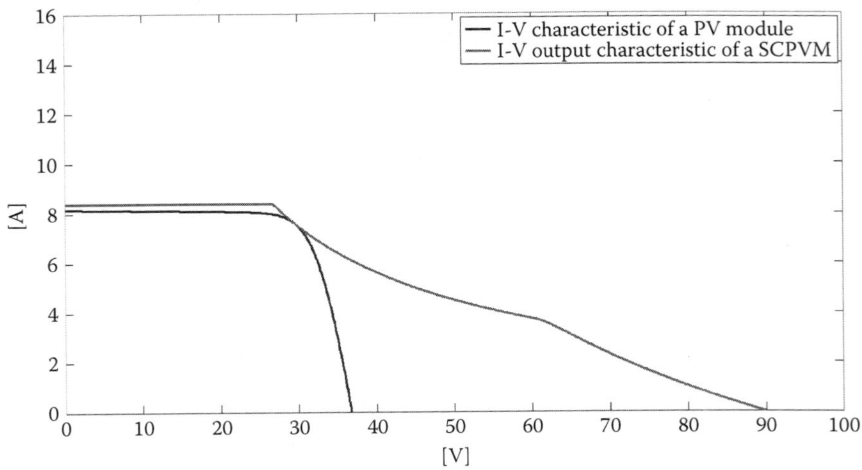

FIGURE 4.51
$G = 1000$ W/m², $T_a = 25$°C; buckboost-based SCPVMs ($V_{ds\ max} = 90$ V, $I_{ds\ max} = 16$ A).

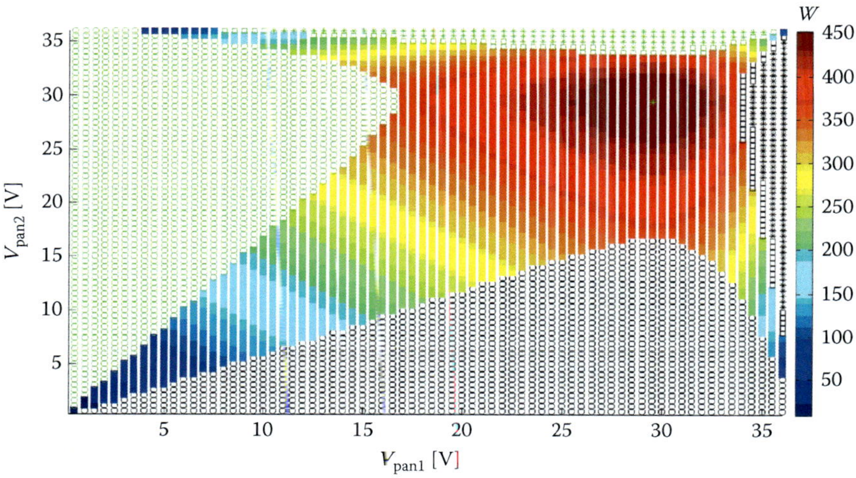

COLOR FIGURE 4.6
(Boost) $G_1 = G_2 = 1000$ W/m², $V_{bulk} = 80$ V, $V_{ds\,max} = 50$ V.

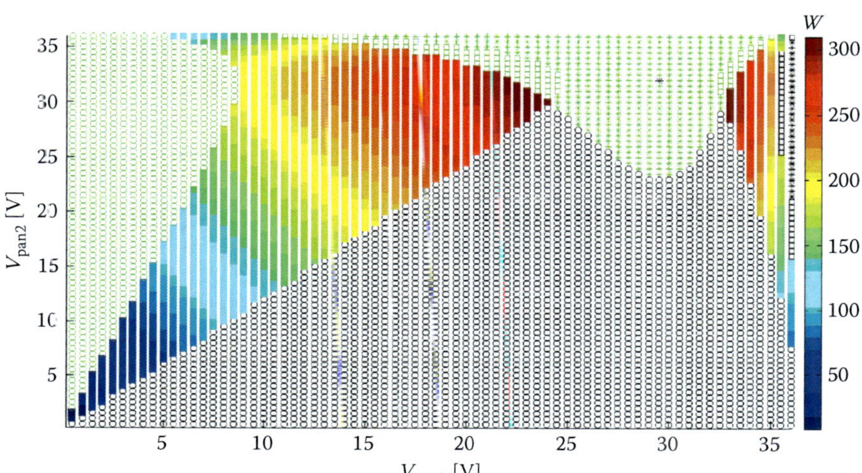

COLOR FIGURE 4.7
(Boost) $G_1 = 1000$ W/m², $G_2 = 500$ W/m², $V_{bulk} = 80$ V, $V_{ds\,max} = 50$ V.

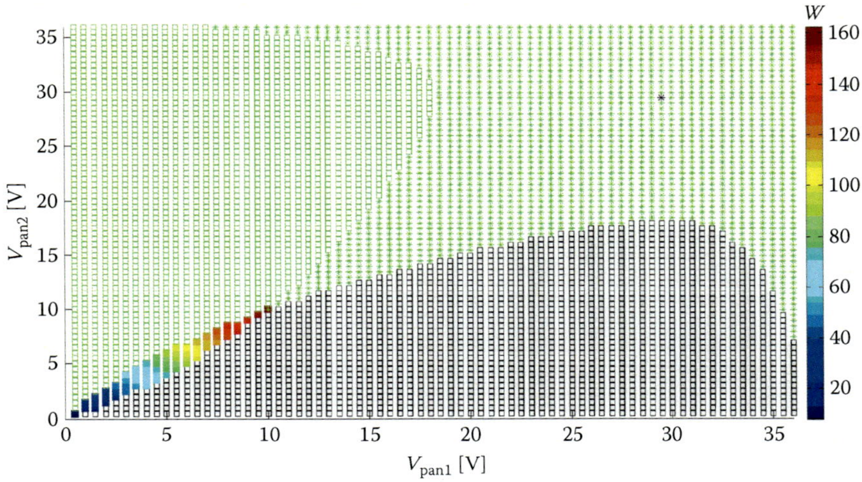

COLOR FIGURE 4.8
(Buckboost) $G_1 = G_2 = 1000$ W/m^2, $V_{bulk} = 80$ V, $V_{ds\,max} = 50$ V, $I_{ds\,max} \to \infty$.

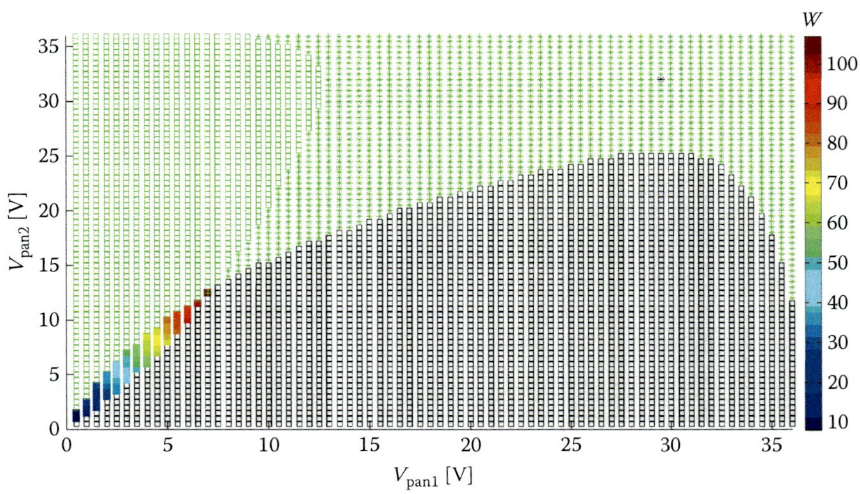

COLOR FIGURE 4.9
(Buckboost) $G_1 = 1000$ W/m^2, $G_2 = 500$ W/m^2, $V_{bulk} = 80$ V, $V_{ds\,max} = 50$ V, $I_{ds\,max} \to \infty$.

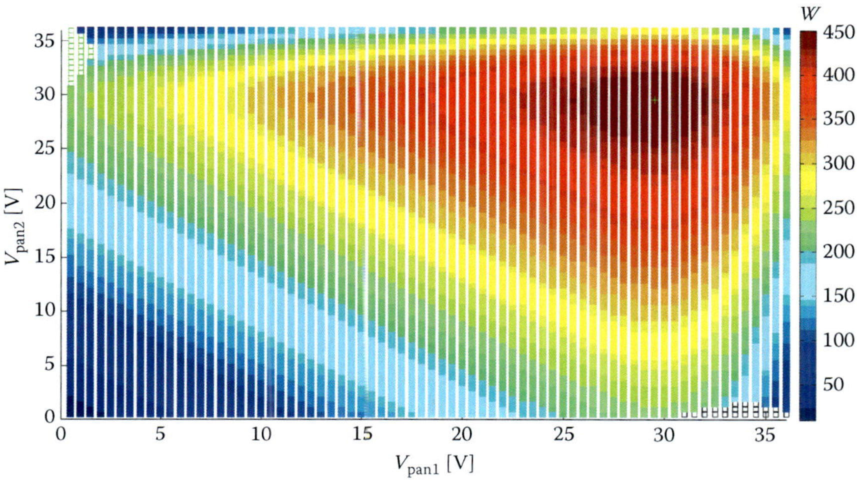

COLOR FIGURE 4.10
(Buckboost) $G_1 = G_2 = 1000$ W/m², $V_{bulk} = 50$ V, $V_{ds\,max} = 80$ V, $I_{ds\,max} \to \infty$.

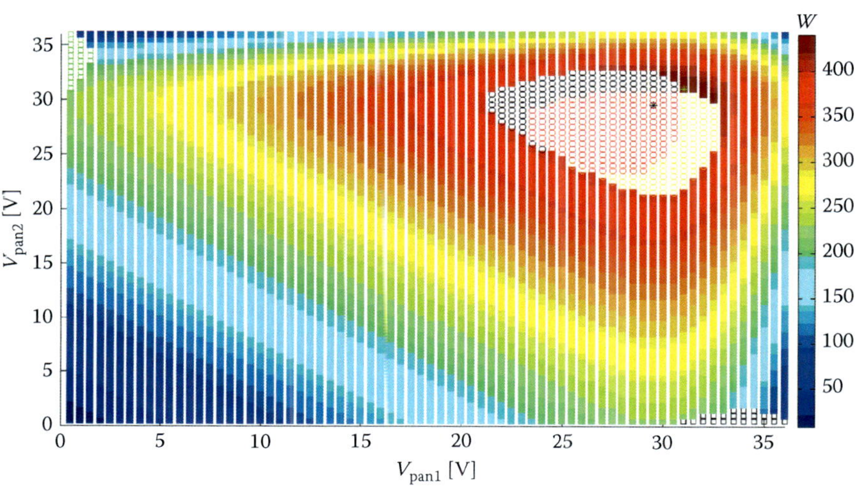

COLOR FIGURE 4.11
(Buckboost) $G_1 = G_2 = 1000$ W/m², $V_{bulk} = 50$ V, $V_{ds\,max} = 80$ V, $I_{ds\,max} = 16$ A.

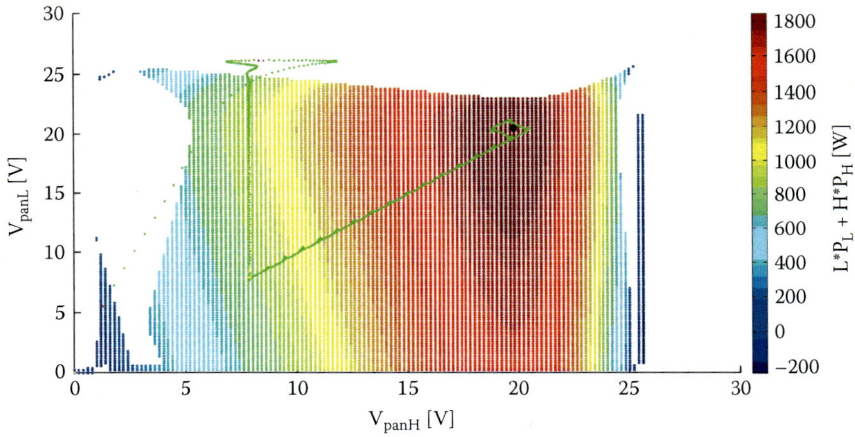

COLOR FIGURE 4.117
Simulation of the system in Figure 4.4 for $NH = 11$ and $G_L = 800$ W/m². (a) Time domain data plotted on the corresponding feasibility map (the black-filled triangle indicates global MPP, the black-filled circle indicates average steady-state operation point of the system).

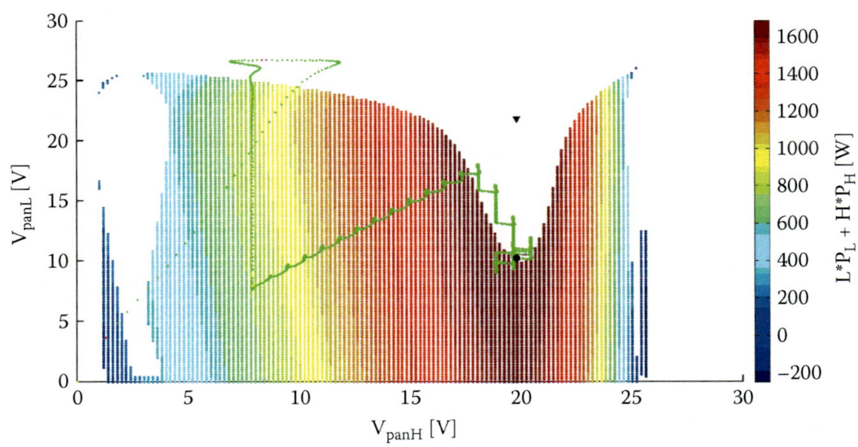

COLOR FIGURE 4.118
Simulation of the system in Figure 4.4 for $NH = 11$ and $G_L = 500$ W/m². (a) Time domain data plotted on the corresponding feasibility map (the black-filled triangle indicates global MPP, the black-filled circle indicates average steady-state operation point of the system).

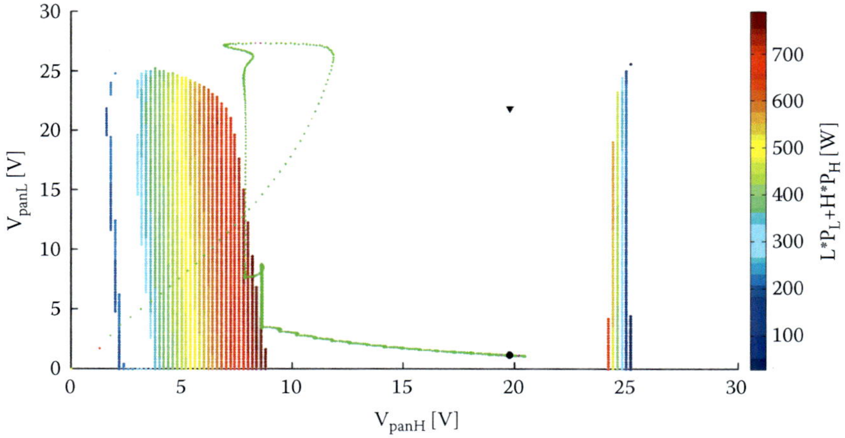

COLOR FIGURE 4.119
Simulation of the system in Figure 4.4 for $NH = 11$ and $G_L = 200$ W/m². (a) Time domain data plotted on the corresponding feasibility map (the black-filled triangle indicates global MPP, the black-filled circle indicates average steady-state operation point of the system).

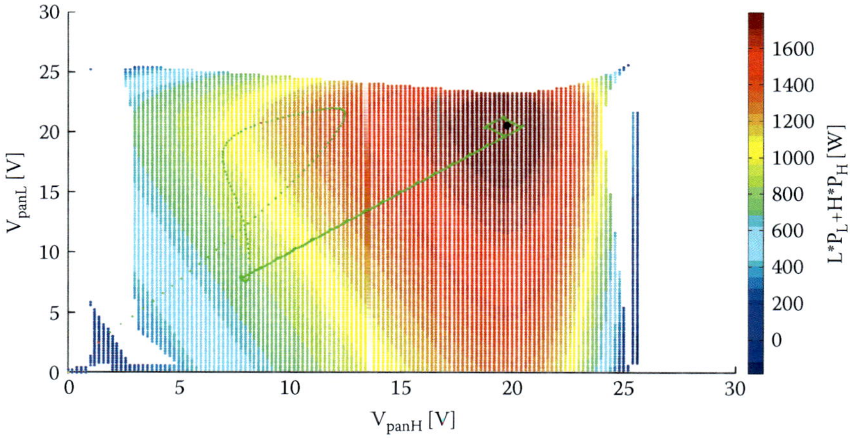

COLOR FIGURE 4.120
Simulation of the system in Figure 4.4 for $NH = 9$ and $G_L = 800$ W/m². (a) Time domain data plotted on the corresponding feasibility map (the black-filled triangle indicates global MPP, the black-filled circle indicates average steady-state operation point of the system).

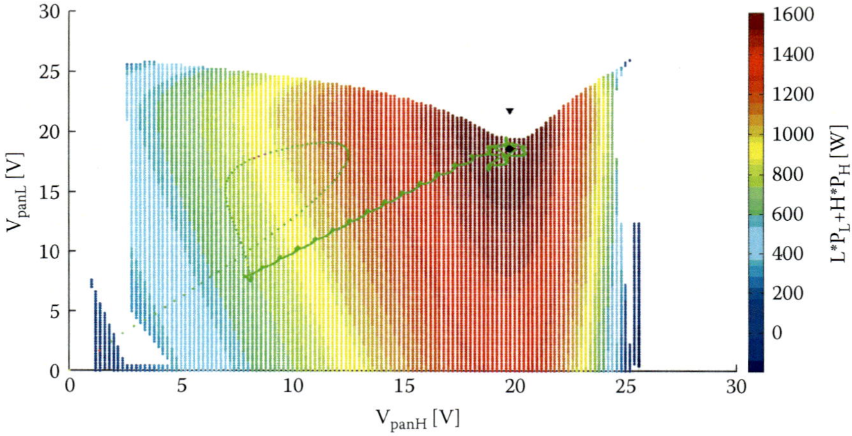

COLOR FIGURE 4.121
Simulation of the system in Figure 4.4 for $NH = 9$ and $G_L = 500$ W/m^2. (a) Time domain data plotted on the corresponding feasibility map (the black-filled triangle indicates global MPP, the black-filled circle indicates average steady-state operation point of the system).

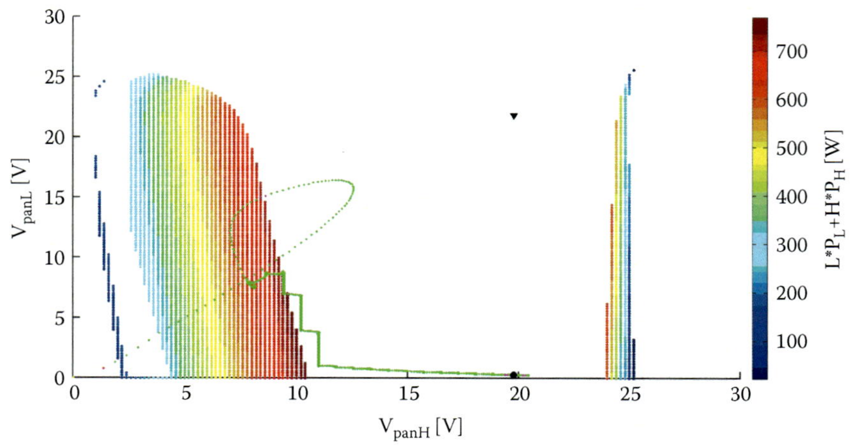

COLOR FIGURE 4.122
Simulation of the system in Figure 4.4 for $NH = 9$ and $G_L = 200$ W/m^2. (a) Time domain data plotted on the corresponding feasibility map (the black-filled triangle indicates global MPP, the black-filled circle indicates average steady-state operation point of the system).

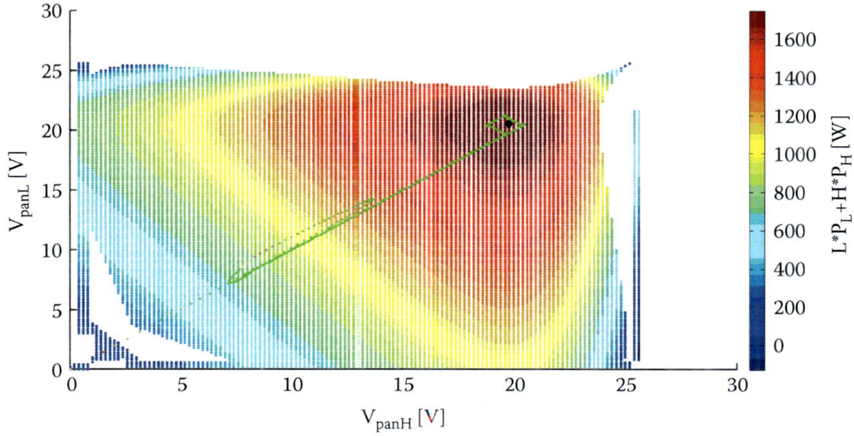

COLOR FIGURE 4.123
Simulation of the system in Figure 4.4 for $NH = 7$ and $G_L = 800$ W/m². (a) Time domain data plotted on the corresponding feasibility map (the black-filled triangle indicates global MPP, the black-filled circle indicates average steady-state operation point of the system).

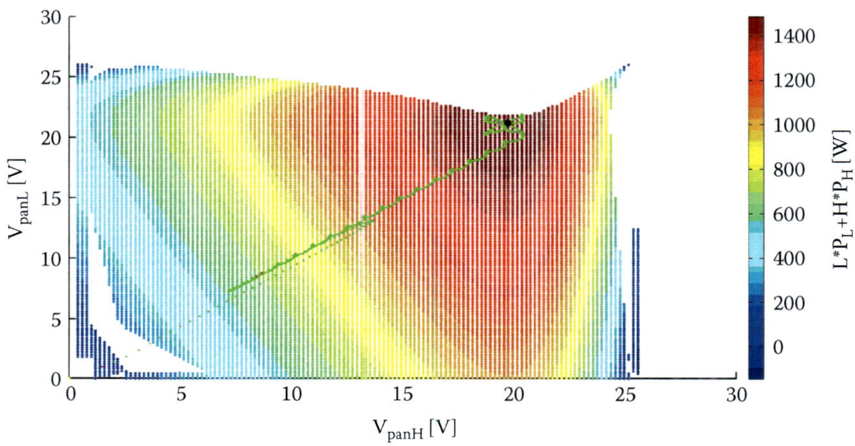

COLOR FIGURE 4.124
Simulation of the system in Figure 4.4 for $NH = 7$ and $G_L = 500$ W/m². (a) Time domain data plotted on the corresponding feasibility map (the black-filled triangle indicates global MPP, the black-filled circle indicates average steady-state operation point of the system).

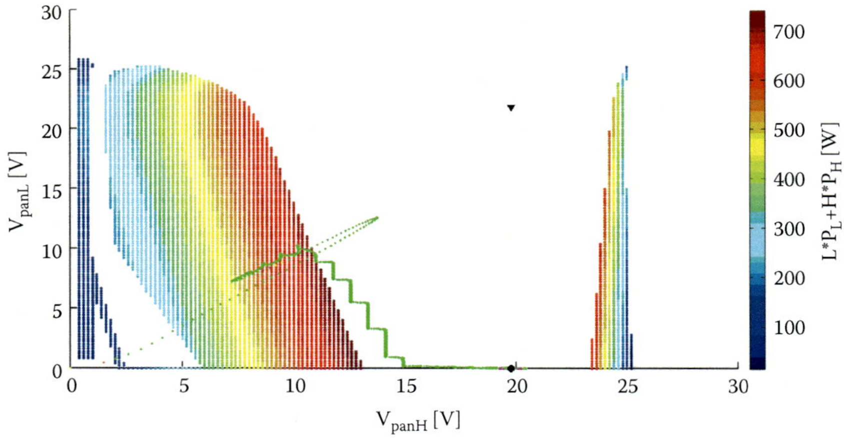

COLOR FIGURE 4.125
Simulation of the system in Figure 4.4 for $NH = 7$ and $G_L = 200$ W/m^2. (a) Time domain data plotted on the corresponding feasibility map (the black-filled triangle indicates global MPP, the black-filled circle indicates average steady-state operation point of the system).

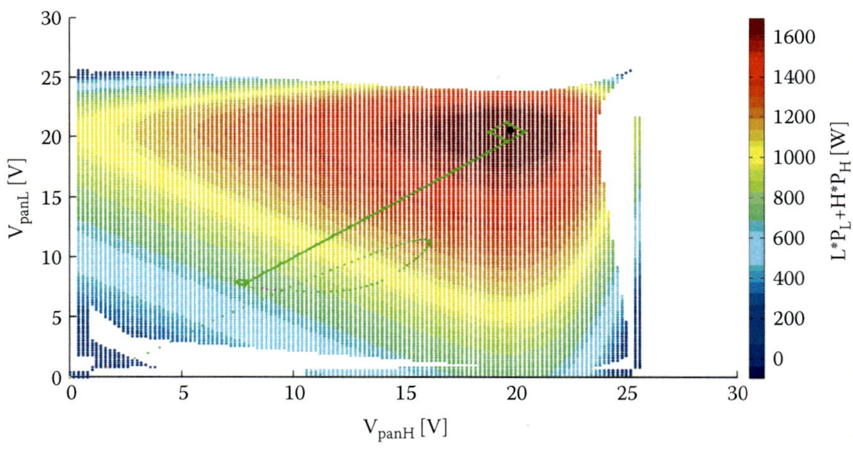

COLOR FIGURE 4.126
Simulation of the system in Figure 4.4 for $NH = 5$ and $G_L = 800$ W/m^2. (a) Time domain data plotted on the corresponding feasibility map (the black-filled triangle indicates global MPP, the black-filled circle indicates average steady-state operation point of the system).

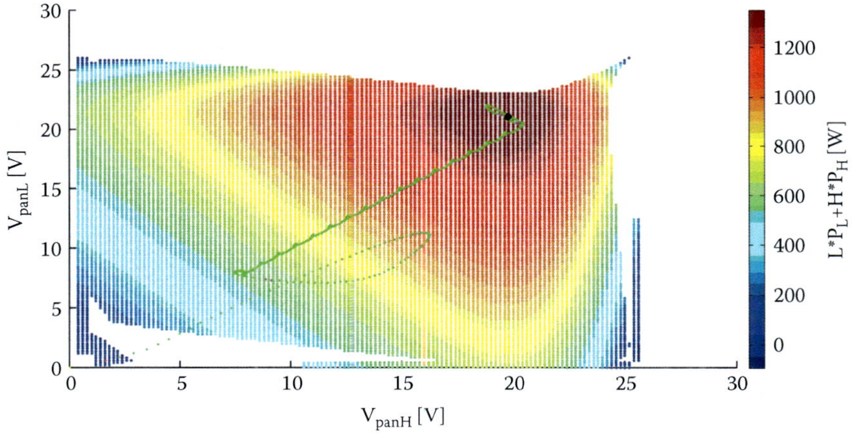

COLOR FIGURE 4.127
Simulation of the system in Figure 4.4 for $NH = 5$ and $G_L = 500$ W/m^2. (a) Time domain data plotted on the corresponding feasibility map (the black-filled triangle indicates global MPP, the black-filled circle indicates average steady-state operation point of the system).

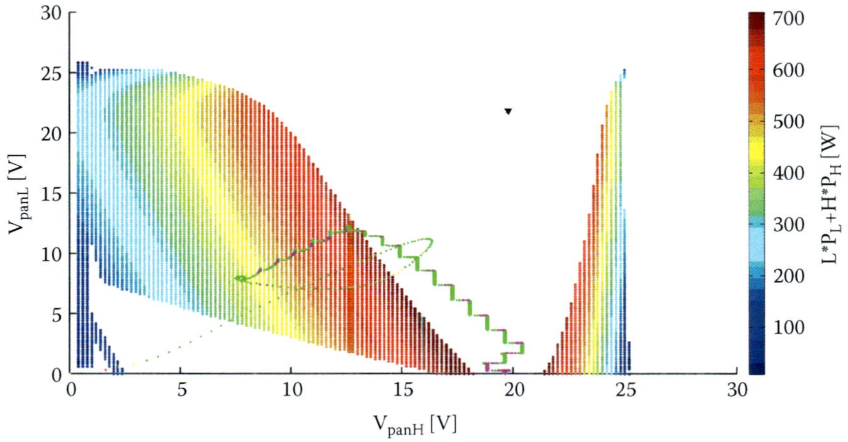

COLOR FIGURE 4.128
Simulation of the system in Figure 4.4 for $NH = 5$ and $G_L = 200$ W/m^2. (a) Time domain data plotted on the corresponding feasibility map (the black-filled triangle indicates global MPP, the black-filled circle indicates average steady-state operation point of the system).

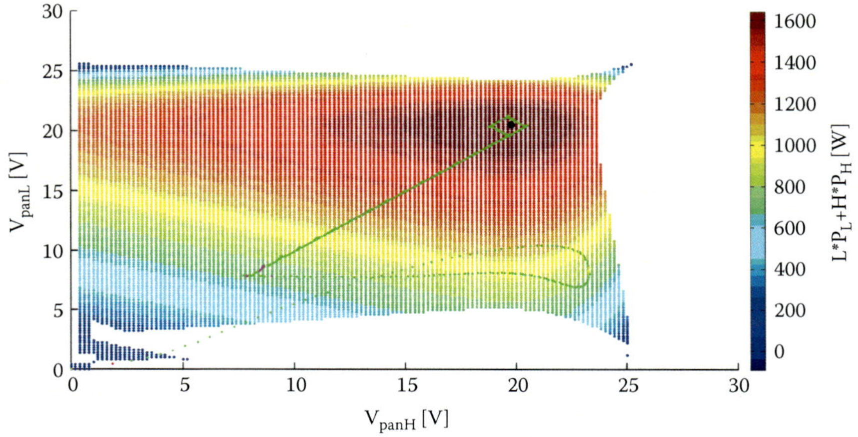

COLOR FIGURE 4.129

Simulation of the system in Figure 4.4 for $NH = 3$ and $G_L = 800$ W/m^2. (a) Time domain data plotted on the corresponding feasibility map (the black-filled triangle indicates global MPP, the black-filled circle indicates average steady-state operation point of the system). (b) PV panel voltage. (c) DC/DC MPPT converter duty cycle.

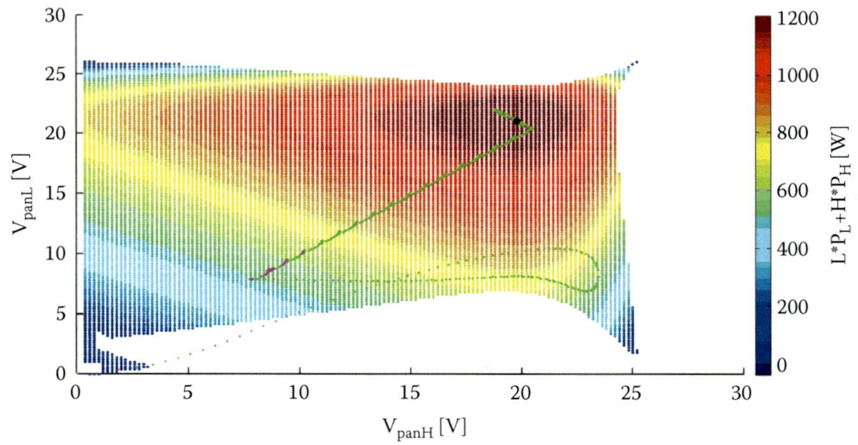

COLOR FIGURE 4.130

Simulation of the system in Figure 4.4 for $NH = 3$ and $G_L = 500$ W/m^2. (a) Time domain data plotted on the corresponding feasibility map (the black-filled triangle indicates global MPP, the black-filled circle indicates average steady-state operation point of the system).

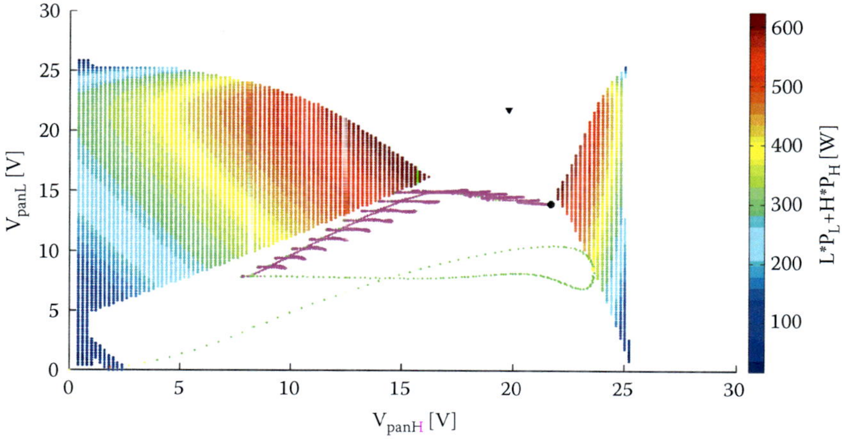

COLOR FIGURE 4.131
Simulation of the system in Figure 4.4 for $NH = 3$ and $G_L = 200$ W/m². (a) Time domain data plotted on the corresponding feasibility map (the black-filled triangle indicates global MPP, the black-filled circle indicates average steady-state operation point of the system).

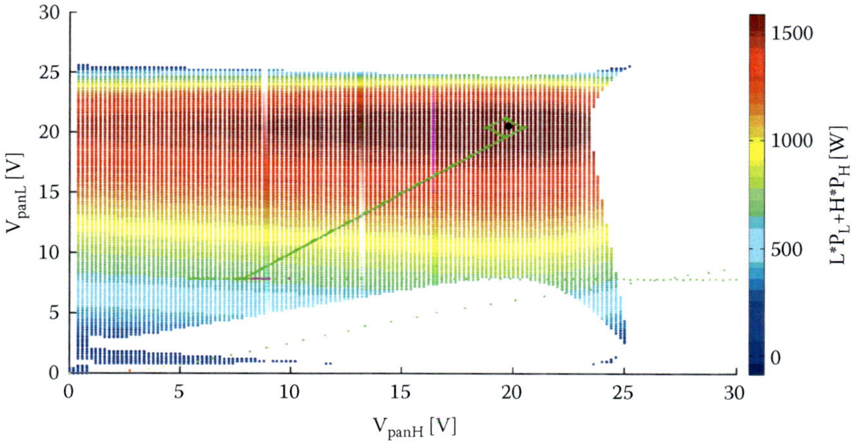

COLOR FIGURE 4.132
Simulation of the system in Figure 4.4 for $NH = 1$ and $G_L = 800$ W/m². (a) Time domain data plotted on the corresponding feasibility map (the black-filled triangle indicates global MPP, the black-filled circle indicates average steady-state operation point of the system).

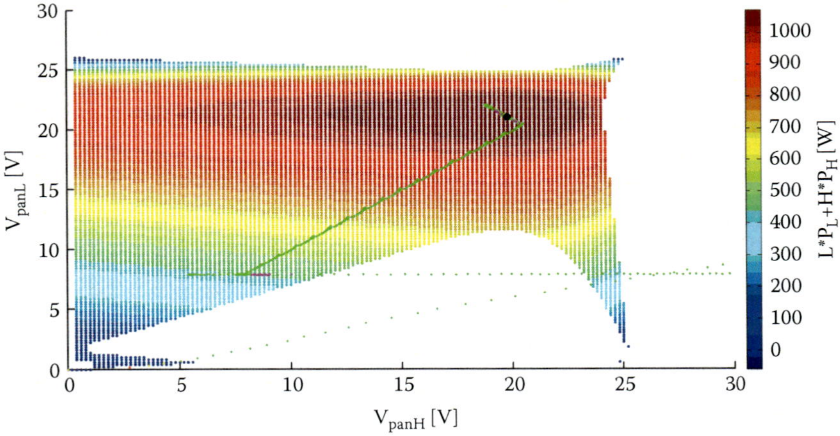

COLOR FIGURE 4.133

Simulation of the system in Figure 4.4 for $NH = 1$ and $G_L = 500$ W/m^2. (a) Time domain data plotted on the corresponding feasibility map (the black-filled triangle indicates global MPP, the black-filled circle indicates average steady-state operation point of the system).

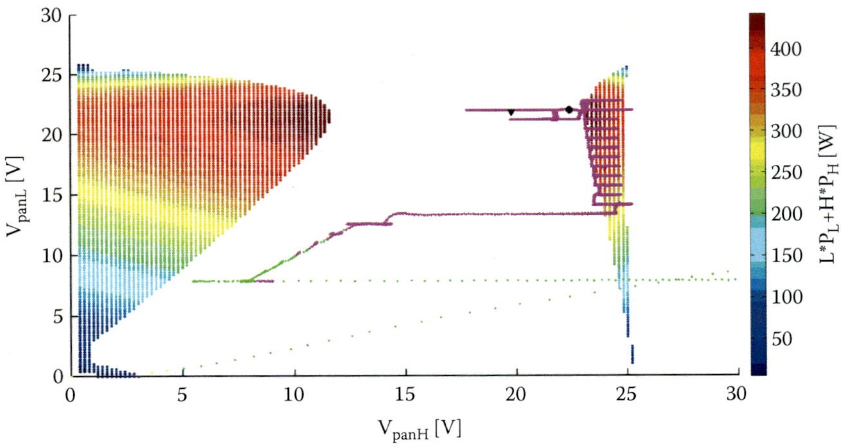

COLOR FIGURE 4.134

Simulation of the system in Figure 4.4 for $NH = 1$ and $G_L = 200$ W/m^2. (a) Time domain data plotted on the corresponding feasibility map (the black-filled triangle indicates global MPP, the black-filled circle indicates average steady-state operation point of the system).

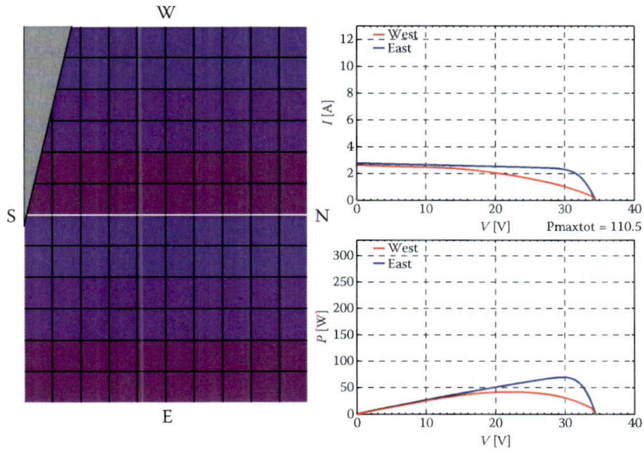

(a)

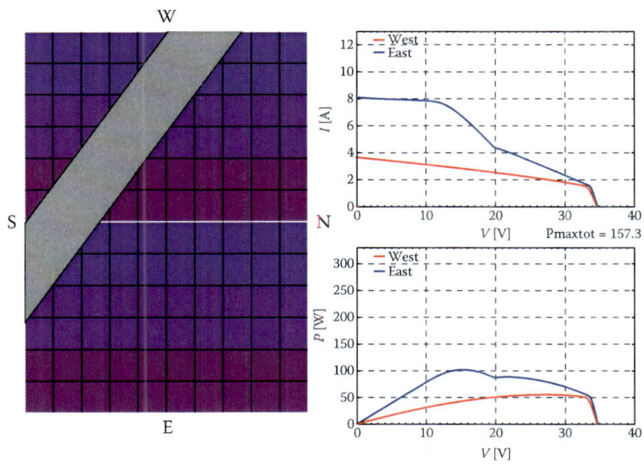

(b)

COLOR FIGURE 5.5
Positions of the shadow in six times cf the day, with related P vs. V plots of PV panels. (a) 7:30 a.m., S = 333 W/m^2. (b) 9:30 a.m., S = 908 W/m^2. (c) 11:30 a.m., S = 1112 W/m^2.

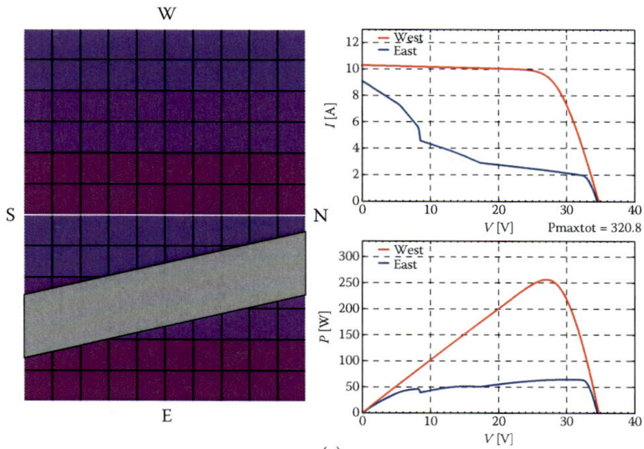

(c)

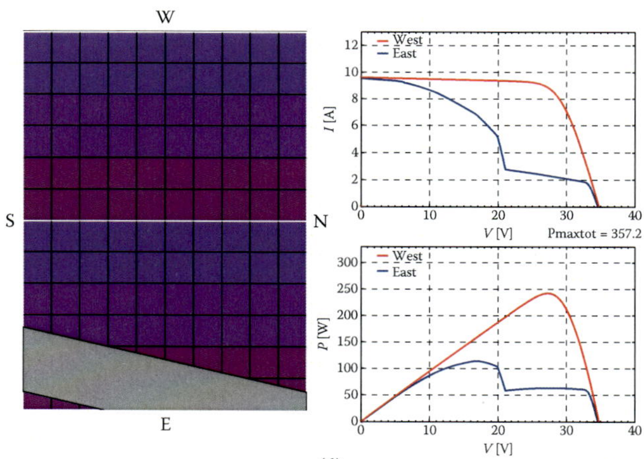

(d)

COLOR FIGURE 5.5
Positions of the shadow in six times of the day, with related P vs. V plots of PV panels. (c) 11:30 a.m., S = 1112 W/m^2. (d) 12:30 a.m., S = 1050 W/m^2. (e) 2:30 p.m., S = 641 W/m^2. (f) 4:30 p.m., S = 163 W/m^2.

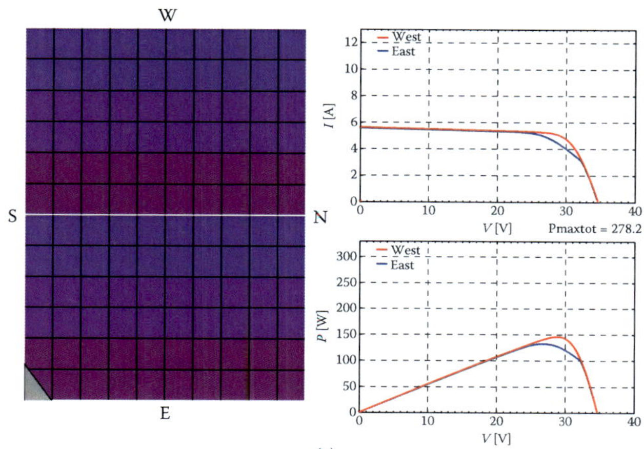

(e)

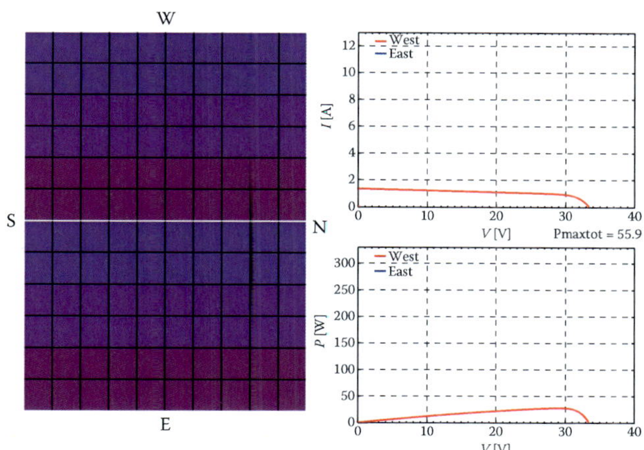

(f)

COLOR FIGURE 5.5
Positions of the shadow in six times of the day, with related P vs. V plots of PV panels. (e) 2:30 p.m., S = 641 W/m². (f) 4:30 p.m., S = 163 W/m².

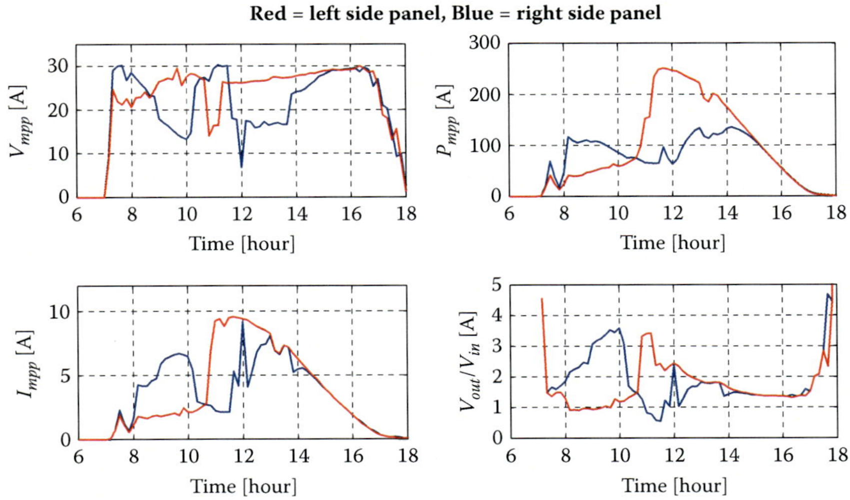

COLOR FIGURE 5.6
MPP voltage, current, and power and input-output voltage conversion ratio: red = west side panel, blue = east side panel.

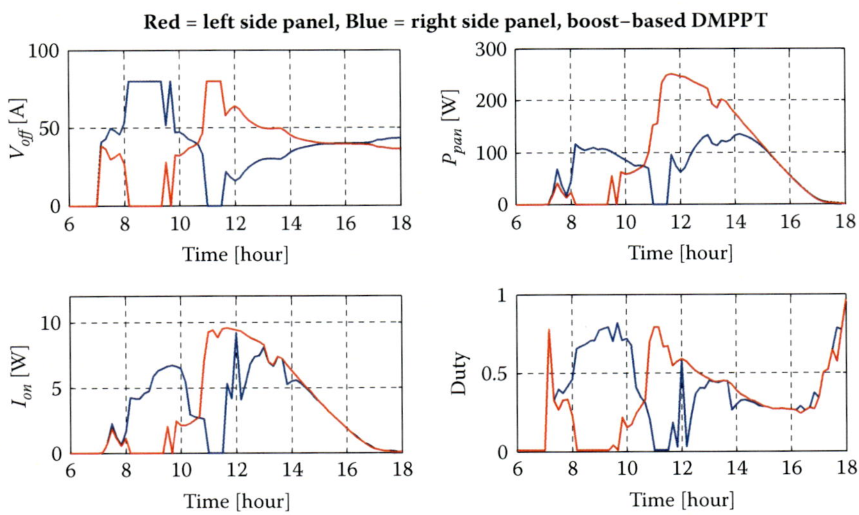

COLOR FIGURE 5.7
Boost DMPPT converter voltage and current stresses: red = west side panel, blue = east side panel.

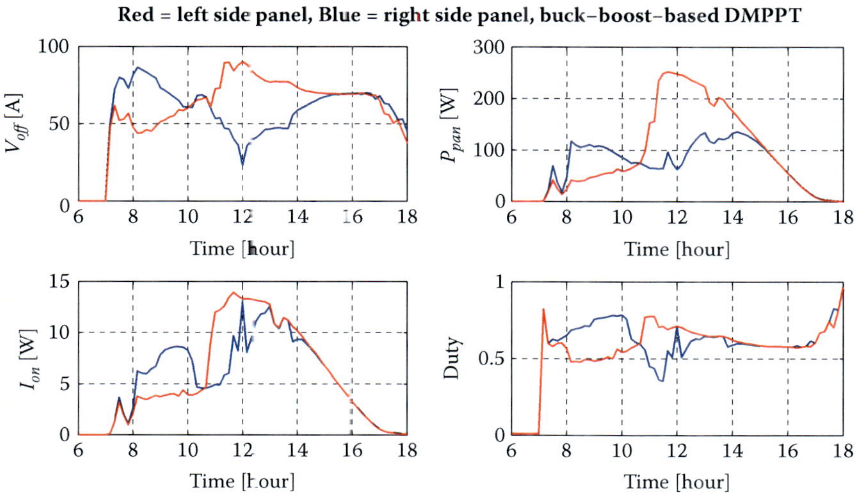

COLOR FIGURE 5.8
Buckboost DMPPT converter voltage and current stresses: red = west side panel, blue = east side panel.

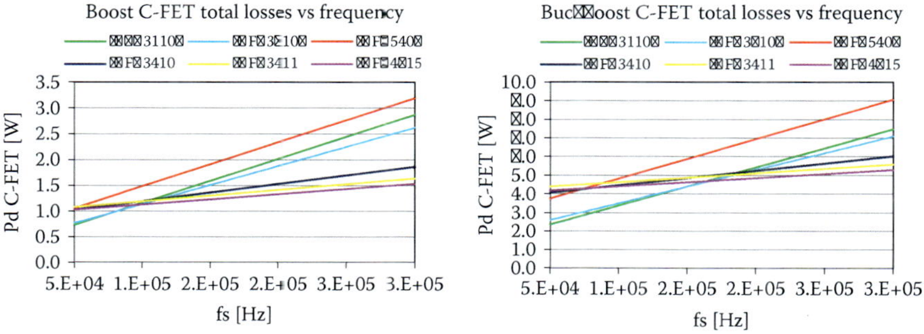

COLOR FIGURE 5.22
C-FET losses vs. switching frequency.

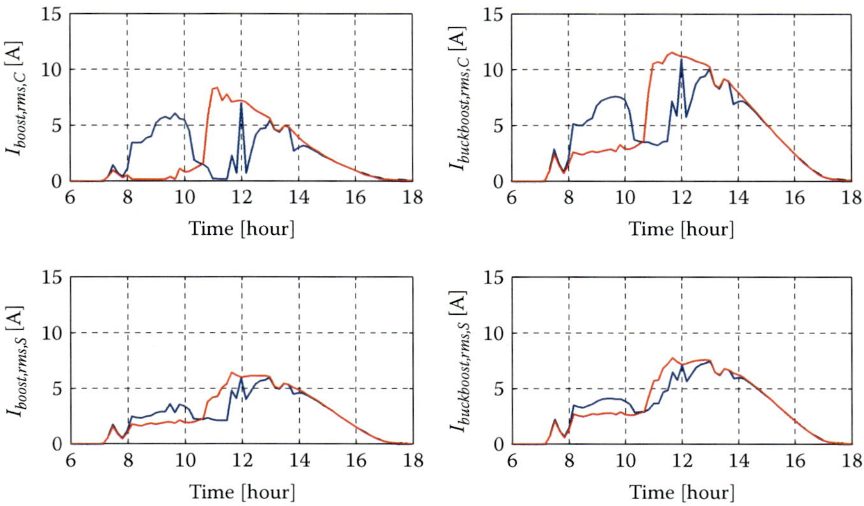

COLOR FIGURE 5.23
Rms currents for C-FET and S-FET for boost and buckboost topologies: red line = east side PV panel, blue line = west side PV panel.

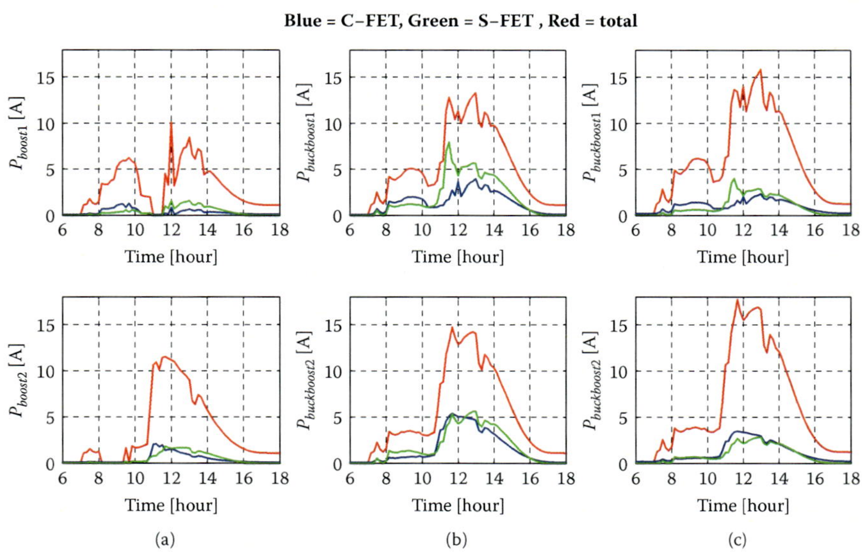

COLOR FIGURE 5.24
Power dissipation for boost and buckboost topologies. (a) Boost. (b) Buckboost single FET. (c) Buckboost double FET.

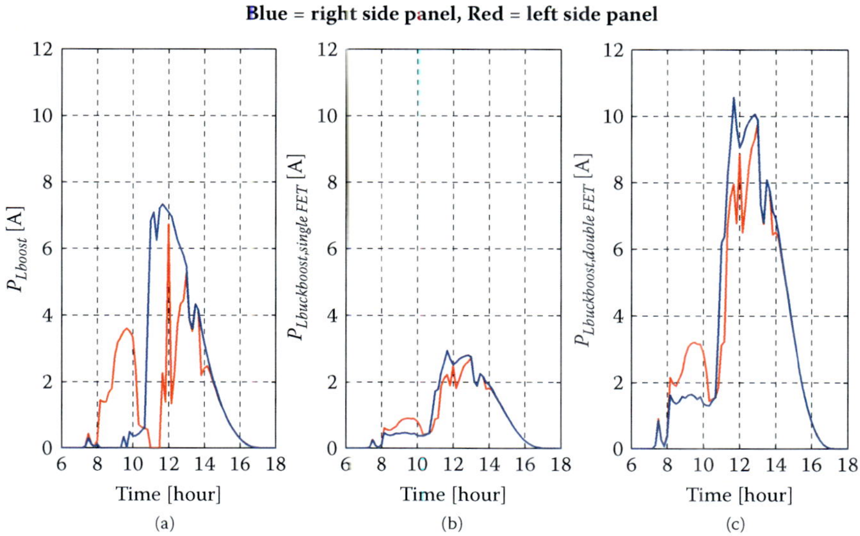

COLOR FIGURE 5.25
Inductor power dissipation for boost and buckboost topologies: red line = east side PV panel, blue line = west side PV panel. (a) Boost. (b) Buckboost single FET. (c) Buckboost double FET.

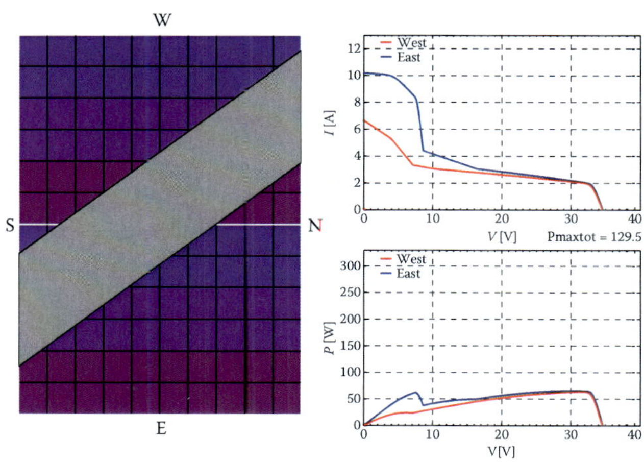

COLOR FIGURE 5.26
PV panels subjected to a shadow twice larger than the one of Figure 5.5.

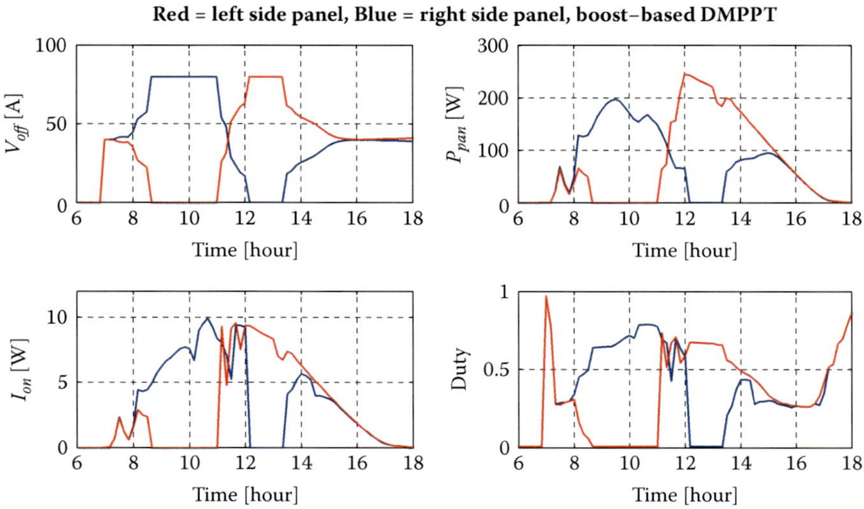

COLOR FIGURE 5.28
Voltage and current stresses, power extracted from the panels, and converter duty cycle for boost topology: red line = east side PV panel, blue line = west side PV panel.

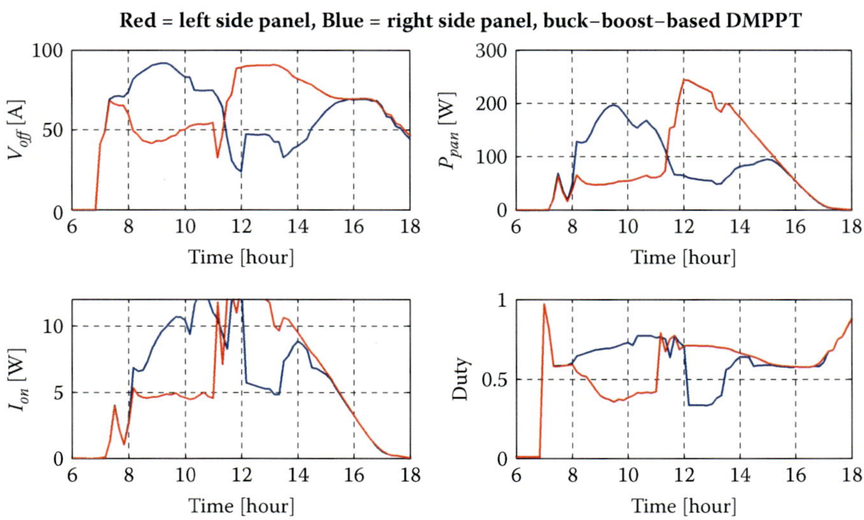

COLOR FIGURE 5.29
Voltage and current stresses, power extracted from the panels, and converter duty cycle for buckboost topology: red line = east side PV panel, blue line = west side PV panel.

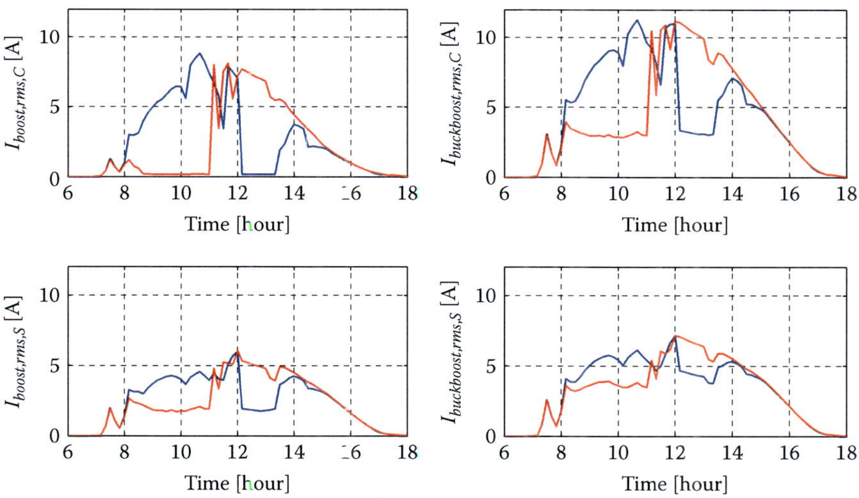

COLOR FIGURE 5.30
Rms currents for C-FET and S-FET for boost and buckboost topologies: red line = east side PV panel, blue line = west side PV panel.

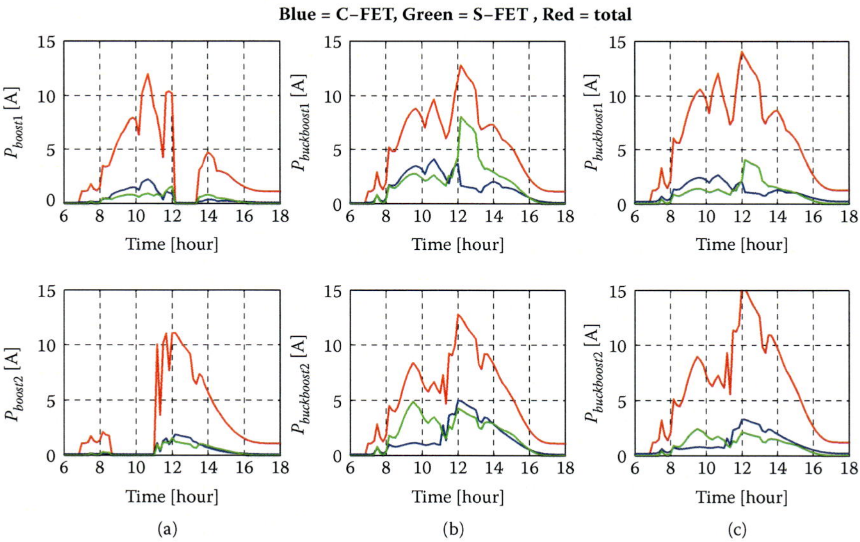

COLOR FIGURE 5.31
Power dissipation for boost and buckboost topologies.

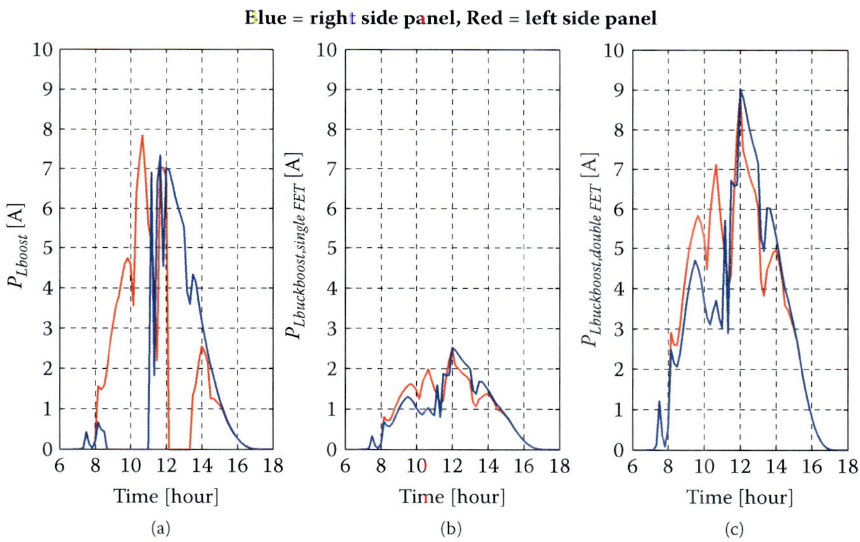

COLOR FIGURE 5.32
Inductors' power dissipation for boost and buckboost topologies.

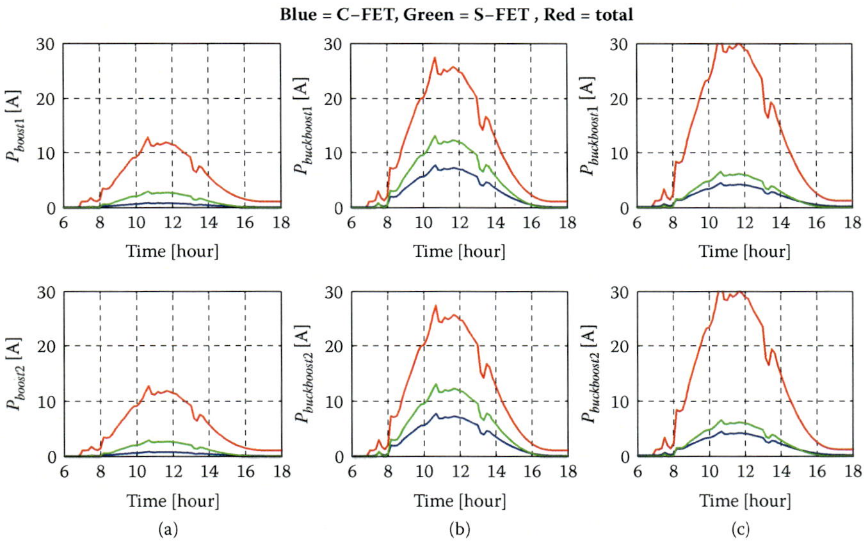

COLOR FIGURE 5.37
Power dissipation for boost and buckboost topologies.

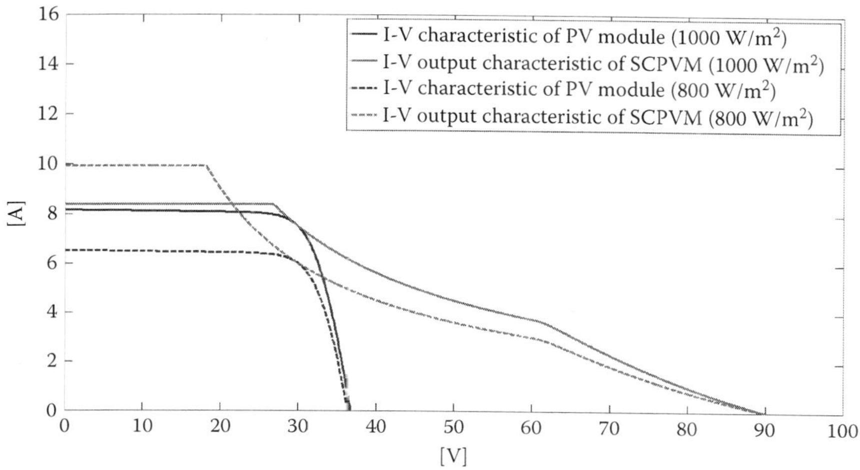

FIGURE 4.52
$G_1 = 1000$ W/m², $G_2 = 800$ W/m², $T_a = 25$°C; buckboost-based SCPVMs ($V_{ds\,max} = 90$ V, $I_{ds\,max} = 16$ A).

Figures 4.51 to 4.63 refer to the case of lossless buckboost MPPT converters characterized by $V_{ds\,max} = 60$ V, $I_{ds\,max} = 16$ A.

In Figure 4.52, the I-V characteristics of two SW225 PV modules respectively operating at $G_1 = 1000$ W/m², $T_a = 25$°C, and at $G_2 = 800$ W/m², $T_a = 25$°C are reported together with the corresponding I-V output characteristics of the two associated buckboost-based SCPVMs.

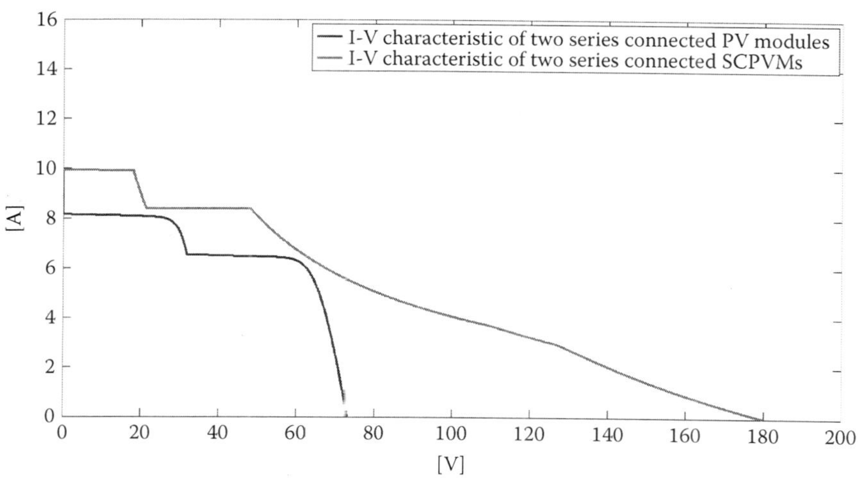

FIGURE 4.53
$G_1 = 1000$ W/m², $G_2 = 800$ W/m², $T_a = 25$°C; buckboost-based SCPVMs ($V_{ds\,max} = 90$ V, $I_{ds\,max} = 16$ A).

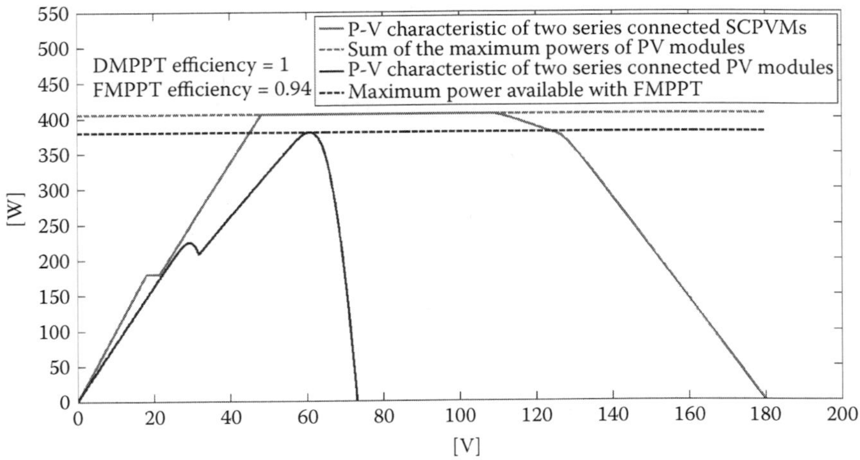

FIGURE 4.54
$G_1 = 1000$ W/m², $G_2 = 800$ W/m², $T_a = 25°C$; buckboost-based SCPVMs ($V_{ds\,max} = 90$ V, $I_{ds\,max} = 16$ A).

In Figure 4.53, the I-V characteristic of the series connection of two PV modules ($G_1 = 1000$ W/m², $G_2 = 800$ W/m², $T_a = 25°C$) and the I-V characteristic of the series connection of the associated buckboost-based SCPVMs are shown.

Finally in Figure 4.54, the P-V characteristic of the series connection of the two PV modules ($G_1 = 1000$ W/m², $G_2 = 800$ W/m², $T_a = 25°C$) and the P-V characteristic of the series connection of the associated buckboost-based SCPVMs are reported. It is $\eta_{FMPPT} = 0.94$, while $\eta_{DMPPT} = 1$ for 48.6 V $< V_{bulk} <$ 109 V.

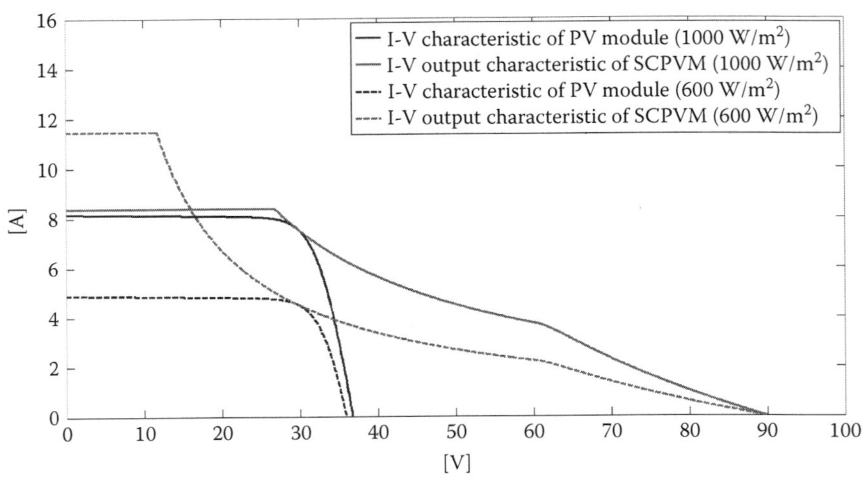

FIGURE 4.55
$G_1 = 1000$ W/m², $G_2 = 600$ W/m², $T_a = 25°C$; buckboost-based SCPVMs ($V_{ds\,max} = 90$ V, $I_{ds\,max} = 16$ A).

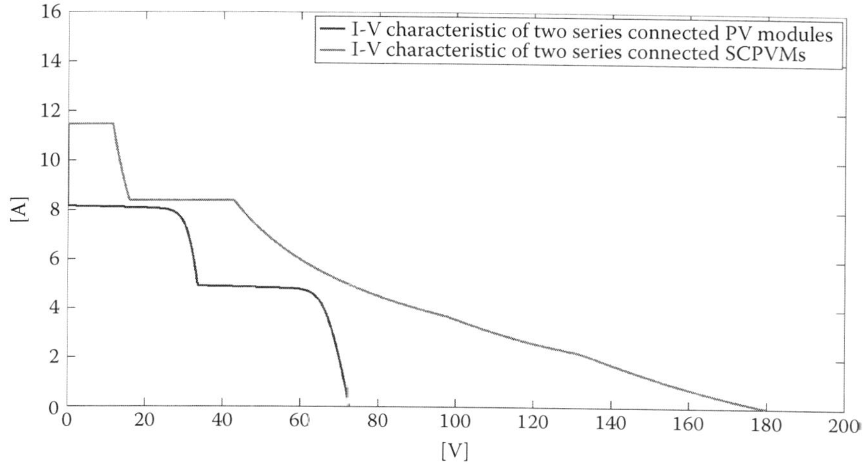

FIGURE 4.56
$G_1 = 1000$ W/m², $G_2 = 600$ W/m², $T_a = 25$°C; buckboost-based SCPVMs ($V_{ds\,max} = 90$ V, $I_{ds\,max} = 16$ A).

Similar considerations also hold for Figures 4.55–4.57, which refer to the case $G_1 = 1000$ W/m², $G_2 = 600$ W/m², and $T_a = 25$°C.

In such a case it is $\eta_{FMPPT} = 0.81$; instead, $\eta_{DMPPT} = 1$ for 43 V < V_{bulk} < 97 V.

Figures 4.58–4.60 refer instead to the case $G_1 = 1000$ W/m², $G_2 = 400$ W/m², and $T_a = 25$°C. In such a case it is $\eta_{FMPPT} = 0.72$; once again, $\eta_{DMPPT} = 1$ for 37 V < V_{bulk} < 84 V.

Eventually Figures 4.61–4.63 refer to the case $G_1 = 1000$ W/m², $G_2 = 200$ W/m², and $T_a = 25$°C. In such a case it is $\eta_{FMPPT} = 0.84$; $\eta_{DMPPT} = 1$ for 32 V < V_{bulk} < 72 V.

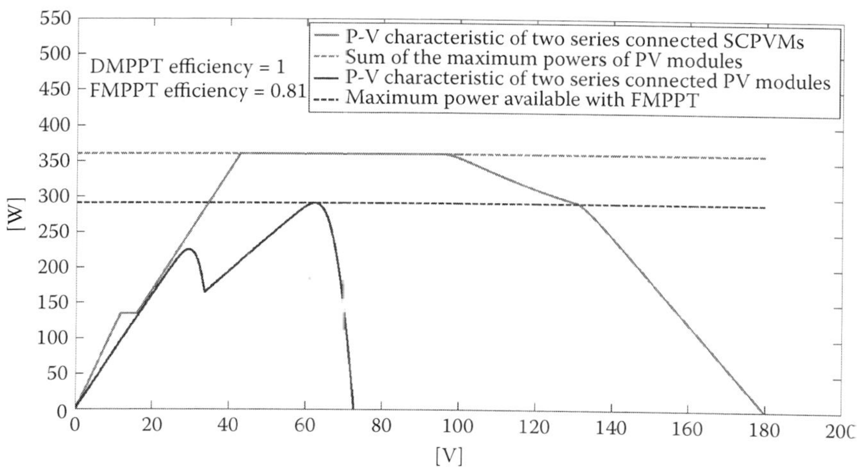

FIGURE 4.57
$G_1 = 1000$ W/m², $G_2 = 600$ W/m², $T_a = 25$°C; buckboost-based SCPVMs ($V_{ds\,max} = 90$ V, $I_{ds\,max} = 16$ A).

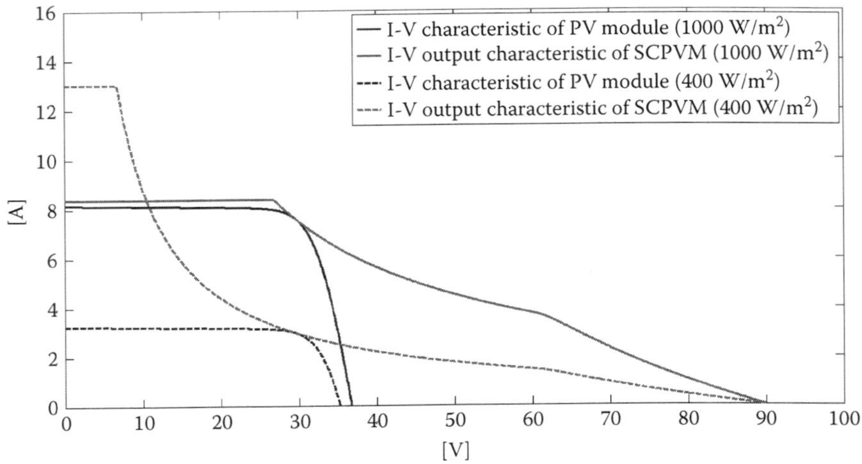

FIGURE 4.58
$G_1 = 1000$ W/m², $G_2 = 400$ W/m², $T_a = 25$°C; buckboost-based SCPVMs ($V_{ds\,max} = 90$ V, $I_{ds\,max} = 16$ A).

By analyzing Figures 4.51–4.63 it is evident that if we consider switching devices characterized by finite voltage and current ratings, in order to get $\eta_{DMPPT} = 1$ it is necessary that the bulk voltage belongs to a range of voltages whose position and amplitude depend on the operating conditions, in terms of irradiance and temperature, of the PV modules. If the bulk voltage does not belong to the above optimal range, the efficiency of DMPPT can be consistently lower than the efficiency obtainable with the simple FMPPT.

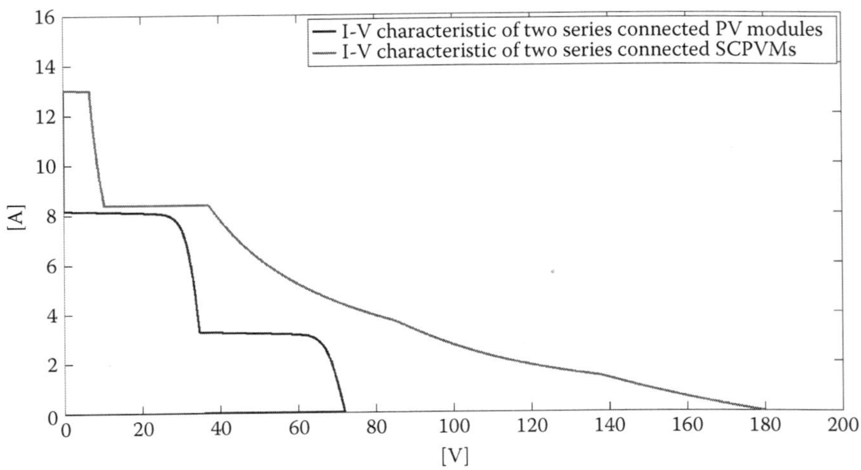

FIGURE 4.59
$G_1 = 1000$ W/m², $G_2 = 400$ W/m², $T_a = 25$°C; buckboost-based SCPVMs ($V_{ds\,max} = 90$ V, $I_{ds\,max} = 16$ A).

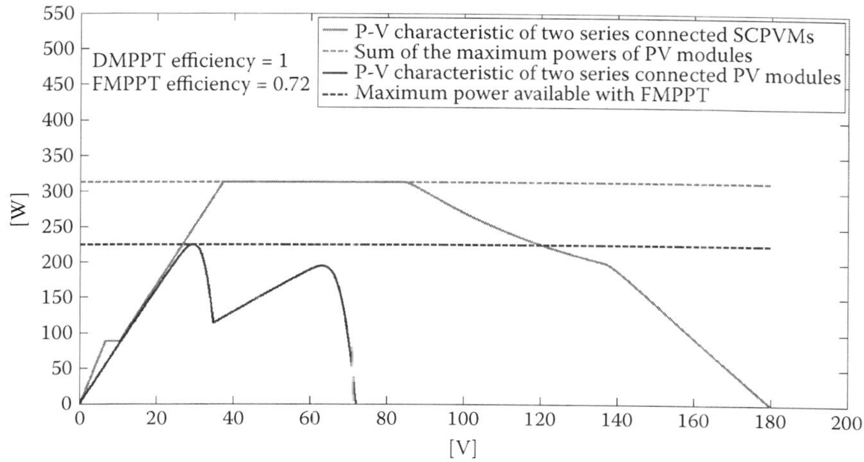

FIGURE 4.60
$G_1 = 1000\ W/m^2$, $G_2 = 400\ W/m^2$, $T_a = 25°C$; buckboost-based SCPVMs ($V_{ds\,max} = 90$ V, $I_{ds\,max} = 16$ A).

4.4 Optimal Operating Range of the DC Inverter Input Voltage

Until now the simple case of only two SCPVMs, with the output ports connected in series, has been considered. But grid-connected inverters' input voltage usually ranges from 180 to 500 V. Therefore a number of PV modules are usually connected in series to supply the inverter with an input voltage within its operating range, and identical strings are then connected in

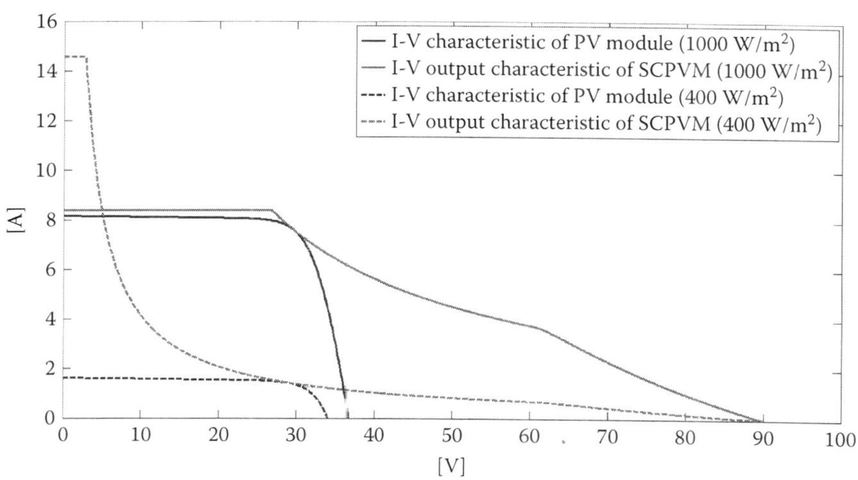

FIGURE 4.61
$G_1 = 1000\ W/m^2$, $G_2 = 200\ W/m^2$, $T_a = 25°C$; buckboost-based SCPVMs ($V_{ds\,max} = 90$ V, $I_{ds\,max} = 16$ A).

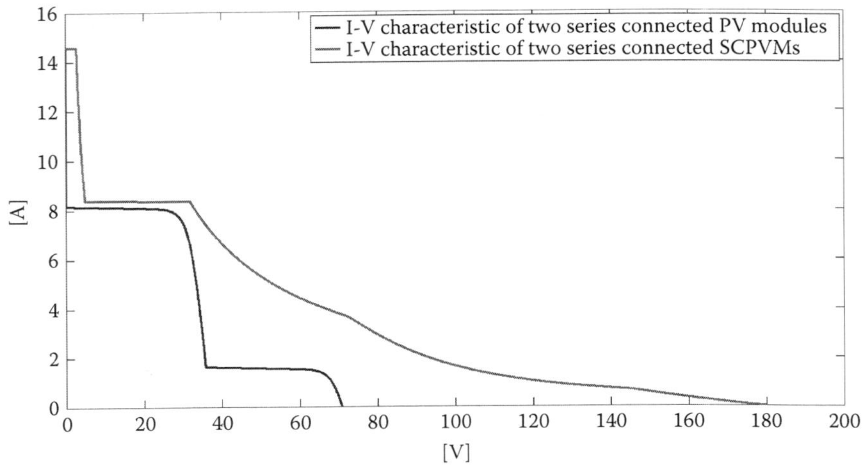

FIGURE 4.62
$G_1 = 1000$ W/m², $G_2 = 200$ W/m², $T_a = 25$°C; buckboost-based SCPVMs ($V_{ds\,max} = 90$ V, $I_{ds\,max} = 16$ A).

parallel to achieve the desired output power. For these reasons, a system composed by parallel strings of a number N of SCPVMs connected in series will be considered in the following (Figure 4.4).

In particular, for simplicity of computation and ease of graphical representation of results, a string made of NH SCPVMs operating under irradiance level G_H and of NL SCPVMs operating under irradiance level G_L will be considered. Of course, it is $N = NH + NL$. The case $N = 11$ will be studied in detail.

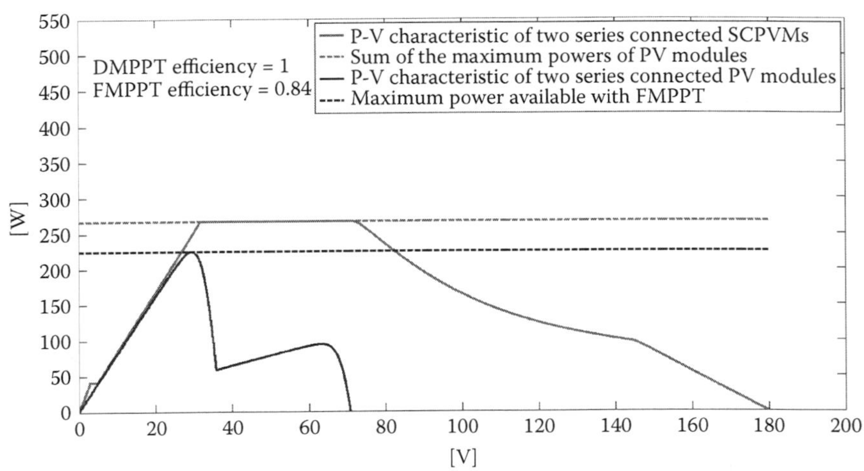

FIGURE 4.63
$G_1 = 1000$ W/m², $G_2 = 200$ W/m², $T_a = 25$°C; buckboost-based SCPVMs ($V_{ds\,max} = 90$ V, $I_{ds\,max} = 16$ A).

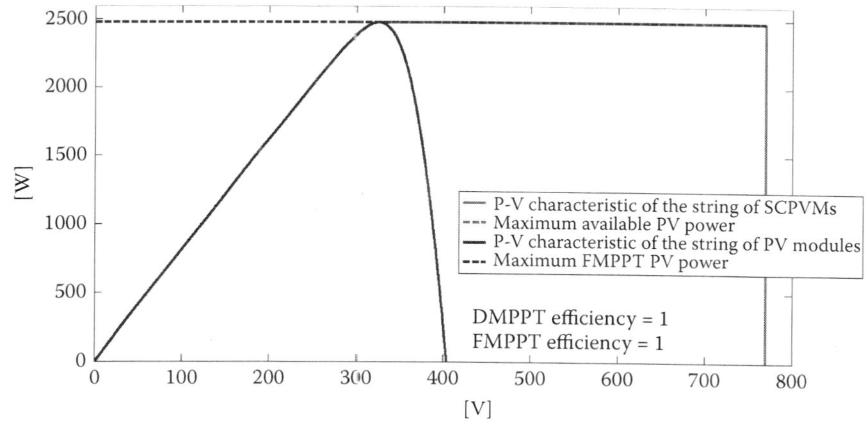

FIGURE 4.64
(Boost-based SCPVMs) $G_H = 1000$ W/m^2, $G_L = 200$ W/m^2, $V_{ds\,max} = 70$ V, $NH = 11$, $NL = 0$.

Figures 4.64–4.74 refer to the case of N boost-based lossless SCPVMs with $G_H = 1000$ W/m^2, $G_L = 200$ W/m^2, and $V_{ds\,max} = 70$ V. For each considered case the maximum value obtainable for the FMPPT efficiency and for the DMPPT efficiency has been provided. It is evident that, in all the cases, the maximum value obtainable for the DMPPT efficiency is lower than 1. Moreover, only in two cases (Figures 4.73 and 4.74) the maximum value obtainable for the DMPPT efficiency is higher than the maximum value obtainable for the FMPPT efficiency; in all the other cases the maximum values of the two efficiencies are equal.

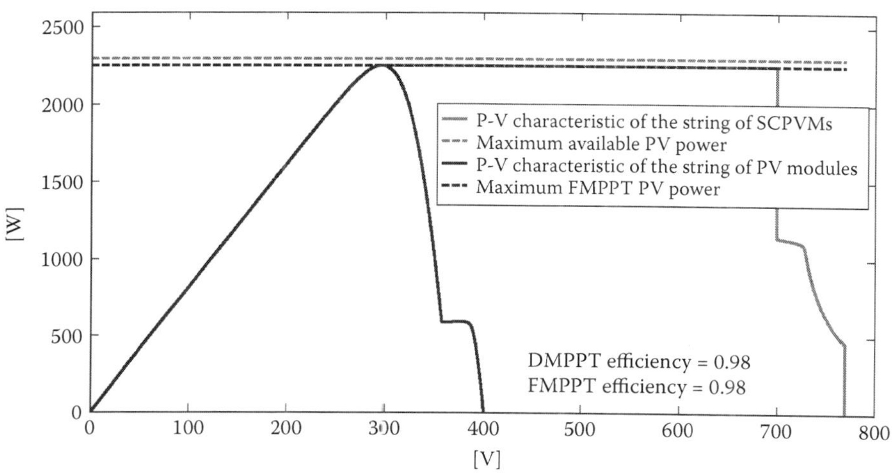

FIGURE 4.65
(Boost-based SCPVMs) $G_H = 1000$ W/m^2, $G_L = 200$ W/m^2, $V_{ds\,max} = 70$ V, $NH = 10$, $NL = 1$.

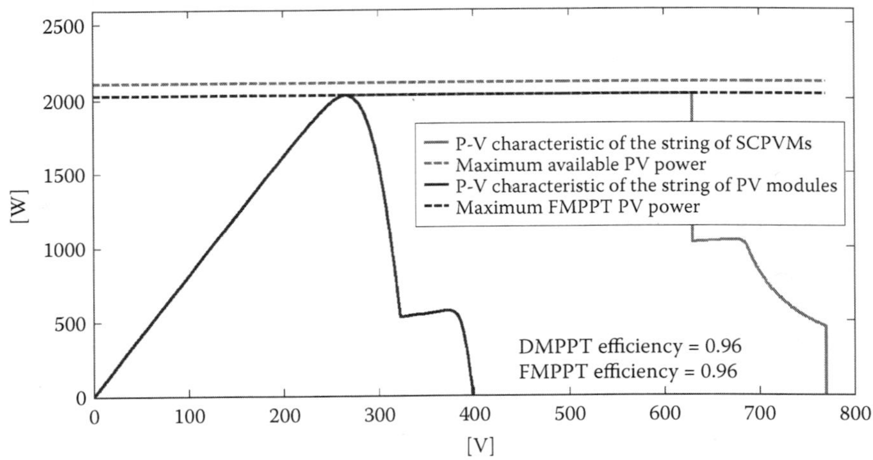

FIGURE 4.66
(Boost-based SCPVMs) $G_H = 1000$ W/m², $G_L = 200$ W/m², $V_{ds\,max} = 70$ V, $NH = 9$, $NL = 2$.

Figures 4.75–4.84 refer instead to the case of boost-based lossless SCPVMs with $G_H = 1000$ W/m², $G_L = 500$ W/m², and $V_{ds\,max} = 70$ V. In this case, the maximum value obtainable for the DMPPT efficiency is always equal to 1, and it is always higher than the maximum value obtainable for the FMPPT efficiency.

Figures 4.64–4.84 clearly show that, of course, in order to be able to get the maximum value of the DMPPT efficiency, the bulk voltage must belong to an optimal range; the position and amplitude of such an optimal range are not fixed, but they change case by case. From this point of view, this means that

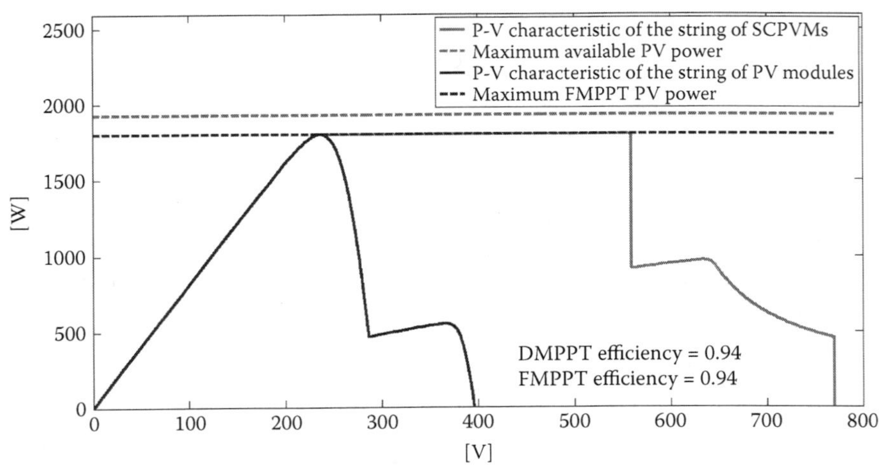

FIGURE 4.67
(Boost-based SCPVMs) $G_H = 1000$ W/m², $G_L = 200$ W/m², $V_{ds\,max} = 70$ V, $NH = 8$, $NL = 3$.

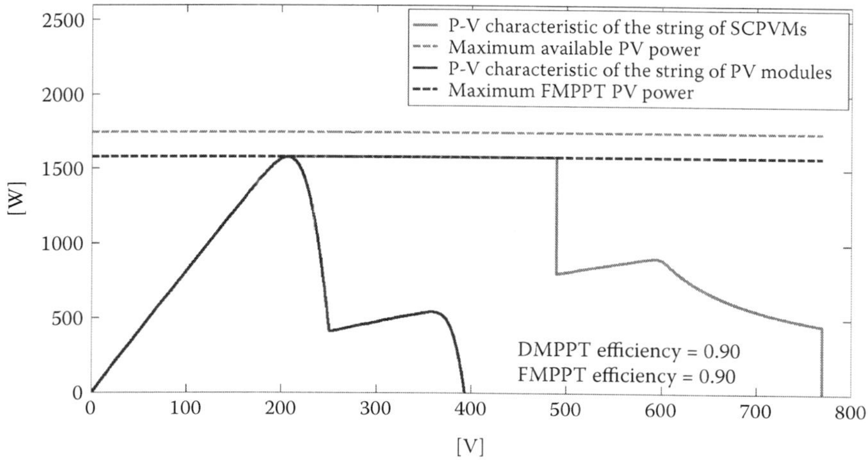

FIGURE 4.68
(Boost-based SCPVMs) $G_H = 1000$ W/m², $G_L = 200$ W/m², $V_{ds\,max} = 70$ V, $NH = 7$, $NL = 4$.

in order to get profit from DMPPT, the inverter MPPT input voltage range should be as large as possible.

In Table 4.1 the main characteristics in terms of efficiency, maximum allowed input power, and MPPT voltage range have been reported for a number of commercial PV inverters available on the market. If we refer, as an example, to the case shown in Figure 4.75 (G_H = 1000 W/m², G_L = 500 W/m², $V_{ds\,max} = 70$ V, $NH = 10$, $NL = 1$) the optimal range for the bulk voltage in

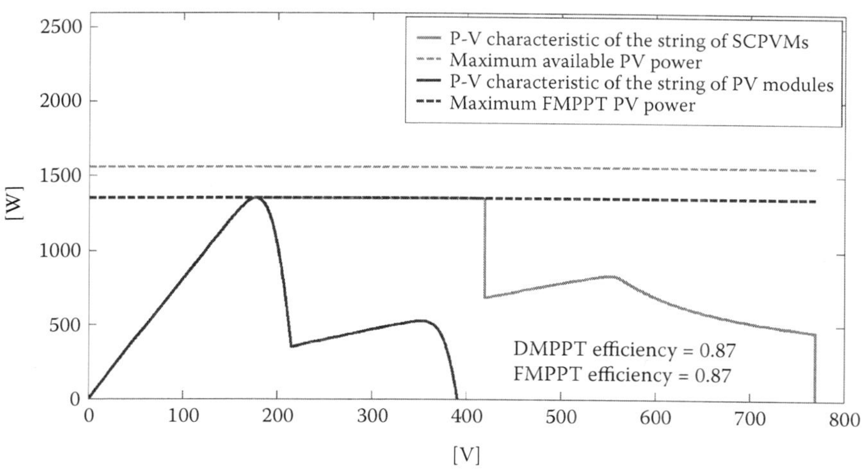

FIGURE 4.69
(Boost-based SCPVMs) $G_H = 1000$ W/m², $G_L = 200$ W/m², $V_{ds\,max} = 70$ V, $NH = 6$, $NL = 5$.

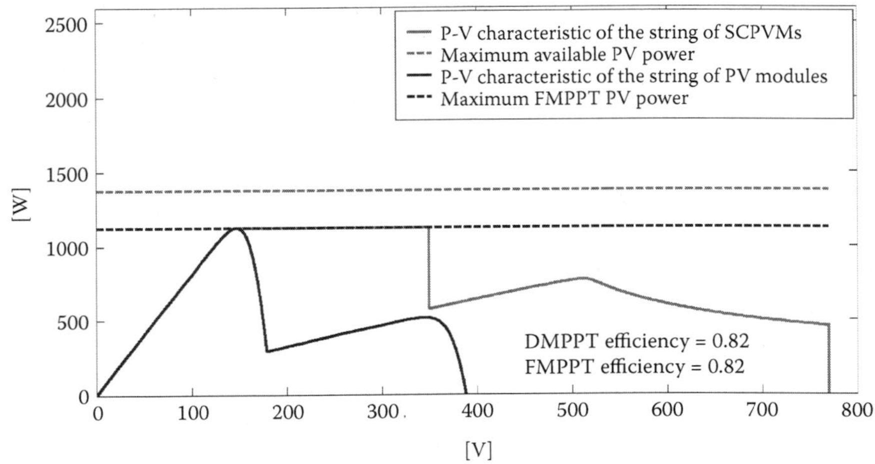

FIGURE 4.70
(Boost-based SCPVMs) $G_H = 1000$ W/m², $G_L = 200$ W/m², $V_{ds\,max} = 70$ V, $NH = 5$, $NL = 6$.

order to get $\eta_{DMPPT} = 1$ is (617 V, 733 V). The only inverters in Table 4.1 able to extract the DMPPT maximum power are the following: AROS (SIRIO 4600P), FRONIUS (IG TL3, IG TL4, IG TL5), MITSUBISHI ELECTRIC (PV-PNS03A TL-IT, PV-PNS04A TL-IT, PV-PNS04A TL2-IT, PV-PNS06A TL-IT), and SOCOMEC (SUNSYS 6065).

Figures 4.85–4.95 refer to the case of N buckboost-based lossless SCPVMs with $G_H = 1000$ W/m², $G_L = 200$ W/m², $V_{ds\,max} = 70$ V, and $I_{ds\,max} = 16$ A. Figures 4.96–4.105 refer instead to the case of buckboost-based lossless SCPVMs with $G_H = 1000$

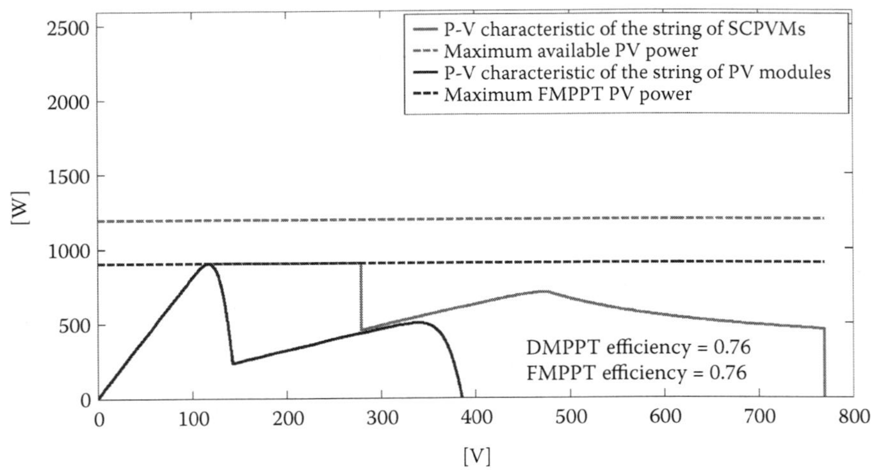

FIGURE 4.71
(Boost-based SCPVMs) $G_H = 1000$ W/m², $G_L = 200$ W/m², $V_{ds\,max} = 70$ V, $NH = 4$, $NL = 7$.

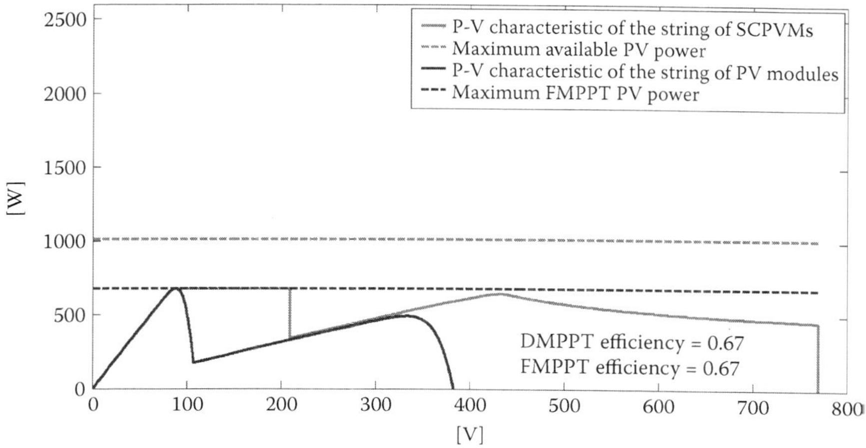

FIGURE 4.72
(Boost-based SCPVMs) $G_H = 1000$ W/m², $G_L = 200$ W/m², $V_{ds\ max} = 70$ V, $NH = 3$, $NL = 8$.

W/m², $G_L = 500$ W/m², $V_{ds\ max} = 70$ V, and $I_{ds\ max} = 16$ A. The maximum value obtainable for the DMPPT efficiency is always equal to 1, and it is always higher than the maximum value obtainable for the FMPPT efficiency.

Also in this case, of course, in order to be able to get the maximum value of the DMPPT efficiency, the bulk voltage must belong to an optimal range; the position and amplitude of such an optimal range are not fixed, but they change case by case.

If we refer, as an example, to the case shown in Figure 4.95 ($G_H = 1000$ W/m², $G_L = 200$ W/m², $V_{ds\ max} = 70$ V, $I_{ds\ max} = 16$ A, $NH = 1$, $NL = 10$), the optimal

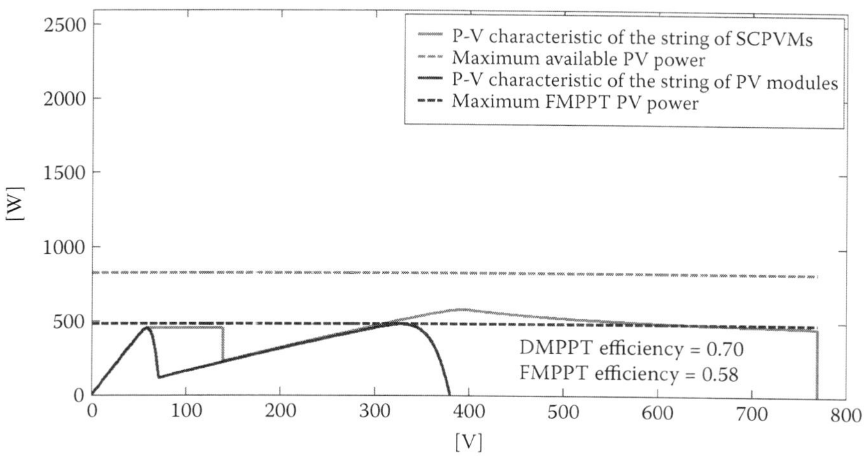

FIGURE 4.73
(Boost-based SCPVMs) $G_H = 1000$ W/m², $G_L = 200$ W/m², $V_{ds\ max} = 70$ V, $NH = 2$, $NL = 9$.

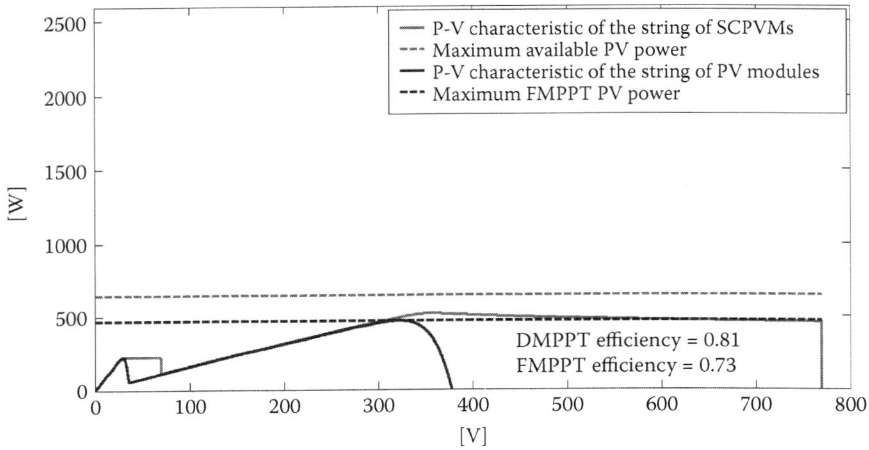

FIGURE 4.74
(Boost-based SCPVMs) $G_H = 1000$ W/m², $G_L = 200$ W/m², $V_{ds\ max} = 70$ V, $NH = 1$, $NL = 10$.

range for the bulk voltage in order to get $\eta_{DMPPT} = 1$ is (77 V, 115 V). The only inverters in the list of Table 4.1 able to extract the DMPPT maximum power are SMA (SUNNY BOY SB 1200-IT) and ELETTRONICA SANTERNO (SUNWAY-M120 3K, SUNWAY-M120 2.5K, SUNWAY-M120 1.8K).

On the basis of the results shown above, it is possible to state that the effective value of η_{DMPPT} depends on the number and type of PV modules and their operating conditions (irradiance level and temperature value), on the conversion ratio of the adopted MPPT DC/DC converters, on the voltage and current ratings of the adopted devices, and on the allowed MPPT range of the inverter input voltage.

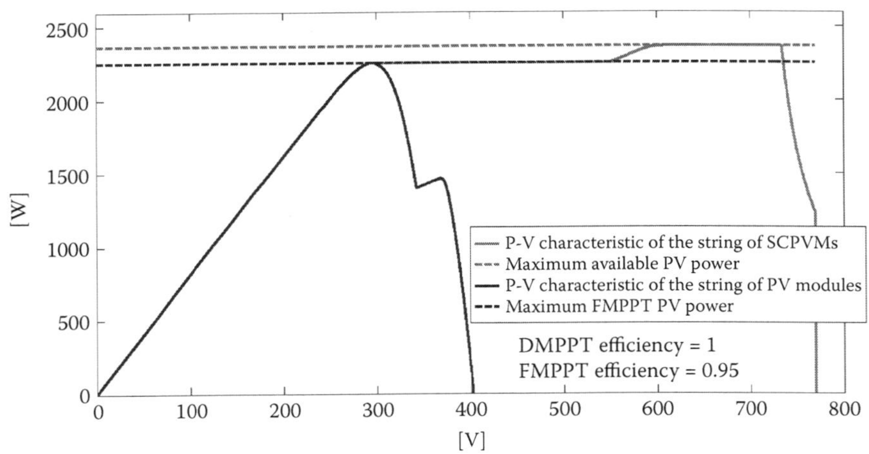

FIGURE 4.75
(Boost-based SCPVMs) $G_H = 1000$ W/m², $G_L = 500$ W/m², $V_{ds\ max} = 70$ V, $NH = 10$, $NL = 1$.

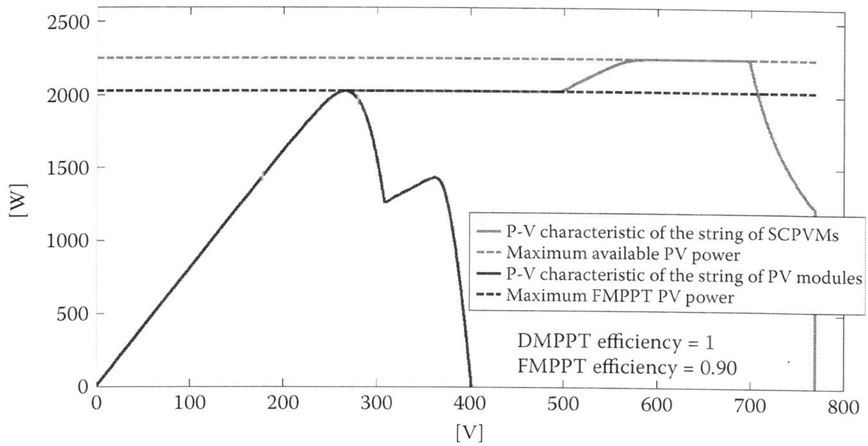

FIGURE 4.76
(Boost-based SCPVMs) $G_H = 1000$ W/m^2, $G_L = 500$ W/m^2, $V_{ds\,max} = 70$ V, $NH = 9$, $NL = 2$.

4.5 AC Analysis of a PV Array with DMPPT

This section is devoted to the AC analysis of an array constituted by a given number of SCPVMs. The aim is to study dynamic behavior and stability of a whole array of SCPVMs. In particular, the impact of system parameters on effectiveness and stability of the DMPPT technique will be analyzed [79]. As a first-order approximation, it is possible to model the DC/AC conversion

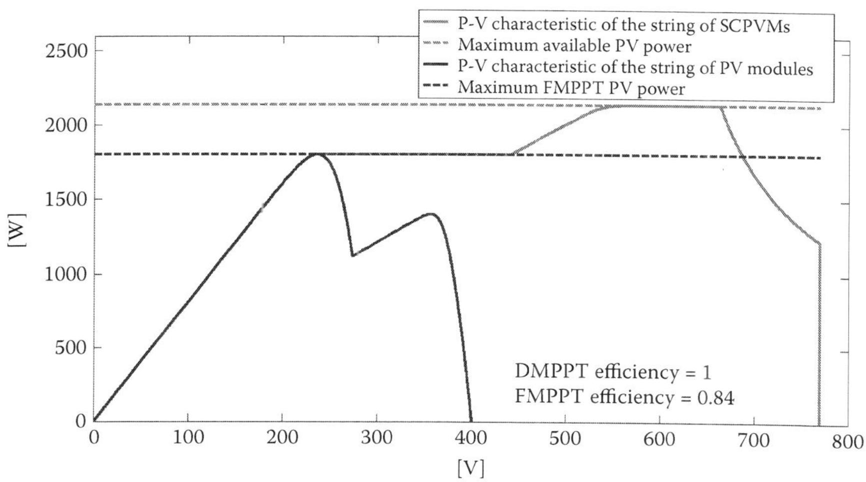

FIGURE 4.77
(Boost-based SCPVMs) $G_H = 1000$ W/m^2, $G_L = 500$ W/m^2, $V_{ds\,max} = 70$ V, $NH = 8$, $NL = 3$.

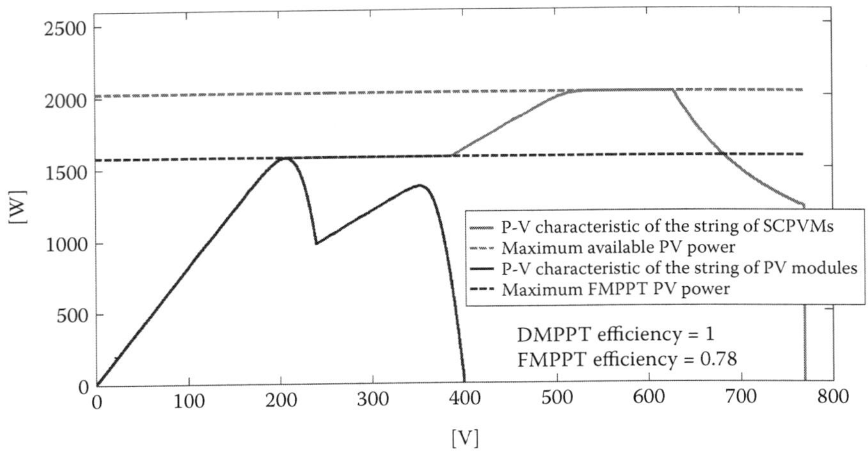

FIGURE 4.78
(Boost-based SCPVMs) $G_H = 1000$ W/m^2, $G_L = 500$ W/m^2, $V_{ds\,max} = 70$ V, $NH = 7$, $NL = 4$.

stage as a voltage source V_{bulk} with a series resistance R_{bulk}. Indeed, a PV inverter is capable of sinking whatever current in a certain range while keeping its input voltage regulated to a fixed average value [73–76]. This assumption greatly simplifies system analysis because, as long as the value of R_{bulk} is small compared to the output impedance of a string of SCPVMs, each string forms an independent loop with the equivalent model of the DC/AC conversion stage, and the analysis of the system of Figure 4.1 can be simplified by resorting to the analysis of a single string of N SCPVMs (Figure 4.106).

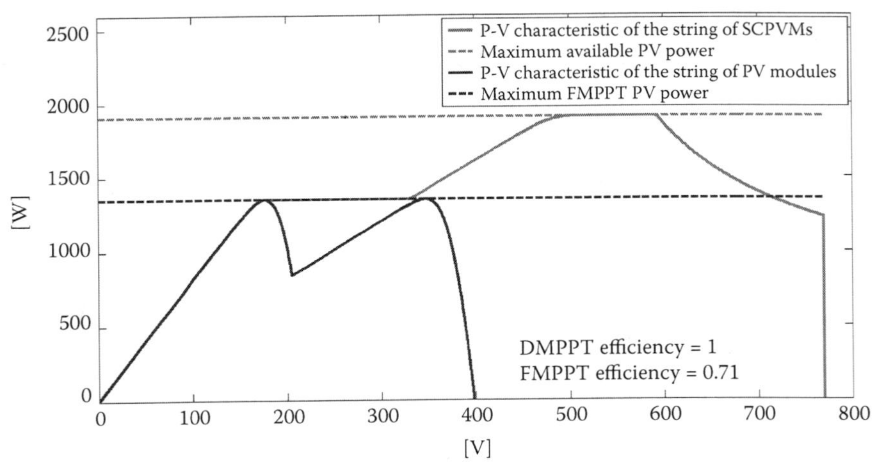

FIGURE 4.79
(Boost-based SCPVMs) $G_H = 1000$ W/m^2, $G_L = 500$ W/m^2, $V_{ds\,max} = 70$ V, $NH = 6$, $NL = 5$.

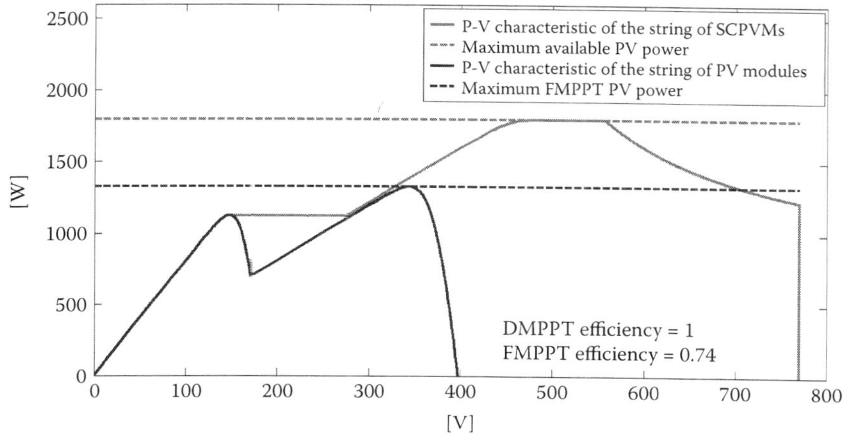

FIGURE 4.80
(Boost-based SCPVMs) $G_H = 1000$ W/m^2, $G_L = 500$ W/m^2, $V_{ds\,max} = 70$ V, $NH = 5$, $NL = 6$.

Without any loss of generality, a boost topology with synchronous rectification will be considered in the following. The MPPT of an SCPVM can be achieved by means of several standard MPPT techniques. A wide variety of them have been proposed in the literature, some of the most recent being described in [9–35]. In the following, the perturb and observe (P&O) technique is considered due to its high performance and simple and cost-effective implementation [9]. In a system like that shown in Figure 4.106, every disturbance introduced on the output voltage, by the inverter or by other SCPVMs, directly propagates on the PV modules' output voltage. This

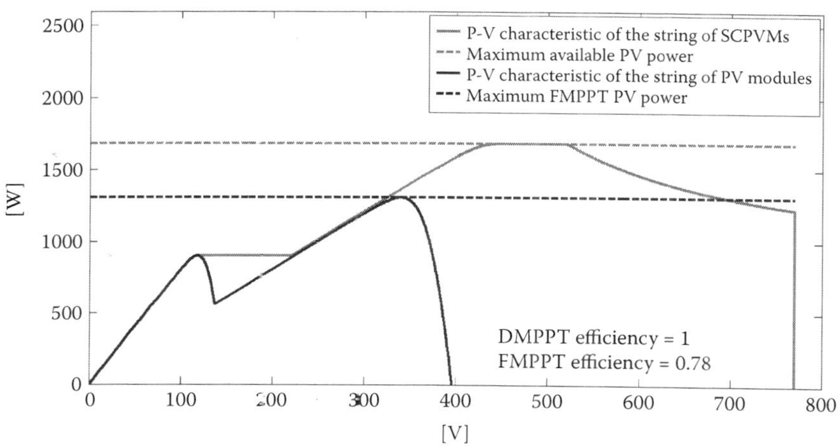

FIGURE 4.81
(Boost-based SCPVMs) $G_H = 1000$ W/m^2, $G_L = 500$ W/m^2, $V_{ds\,max} = 70$ V, $NH = 4$, $NL = 7$.

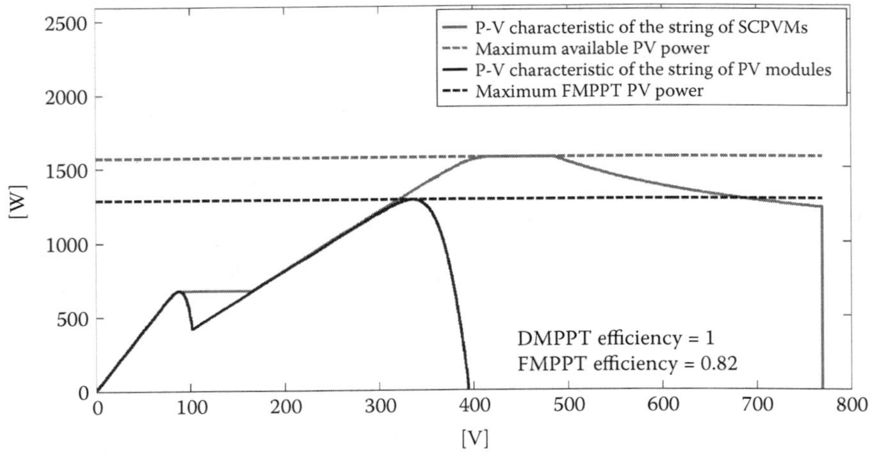

FIGURE 4.82
(Boost-based SCPVMs) $G_H = 1000$ W/m^2, $G_L = 500$ W/m^2, $V_{ds\,max} = 70$ V, $NH = 3$, $NL = 8$.

may lead to instability or dynamic performance degradation [47]. Therefore, as discussed in Chapter 2 (Section 2.4), a boost converter with linear input voltage feedback control is considered (Figure 4.107). In such a system, the P&O control variable is the reference voltage V_{ref}. If the control loop is fast enough, disturbances in the output loop of the system do not significantly affect the operation of a single SCPVM [47, 76]. The implementation of a dynamic duty cycle limitation prevents output overvoltages when deep mismatching conditions occur [79].

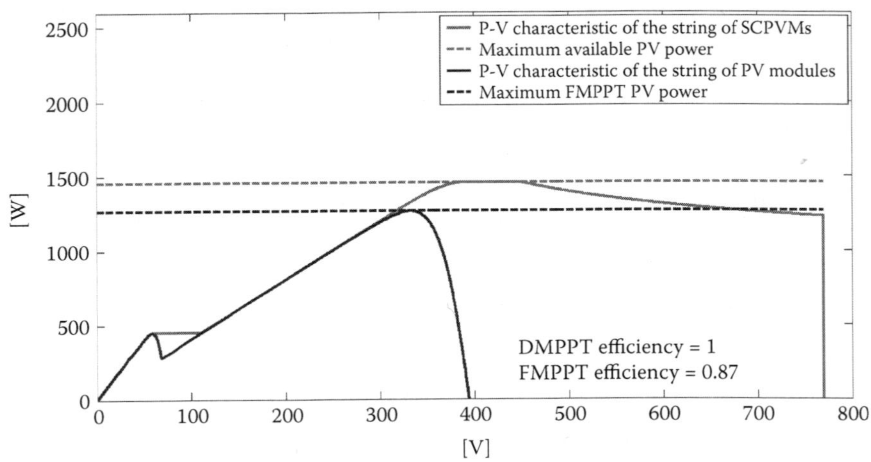

FIGURE 4.83
(Boost-based SCPVMs) $G_H = 1000$ W/m^2, $G_L = 500$ W/m^2, $V_{ds\,max} = 70$ V, $NH = 2$, $NL = 9$.

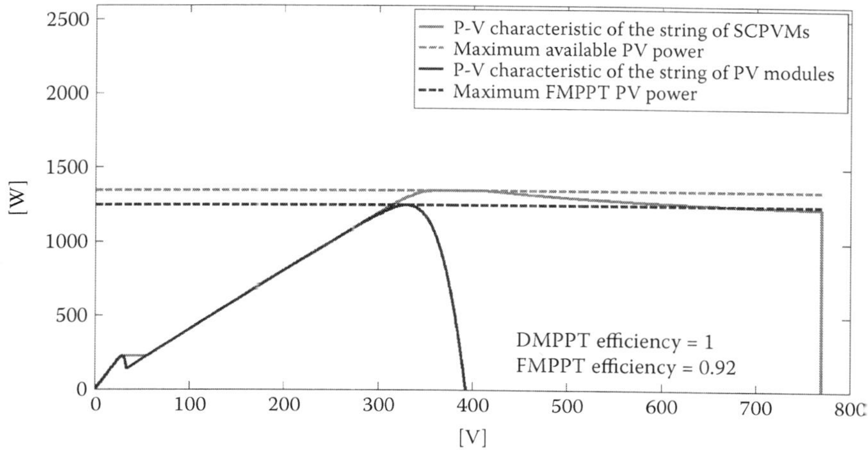

FIGURE 4.84
(Boost-based SCPVMs) $G_H = 1000\ \text{W/m}^2$, $G_L = 500\ \text{W/m}^2$, $V_{ds\,max} = 70$ V, $NH = 1$. $NL = 10$.

In conclusion the boost converter has two control loops. The inner control loop is constituted by a linear regulator for input voltage feedback control. The outer control loop is constituted by a MPPT P&O controller; its dynamics is slower by at least one order of magnitude compared to the dynamics of the controlled converter on which it operates. As in the case of the DC

TABLE 4.1

Main Characteristics of PV Inverters Available on the Market

Inverter Manufacturer		$P_{in}\,[W_p]$	η		MPPT Voltage Range (V)
			η_{max}	$\eta_{european}$	
SAMIL POWER	SolarRiver SR3K3TLA1	3480	97.4%	96.5%	
	SolarRiver SR4K4TLA1	4580	97.6%	97.1%	
	SolarRiver SR4K4TLA1/PT	4000	97.6%	97.1%	200–500
	SolarRiver SR5KTLA1	5200	97.6%	96.8%	
SMA	(SUNNY BOY) SB 1200-IT	1320	92.1%	90.9%	100–320
	SB 1200-IT	1850	93.5%	91.8%	139–320
	SB 2100-IT	2200	96.0%	95.2%	125–480
	SB 2500-IT	2700	94.1%	93.2%	224–480
	SB 3000-IT	3200	95.0%	93.6%	264–480
	SB 3800-IT	4040	95.6%	94.7%	200–400
	SB 3300HLTC-IT	3440	96.0%	94.6%	125–600

(*Continued*)

TABLE 4.1 (CONTINUED)

Inverter Manufacturer		$P_{in}\left[W_p\right]$	η		MPPT Voltage Range (V)
			η_{max}	$\eta_{european}$	
	SB 4000HLTC-IT	4200	97.0%	97.4%	125-440
	SB 5000HLTC-IT	5300	97.0%	96.4%	125–440
	(SUNNY MINI CENTRAL) SMC 5000A-IT	5750	96.1%	95.2%	246–480
	SMC 6000A-IT	6300	96.1%	95.2%	246–480
POWER ONE	AURORA PVI-2000-OUTD-IT	2300	96%	95%	210–530
	AURORA PVI-6000-OUTD	6900	97%	96.4%	180–530
ELETTRONICA SANTERNO	SUNWAY-M120 3K	3800	20% pin >90%	75% pin >93%	
	SUNWAY-M120 2.5K	3400	>90%	>93%	99–170
	SUNWAY-M120 1.8K	2300	>90%	>93%	
ECOJOULE	EJ-APOLLO4	5600	96.9%	95.9%	330–600
	EJ-APOLLO5	6400	96.9%	96.1%	330–600
	EJ-APOLLO 2 PLUS	2800	96.7%	95.4%	200–600
	EJ-APOLLO 3 PLUS	4200	96.8%	96.0%	
	EJ-APOLLO 4 PLUS	5600	96.9%	95.9%	
	EJ-APOLLO 5 PLUS	6400	96.9%	96.1%	
KAKO	POWADOR 2002	2000	95.9%	95.3%	125–510
	POWADOR 3002	3000	96.0%	95.4%	200–510
	POWADOR 4202	4200	95.9%	95.1%	
	POWADOR 5002	5000	95.9%	95.3%	200–510
	POWADOR 6002	6000	95.9%	95.3%	
	POWADOR 2500 xi	3200	96.4%	95.8%	
	POWADOR 3600 xi	4400	96.4%	95.8%	
	POWADOR 4000 xi	5250	96.3%	95.8%	350–600
	POWADOR 4500 xi	6000	96.3%	95.3%	
	POWADOR 5000 xi	6800	96.3%	95.3%	

TABLE 4.1 (CONTINUED)

Inverter Manufacturer		$P_{in}\left[W_p\right]$	η		MPPT Voltage Range (V)
			η_{max}	$\eta_{european}$	
	POWADOR supreme 4000	3200	96.4%	95.8%	350–510
AROS	SIRIO 1500	1900	96.3%	95.0%	200–405
	SIRIO 2000	2500	96.5%	95.1%	250–405
	SIRIO 2800	3500	97.1%	96.0%	250–405
	SIRIO 3100	3900	96.1%	95.3%	190–450
	SIRIO 4000	5000	96.2%	95.7%	250–450
	SIRIO 4000P	5000	96.2%	95.7%	250–450
	SIRIO 4600P	5700	96.2%	95.2%	190–700
	SIRIO 6000P	7500	97.6%	96.7%	230–500
FRONIUS	IG 15	2000	94.2%	91.4%	150–400
	IG 20	2700	94.3%	92.3%	
	IG 30	3600	94.3%	92.7%	
	IG 60 HW	6700	94.3%	93.5%	
	IG PLUS 35V	3700	95.7%	95%	230–500
	IG PLUS 50V	4200	95.7%	95%	
	IG TL3	3130	97.7%	97.1%	
	IG TL4	4190	97.7%	97.3%	350–700
	IG TL5	5250	97.7%	97.3%	
MITSUBISHI ELECTRIC	PV-PNS03A TL-IT	3250	96.1%	94.6%	160–650
	PV-PNS04A TL-IT	4300	96.2%	95.1%	190–650
	PV-PNS04A TL2-IT	3250	96.2%	95.4%	260–650
	PV-PNS06A TL-IT	6000	96.1%	94.6%	160–650
GEFRAN	RADIUS 2K	2200	95.4%	94.4%	210–400
	RADIUS 3.6K	3900	95.4%	94.4%	
SOCOMEC	SUNSYS 2254	2200	95.4%	94.4%	210–400
	SUNSYS 2265	2200	95.4%	94.4%	
	SUNSYS 3954	3900	95.4%	94.4%	
	SUNSYS 3965	3900	95.4%	94.4%	
	SUNSYS 6065	6000	98.0%	97.5%	310–800
ECOJOULE	EJ-APOLLO2	2800	96.7%	95.4%	330–600
	EJ-APOLLO3	4200	96.8%	96.0%	

analysis, a string made of NH SCPVMs operating under irradiance level G_H and of NL SCPVMs operating under irradiance level G_L will be considered (Figure 4.108). It is $N = NH + NL$. It will be assumed that all the NL SCPVMs are identical and that their MPPT controllers are synchronized. This assumption simplifies system analysis and simulation. Simulations including

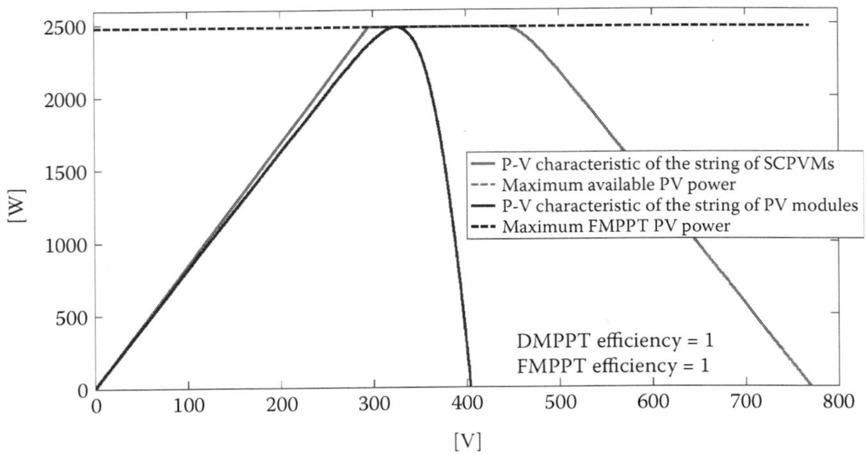

FIGURE 4.85
(Buckboost-based SCPVMs) $G_H = 1000$ W/m^2, $G_L = 200$ W/m^2, $V_{ds\,max} = 70$ V, $I_{ds\,max} = 16$ A, $NH = 11$, $NL = 0$.

random delay between MPPT controllers have also been carried out, and a negligible difference has been found between these two approaches [79]. The same assumption is also valid for the NH SCPVMs.

Due to the series connection of the output ports of the SCPVMs, the output voltage of a given SCPVM is related to the ratio between its output power and the total output power:

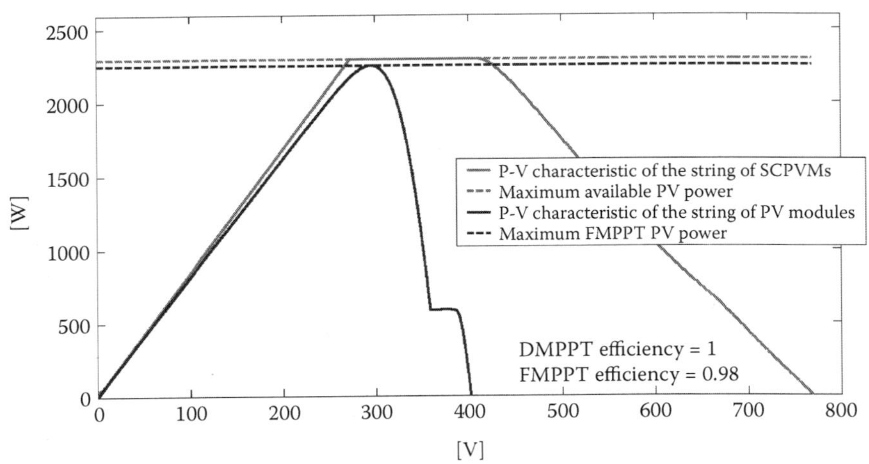

FIGURE 4.86
(Buckboost-based SCPVMs) $G_H = 1000$ W/m^2, $G_L = 200$ W/m^2, $V_{ds\,max} = 70$ V, $I_{ds\,max} = 16$ A, $NH = 10$, $NL = 1$.

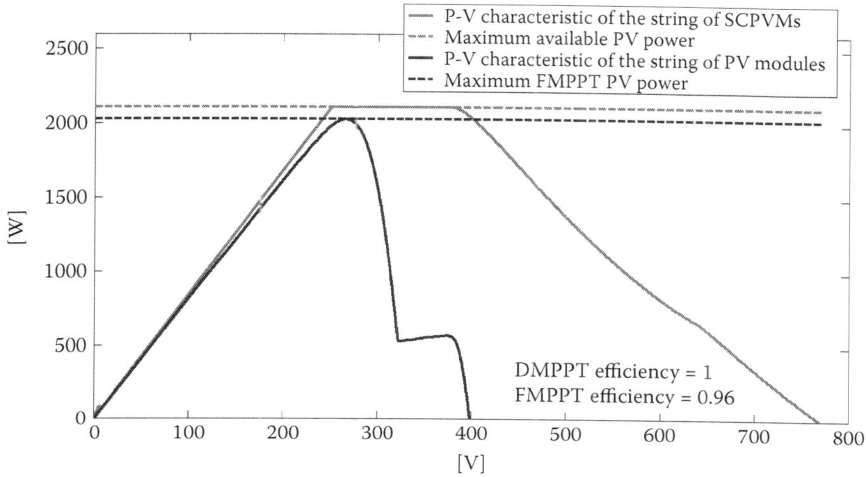

FIGURE 4.87
(Buckboost-based SCPVMs) $G_H = 1000$ W/m², $G_L = 200$ W/m², $V_{ds\,max} = 70$ V, $I_{ds\,max} = 16$ A, $NH = 9$, $NL = 2$.

$$V_{out\ H} = V_{bulk} \frac{P_{outH}}{NH \cdot P_{outH} + NL \cdot P_{outL}} \tag{4.2}$$

$$V_{out\ L} = V_{bulk} \frac{P_{outL}}{NH \cdot P_{outH} + NL \cdot P_{outL}} \tag{4.3}$$

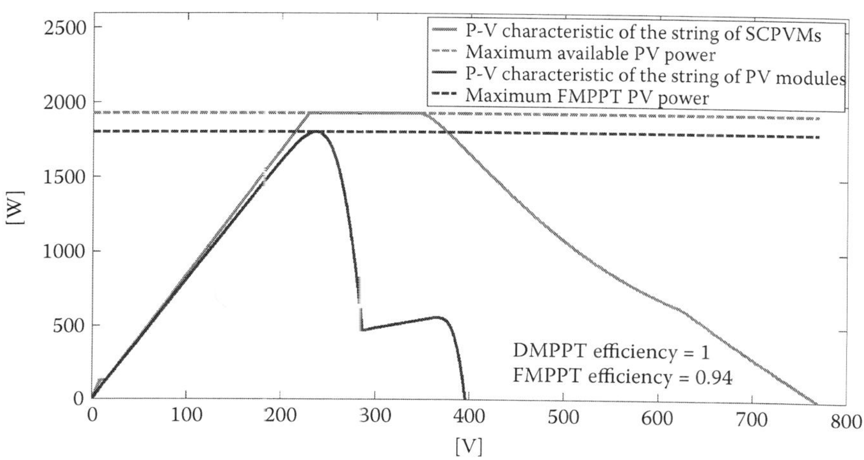

FIGURE 4.88
(Buckboost-based SCPVMs) $G_H = 1000$ W/m², $G_L = 200$ W/m², $V_{ds\,max} = 70$ V, $I_{ds\,max} = 16$ A, $NH = 8$, $NL = 3$.

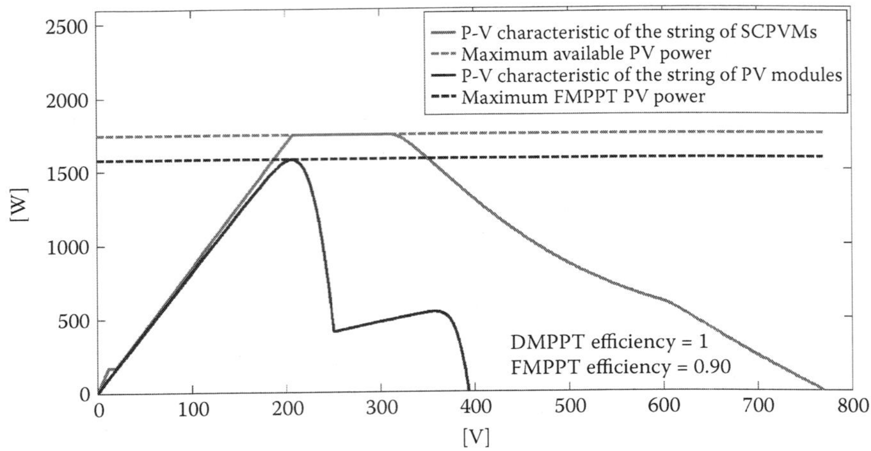

FIGURE 4.89
(Buckboost-based SCPVMs) $G_H = 1000$ W/m², $G_L = 200$ W/m², $V_{ds\,max} = 70$ V, $I_{ds\,max} = 16$ A, $NH = 7$, $NL = 4$.

where P_{outH} and P_{outL} are the output powers of SCPVMs with irradiations G_H and G_L, respectively, and V_{outH} and V_{outL} are the corresponding output voltages. Equations (4.2) and (4.3) highlight that the output voltage of an SCPVM can vary in a wide range due to possible imbalances among powers delivered by modules. If, for example, $P_{outL} \approx 0$ (the NL modules are totally shadowed) while $P_{outH} \neq 0$, from (4.2) and (4.3), we get $V_{outL} \approx 0$ and $V_{outH} = V_{bulk}/NH$. Therefore, if NH is sufficiently low, V_{outH} can become very large, causing

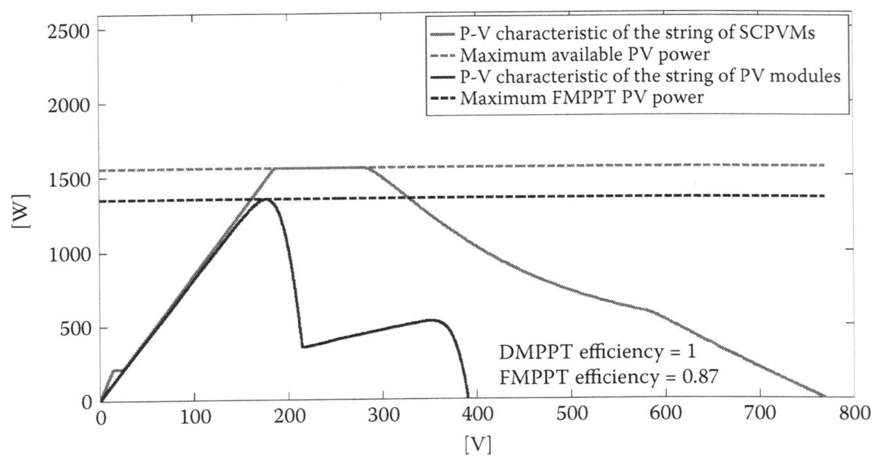

FIGURE 4.90
(Buckboost-based SCPVMs) $G_H = 1000$ W/m², $G_L = 200$ W/m², $V_{ds\,max} = 70$ V, $I_{ds\,max} = 16$ A, $NH = 6$, $NL = 5$.

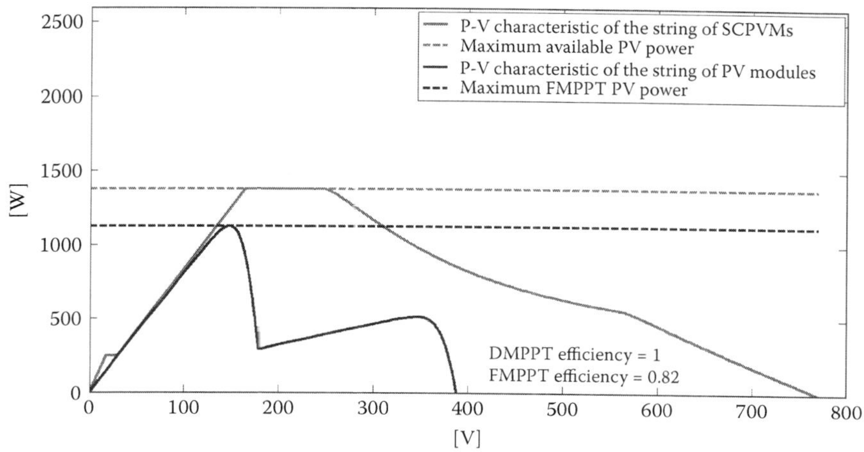

FIGURE 4.91
(Buckboost-based SCPVMs) $G_H = 1000$ W/m², $G_L = 200$ W/m², $V_{ds\,max} = 70$ V, $I_{ds\,max} = 16$ A, $NH = 5$, $NL = 6$.

high switch stresses. Output voltage in a boost converter corresponds to the maximum voltage across the switches and the output capacitor. In order to prevent the output voltage of one or more SCPVMs from exceeding a given maximum value $V_{ds\,max}$, an output voltage limitation technique must be adopted by means of a suitable duty cycle limitation function [79], as shown in Figure 4.107.

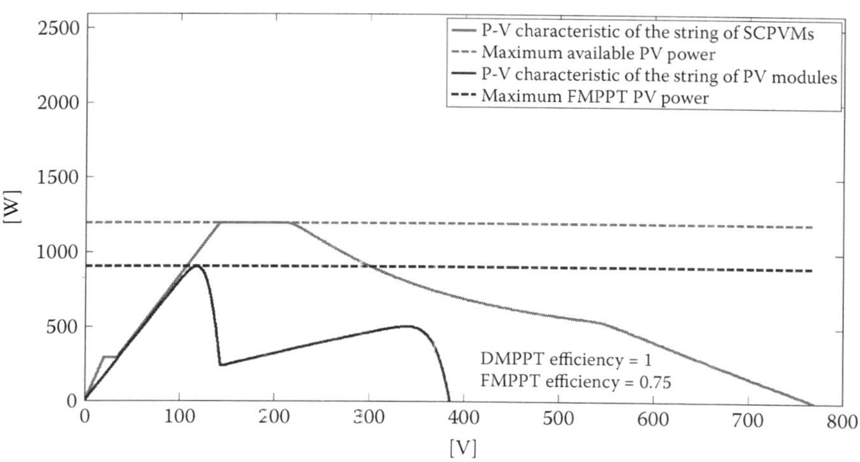

FIGURE 4.92
(Buckboost-based SCPVMs) $G_H = 1000$ W/m², $G_L = 200$ W/m², $V_{ds\,max} = 70$ V, $I_{ds\,max} = 16$ A, $NH = 4$, $NL = 7$.

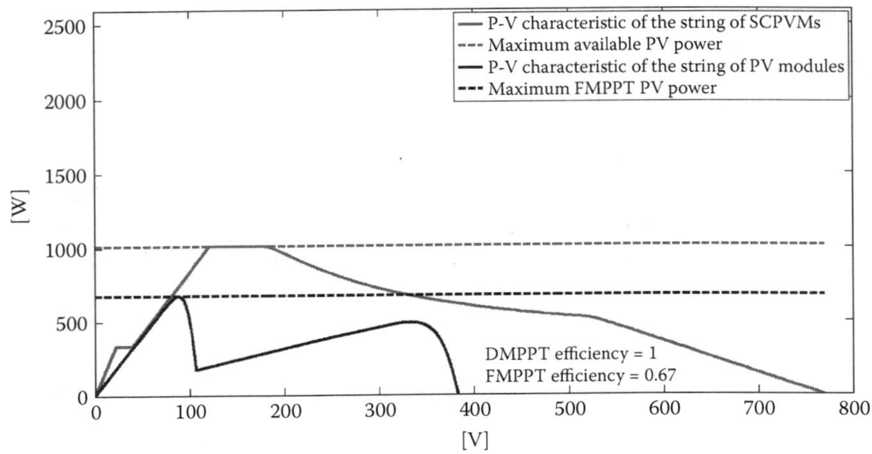

FIGURE 4.93
(Buckboost-based SCPVMs) $G_H = 1000$ W/m^2, $G_L = 200$ W/m^2, $V_{ds\,max} = 70$ V, $I_{ds\,max} = 16$ A, $NH = 3$, $NL = 8$.

4.5.1 AC Model of a Single SCPVM

Figure 4.109 shows a detailed schematic of the SCPVM under study, including circuitry achieving feedback regulation of the input voltage. The corresponding small-signal model is schematized in Figure 4.110.

In the following small letters like v, i, and d are used to indicate large-signal values of voltages, currents, and duty cycles, respectively; capital letters

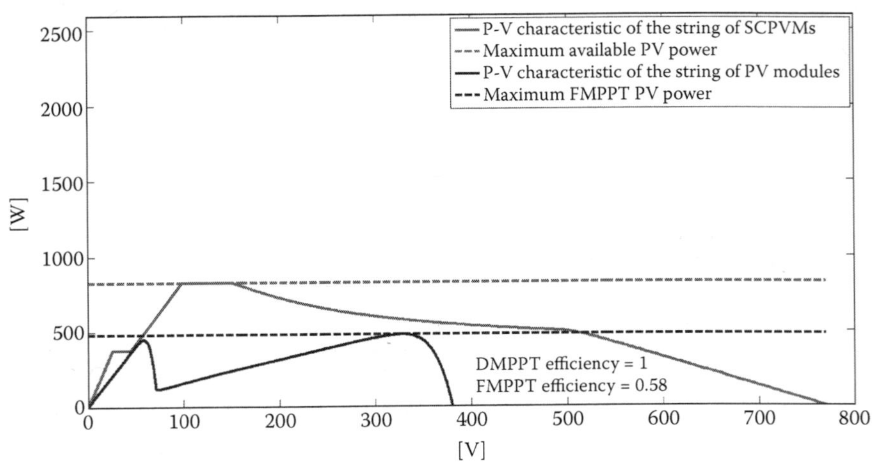

FIGURE 4.94
(Buckboost-based SCPVMs) $G_H = 1000$ W/m^2, $G_L = 200$ W/m^2, $V_{ds\,max} = 70$ V, $I_{ds\,max} = 16$ A, $NH = 2$, $NL = 9$.

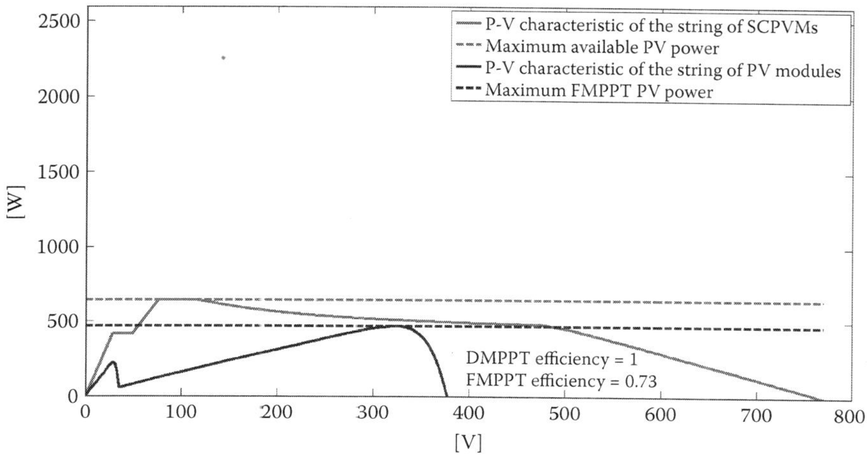

FIGURE 4.95
(Buckboost-based SCPVMs) $G_H = 1000$ W/m², $G_L = 200$ W/m², $V_{ds\,max} = 70$ V, $I_{ds\,max} = 16$ A, $NH = 1$, $NL = 10$.

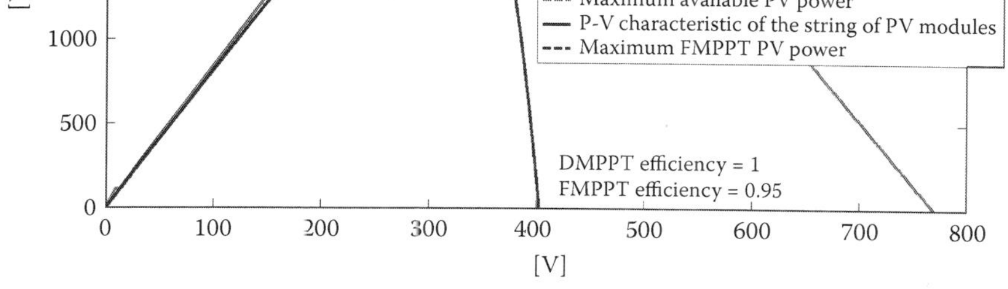

FIGURE 4.96
(Buckboost-based SCPVMs) $G_H = 1000$ W/m², $G_L = 500$ W/m², $V_{ds\,max} = 70$ V, $I_{ds\,max} = 16$ A, $NH = 10$, $NL = 1$.

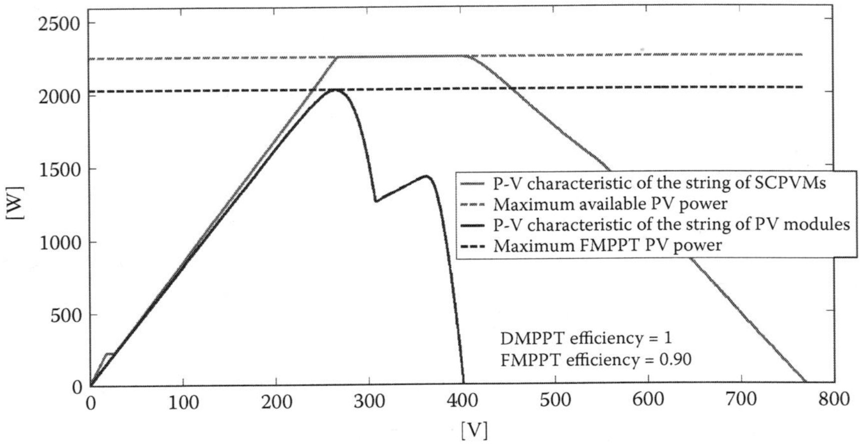

FIGURE 4.97
(Buckboost-based SCPVMs) $G_H = 1000$ W/m^2, $G_L = 500$ W/m^2, $V_{ds\ max} = 70$ V, $I_{ds\ max} = 16$ A, $NH = 9$, $NL = 2$.

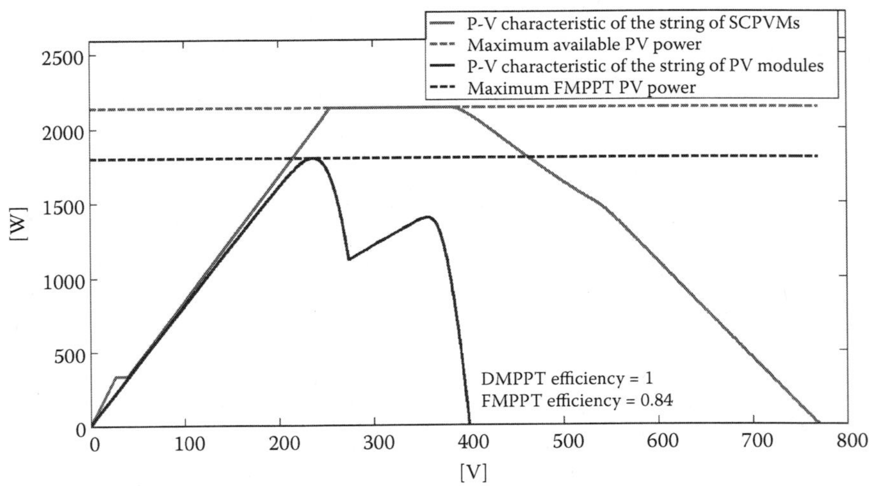

FIGURE 4.98
(Buckboost-based SCPVMs) $G_H = 1000$ W/m^2, $G_L = 500$ W/m^2, $V_{ds\ max} = 70$ V, $I_{ds\ max} = 16$ A, $NH = 8$, $NL = 3$.

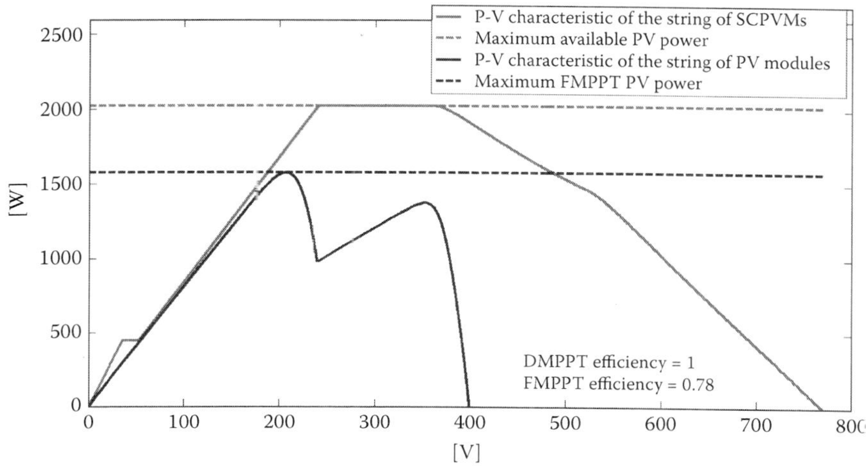

FIGURE 4.99
(Buckboost-based SCPVMs) $G_H = 1000$ W/m², $G_L = 500$ W/m², $V_{ds\ max} = 70$ V, $I_{ds\ max} = 16$ A, $NH = 7$, $NL = 4$.

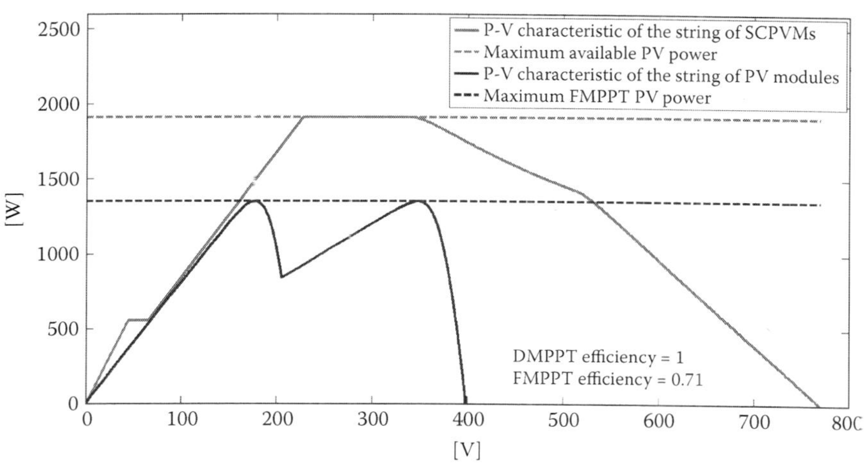

FIGURE 4.100
(Buckboost-based SCPVMs) $G_H = 1000$ W/m², $G_L = 500$ W/m², $V_{ds\ max} = 70$ V, $I_{ds\ max} = 16$ A, $NH = 6$, $NL = 5$.

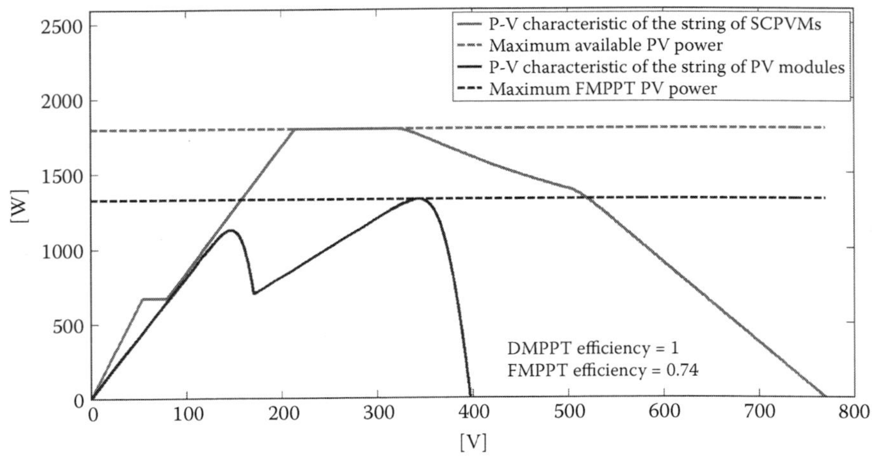

FIGURE 4.101
(Buckboost-based SCPVMs) $G_H = 1000$ W/m^2, $G_L = 500$ W/m^2, $V_{ds\,max} = 70$ V, $I_{ds\,max} = 16$ A, $NH = 5$, $NL = 6$.

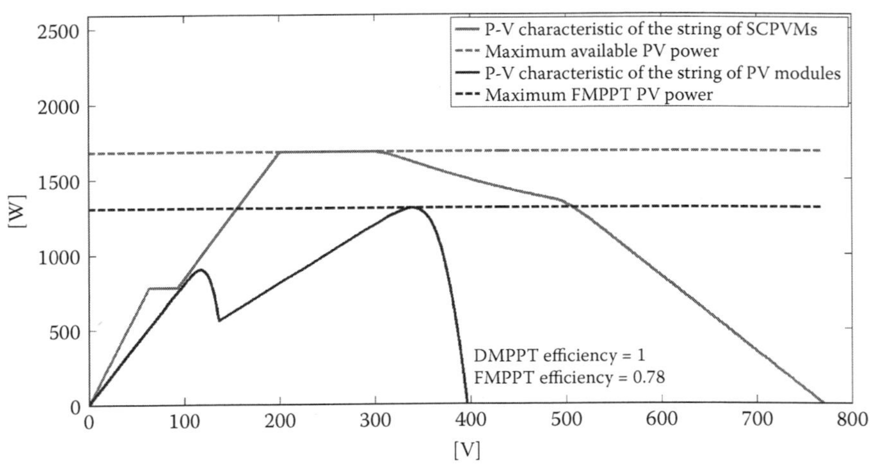

FIGURE 4.102
(Buckboost-based SCPVMs) $G_H = 1000$ W/m^2, $G_L = 500$ W/m^2, $V_{ds\,max} = 70$ V, $I_{ds\,max} = 16$ A, $NH = 4$, $NL = 7$.

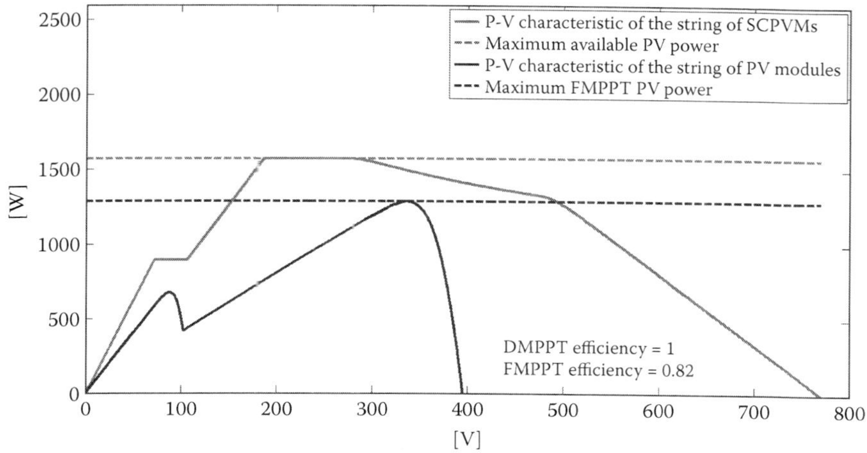

FIGURE 4.103
(Buckboost-based SCPVMs) $G_H = 1000$ W/m^2, $G_L = 500$ W/m^2, $V_{ds\ max} = 70$ V, $I_{ds\ max} = 16$ A, $NH = 3$, $NL = 8$.

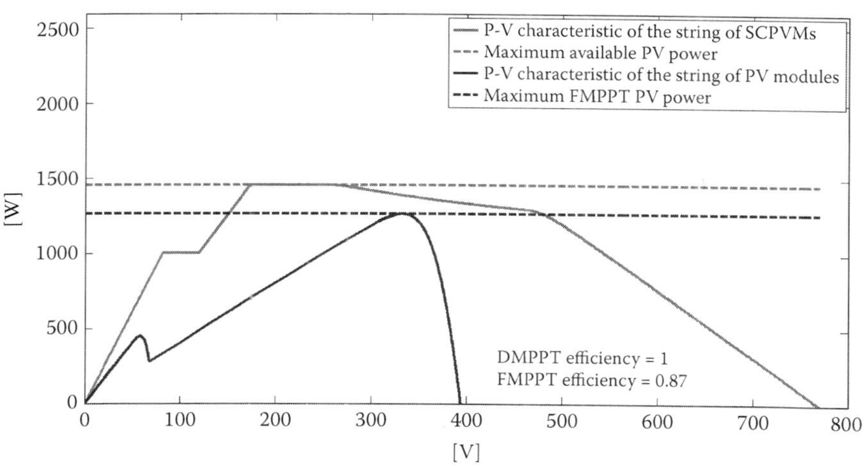

FIGURE 4.104
(Buckboost-based SCPVMs) $G_H = 1000$ W/m^2, $G_L = 500$ W/m^2, $V_{ds\ max} = 70$ V, $I_{ds\ max} = 16$ A, $NH = 2$, $NL = 9$.

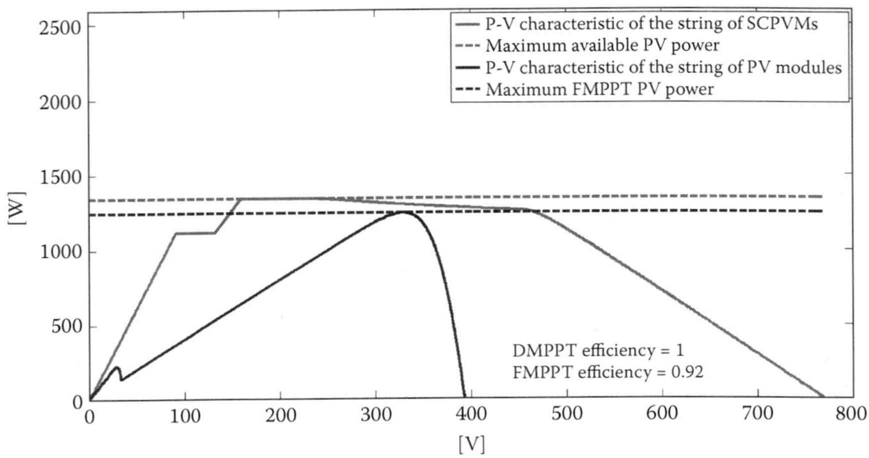

FIGURE 4.105
(Buckboost-based SCPVMs) $G_H = 1000$ W/m^2, $G_L = 500$ W/m^2, $V_{ds\,max} = 70$ V, $I_{ds\,max} = 16$ A, $NH = 1$, $NL = 10$.

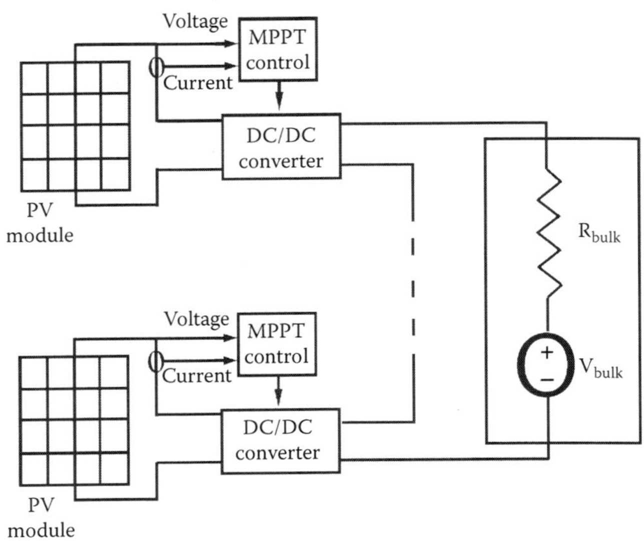

FIGURE 4.106
Simplified model of a string of SCPVMs.

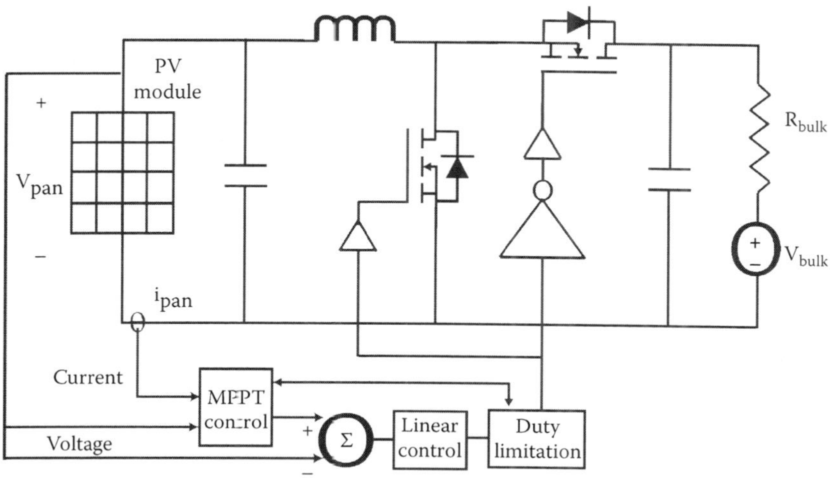

FIGURE 4.107
Schematic of an SCPVM employing a boost converter with synchronous rectification, input voltage feedback control, and dynamic duty cycle limitation.

like *V*, *I*, and *D* are used to indicate DC values of the same variables; and small letters with hat symbols indicate small-signal variations around their steady-state operating point.

4.5.1.1 Open-Loop Transfer Functions of an SCPVM

The following transfer functions can be defined and computed by analyzing the circuit of Figure 4.110:

$$G_{vpan\ d\ OL}(s) = \left.\frac{\hat{v}_{pan}(s)}{\hat{d}(s)}\right|_{\substack{\hat{i}_g(s)=0 \\ \hat{v}_{bulk}(s)=0 \\ open\ loop}} \tag{4.4}$$

$$G_{ibulk\ d\ OL}(s) = \left.\frac{\hat{i}_{bulk}(s)}{\hat{d}(s)}\right|_{\substack{\hat{i}_g(s)=0 \\ \hat{v}_{bulk}(s)=0 \\ open\ loop}} \tag{4.5}$$

$$Z_{vpan\ i_g\ OL}(s) = \left.\frac{\hat{v}_{pan}(s)}{\hat{i}_g(s)}\right|_{\substack{\hat{d}(s)=0 \\ \hat{v}_{bulk}(s)=0 \\ open\ loop}} \tag{4.6}$$

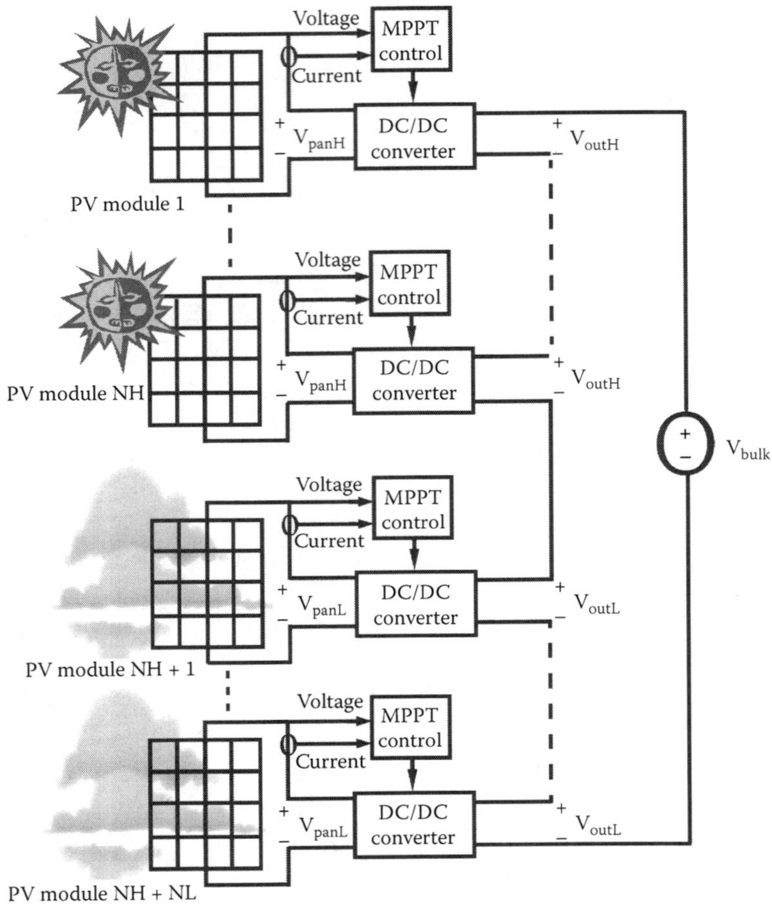

FIGURE 4.108
Simplified model of a string of SCPVMs under uneven irradiation.

$$G_{ibulk\ i_g\ OL}(s) = \left.\frac{\hat{i}_{bulk}(s)}{\hat{i}_g(s)}\right|_{\substack{\hat{d}(s)=0 \\ \hat{v}_{bulk}(s)=0 \\ open\ loop}} \tag{4.7}$$

$$G_{vpan\ v_{bulk}\ OL}(s) = \left.\frac{\hat{v}_{pan}(s)}{\hat{v}_{bulk}(s)}\right|_{\substack{\hat{i}_g(s)=0 \\ \hat{d}(s)=0 \\ open\ loop}} \tag{4.8}$$

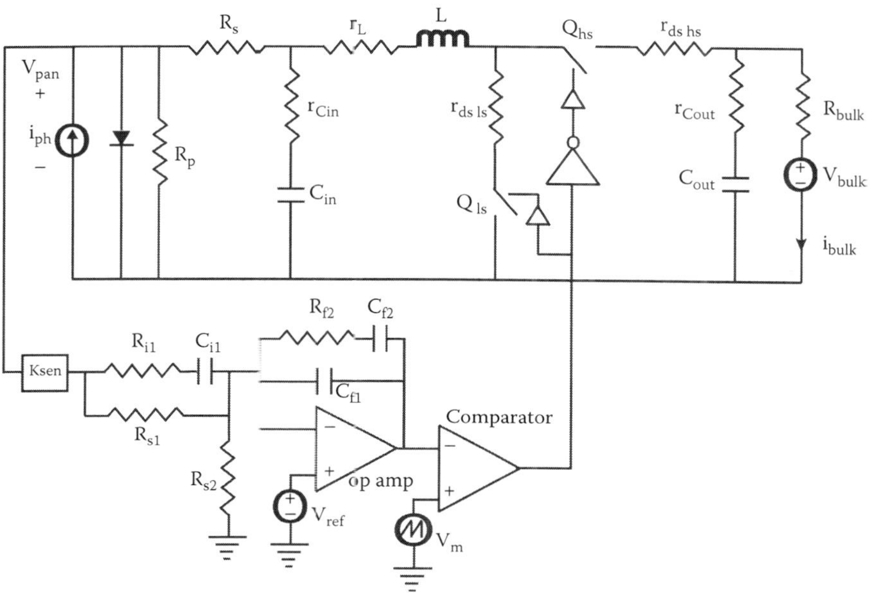

FIGURE 4.109
Schematics of an SCPVM under study (i_{ph}, Rp, and Rs are the parameters of the large-signal model of the PV module; see Chapter 1).

$$Z_{out\ OL}(s) = -\left.\frac{\hat{v}_{bulk}(s)}{\hat{i}_{bulk}(s)}\right|_{\substack{\hat{i}_g(s)=0 \\ \hat{d}(s)=0 \\ open\ loop}} \tag{4.9}$$

Input voltage control is implemented via a standard operational amplifier with negative feedback. Its dynamics is described by the transfer function

$$G_c(s) = \frac{Z_f(s)}{Z_i(s)} \tag{4.10}$$

where

$$Z_f(s) = (R_{f2} + \frac{1}{sC_{f2}})\,||\,\frac{1}{sC_{f1}} \tag{4.11}$$

$$Z_i(s) = (R_{i1} + \frac{1}{sC_{i1}})\,||\,R_{s1} \tag{4.12}$$

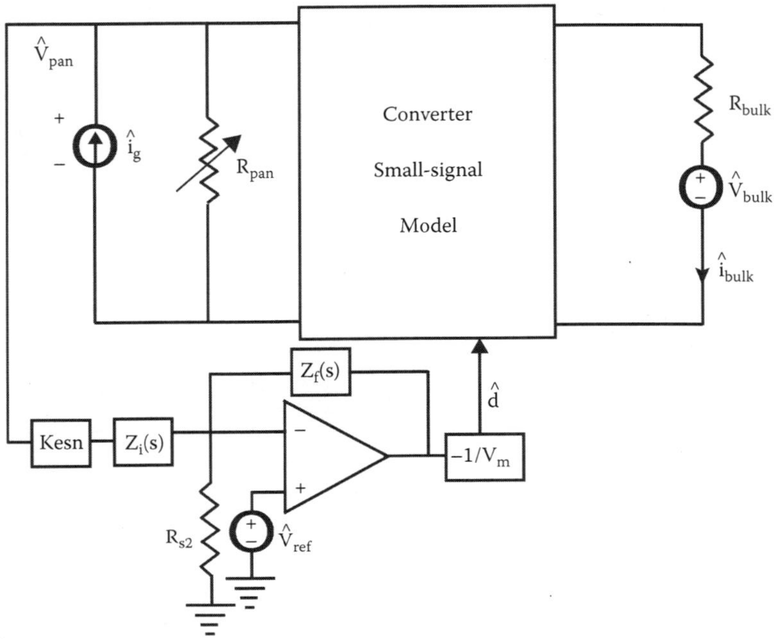

FIGURE 4.110
Small-signal model of the SCPVM.

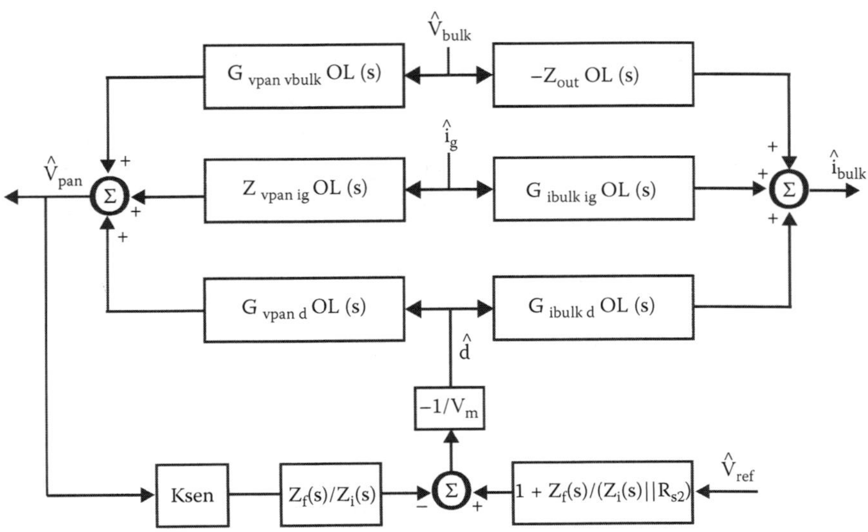

FIGURE 4.111
Block diagram of the small-signal AC model of the SCPVM of Figure 4.110.

Given two parallel connected impedances z_1 and z_2, the symbol $z_1||z_2$ indicates their equivalent impedance $(z_1 \cdot z_2)/(z_1 + z_2)$. The negative sign of the transfer function of the operational amplifier in the inverting topology is considered part of the feedback loop rather than part of the compensation network transfer function according to standard control theory convention. The loop transfer function can then be written as

$$T_c(s) = -G_{vpan\ d\ OL}(s) K_{sen} G_c(s) \frac{1}{V_m} \tag{4.13}$$

Note that the negative sign in (4.13) is introduced by the negative PWM block in the circuit of Figure 4.109. This is due to the fact that the transfer function $G_{vpan\ d\ OL}(s)$ is negative at low frequencies, and therefore it is necessary to introduce a negative sign in the feedback loop to have a positive loop transfer function and thus a stable closed-loop system.

4.5.1.2 Closed-Loop Transfer Functions of an SCPVM

The block diagram of Figure 4.111 schematizes the dynamic interactions between input and output variables of the SCPVM of Figure 4.110 and allows the computation of the closed-loop input-to-output transfer functions of the SCPVM.

$$G_{v_{pan} v_{ref} CL}(s) = \left. \frac{\hat{v}_{pan}(s)}{\hat{v}_{ref}(s)} \right|_{\substack{\hat{i}_g(s)=0 \\ \hat{v}_{bulk}(s)=0 \\ closed\ loop}} = \frac{\left[1 + \dfrac{Z_f(s)}{Z_i(s) R_{s2}}\right] G_{v_{pan}\ d OL}(s) \dfrac{1}{V_m}}{1 + T_c(s)} \tag{4.14}$$

$$G_{i_{bulk}\ v_{ref} CL}(s) = \left. \frac{\hat{i}_{bulk}(s)}{\hat{v}_{ref}(s)} \right|_{\substack{\hat{i}_g(s)=0 \\ \hat{v}_{bulk}(s)=0 \\ closed\ loop}} = \frac{\left[1 + \dfrac{Z_f(s)}{Z_i(s) R_{s2}}\right] G_{i_{bulk}\ d OL}(s) \dfrac{1}{V_m}}{1 + T_c(s)} \tag{4.15}$$

$$G_{v_{pan}\ v_{bulk} CL}(s) = \left. \frac{\hat{v}_{pan}(s)}{\hat{v}_{bulk}(s)} \right|_{\substack{\hat{i}_g(s)=0 \\ \hat{v}_{ref}(s)=0 \\ closed\ loop}} = \frac{G_{v_{pan}\ v_{bulk} OL}(s)}{1 + T_c(s)} \tag{4.16}$$

$$Z_{out\ CL}(s) = -\left. \frac{\hat{v}_{bulk}(s)}{\hat{i}_{bulk}(s)} \right|_{\substack{\hat{i}_g(s)=0 \\ \hat{v}_{ref}(s)=0 \\ closed\ loop}} = \frac{Z_{out\ OL}(s)}{1 - \dfrac{Z_{out\ OL}(s)\ G_{i_{bulk}\ d\ OL}(s) G_{v_{pan} v_{bulk}\ OL}(s) K_{sen} G_c(s) \dfrac{1}{V_m}}{1 + T_c(s)}} \tag{4.17}$$

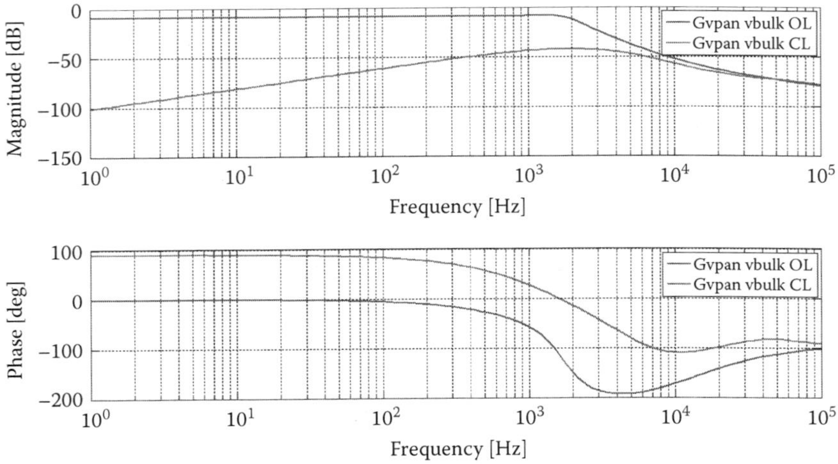

FIGURE 4.112
Output voltage to input voltage transfer function of the circuit of Figure 4.109 in open-loop (OL) and closed-loop (CL) conditions.

$$G_{i_{bulk}\ i_g\ \ CL}(s) = \left.\frac{\hat{i}_{bulk}(s)}{\hat{i}_g(s)}\right|_{\substack{\hat{v}_{bulk}(s)=0 \\ \hat{v}_{ref}(s)=0 \\ closed\ loop}} = G_{i_{bulk}\ i_g\ \ OL}(s) + \frac{G_{i_{bulk}\ d\ OL}(s) Z_{v_{bulk}\ i_g\ \ OL}(s) K_{sen} G_c(s) \frac{1}{V_m}}{1+T_c(s)} \tag{4.18}$$

$$Z_{v_{pan}\ i_g\ \ CL}(s) = \left.\frac{\hat{v}_{pan}(s)}{\hat{i}_g(s)}\right|_{\substack{\hat{v}_{bulk}(s)=0 \\ \hat{v}_{ref}(s)=0 \\ closed\ loop}} = \frac{Z_{v_{pan}\ i_g\ \ OL}(s)}{1+T_c(s)} \tag{4.19}$$

Figure 4.112 shows the Bode plots of the transfer functions $G_{v_{pan}\ v_{bulk}CL}(s)$ and $G_{v_{pan}\ v_{bulk}OL}(s)$. It is worth noting that the output voltage disturbance rejection is obtained by using input voltage feedback control in the SCPVM. Values of the parameters used to plot the Bode diagrams of Figure 4.112 are as follows: $L = 100$ μH, $r_L = 0.082\ \Omega$, $r_{ds\ hs} = 0.021\ \Omega$, $r_{ds\ ls} = 0.013\ \Omega$, $C_{out} = 99$ μF, $r_{Cout} = 0.12\ \Omega$, $C_{in} = 94$ μF, $r_{Cin} = 0.18\ \Omega$, $fs = 160$ kHz, $V_f = 0.6$ V, $R_{s1} = 1$ kΩ, $R_{s2} = 0.39$ kΩ, $R_{i1} = 100\ \Omega$, $C_{i1} = 47$ nF, $C_{f1} = 2.2$ nF, $R_{f2} = 1$ kΩ, $C_{f2} = 47$ nF, $K_{sen} = 0.454$, $V_m = 1.8$ V, irradiance $G = 500$ W/m^2, $V_{pan} = 15$ V, $I_g = 4$ A, $R_{pan} = 84\ \Omega$, $V_{bulk} = 27.5$ V, $R_{bulk} = 1$ mΩ.

4.5.2 Small-Signal Model of a Photovoltaic Array with DMPPT

In the previous sections the model of an SCPVM has been derived. Such a model describes the steady-state and small-signal dynamic behavior of the proposed SCPVM. The purpose of this section is to derive a similar model for an array of SCPVMs operating under mismatched irradiances. As in steady-state DC operating conditions, also at higher frequencies the input port of the PV inverter (in case of grid-connected PV systems) or the battery stack (in case of stand-alone systems) can be modeled as a voltage source. Indeed, common battery chargers and inverters have high-value input capacitors and their input voltage is controlled by a feedback regulator [72–76]. Therefore it is possible to derive a small-signal model of a PV array by considering one string of SCPVMs connected to a voltage source. Transfer functions of the system depicted in Figure 4.108 will be defined and calculated by extending the small-signal model of the SCPVM described previously. Finally, stability considerations will be drawn based on the proposed model. The same approach can be followed to further extend such a model to account for inverter dynamics. Other sources of mismatch among PV modules or operating conditions can be accounted for in a similar way [79].

4.5.2.1 Small-Signal Model of a String of SCPVMs

In this section a small-signal AC model is derived for the system of Figure 4.108. Initially, for the sake of simplicity, the case $NH = NL = 1$ is considered. Subscripts 1 and 2 indicate the first and second SCPVMs, respectively, of the string under study. The input variables are i_{g1}, i_{g2}, V_{ref1}, V_{ref2}, and V_{bulk}. The output variables are v_{pan1} and v_{pan2}. The resulting small-signal model is depicted in Figure 4.113.

Appropriate transfer functions should be defined to relate input variables to output variables of the system in Figure 4.113. The following transfer functions relate v_{pan1} to input variables:

$$G_{v_{pan1}\,v_{ref1}\,SYS}(s) = \left.\frac{\hat{v}_{pan1}(s)}{\hat{v}_{ref1}(s)}\right|_{\substack{\hat{v}_{bulk}(s)=0\\ \hat{v}_{ref2}(s)=0\\ \hat{i}_{g1}(s)=0\\ \hat{i}_{g2}(s)=0}} \tag{4.20}$$

$$G_{v_{pan1}\,v_{ref2}\,SYS}(s) = \left.\frac{\hat{v}_{pan1}(s)}{\hat{v}_{ref2}(s)}\right|_{\substack{\hat{v}_{bulk}(s)=0\\ \hat{v}_{ref1}(s)=0\\ \hat{i}_{g1}(s)=0\\ \hat{i}_{g2}(s)=0}} \tag{4.21}$$

$$Z_{v_{pan1}\,i_{g1}\,SYS}(s) = \left.\frac{\hat{v}_{pan1}(s)}{\hat{i}_{g1}(s)}\right|_{\substack{\hat{v}_{bulk}(s)=0\\ \hat{v}_{ref1}(s)=0\\ \hat{v}_{ref2}(s)=0\\ \hat{i}_{g2}(s)=0}} \tag{4.22}$$

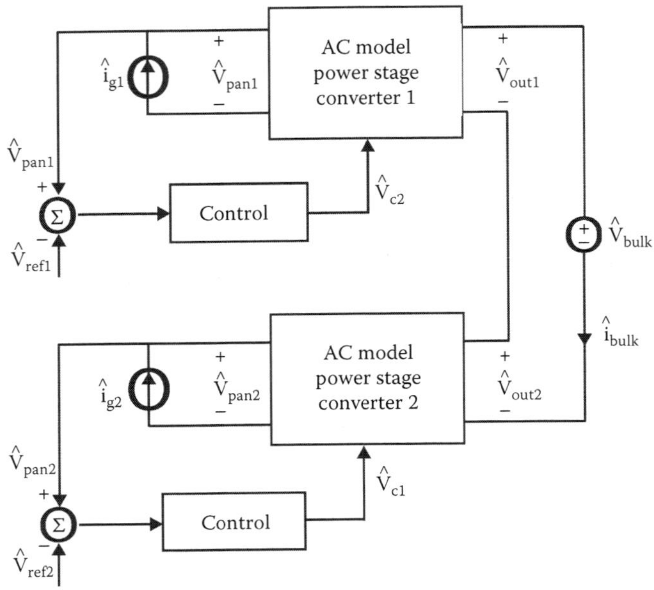

FIGURE 4.113
Small-signal model of a string of SCPVMs ($NH = 1$, $NL = 1$).

$$Z_{v_{pan1}\, i_{g2}\, SYS}(s) = \left. \frac{\hat{v}_{pan1}(s)}{\hat{i}_{g2}(s)} \right|_{\substack{\hat{v}_{bulk}(s)=0 \\ \hat{v}_{ref1}(s)=0 \\ \hat{v}_{ref2}(s)=0 \\ \hat{i}_{g1}(s)=0}} \tag{4.23}$$

$$G_{v_{pan1}\, v_{bulk} SYS}(s) = \left. \frac{\hat{v}_{pan1}(s)}{\hat{v}_{bulk}(s)} \right|_{\substack{\hat{v}_{ref1}(s)=0 \\ \hat{v}_{ref2}(s)=0 \\ \hat{i}_{g1}(s)=0 \\ \hat{i}_{g2}(s)=0}} \tag{4.24}$$

Transfer functions relating v_{pan2} to input variables can be defined similarly. The transfer functions defined above can be derived from the transfer functions of a single SCPVM by applying Middlebrook's extra element theorem [77–79] and Kirchhoff's laws. By applying Middlebrook's theorem we get

$$G_{v_{pan1}\, v_{ref1}\, SYS}(s) = G_{v_{pan}\, v_{ref} CL1}(s) \frac{1 + \dfrac{Z_{out\ CL2}(s)}{Z_{N\ CL1}(s)}}{1 + \dfrac{Z_{out\ CL2}(s)}{Z_{D\ CL1}(s)}} \tag{4.25}$$

where $Z_{out\ CL2}(s)$ is the closed-loop output impedance of SCPVM number 2 and $G_{v_{pan}\ v_{ref}CL1}(s)$ is the V_{ref} to v_{pan} transfer function of the SCPVM 1 as defined in the previous section.

Applying Middlebrook's extra element theorem to the case under study yields the following definitions of impedances $Z_{N\ CL1}(s)$ and $Z_{D\ CL1}(s)$:

$$Z_{N\ CL1}(s) = \left.\frac{\hat{v}_{out\ 1}(s)}{\hat{i}_{out\ 1}(s)}\right|_{\substack{\hat{v}_{pan\ 1}(s)\to 0 \\ \hat{i}_{g\ 1}(s)=0 \\ closed\ loop}} \tag{4.26}$$

$$Z_{D\ CL1}(s) = \left.\frac{\hat{v}_{out\ 1}(s)}{\hat{i}_{out\ 1}(s)}\right|_{\substack{\hat{v}_{ref\ 1}(s)=0 \\ \hat{i}_{g\ 1}(s)=0 \\ closed\ loop}} \tag{4.27}$$

For the system in Figure 4.113 it is $Z_{D\ CL1}(s) = Z_{out\ CL1}(s)$ and

$$Z_{N\ CL1}(s) = \frac{Z_{out\ CL\ 1}(s)}{1 + \dfrac{Z_{out\ CL1}(s)}{1+T_{c1}(s)}\dfrac{G_{v_{pan}\ v_{bulk}\ OL\ 1}(s)G_{i_{bulk}\ d\ OL\ 1}(s)}{G_{v_{pan}\ d\ OL\ 1}(s)}} \tag{4.28}$$

Moreover, it is

$$G_{v_{pan1}\ v_{ref2}\ SYS}(s) = -\frac{G_{i_{bulk}\ v_{ref}\ CL\ 2}(s)G_{v_{pan}\ v_{bulk}\ CL\ 1}(s)Z_{out\ CL\ 1}(s)}{1 + \dfrac{Z_{out\ CL\ 1}(s)}{Z_{out\ CL\ 2}(s)}} \tag{4.29}$$

$$Z_{v_{pan1}\ g1\ SYS}(s) = Z_{v_{pan}\ i_g\ CL1}(s)\frac{1 + \dfrac{Z_{out\ CL\ 2}(s)}{Z_{N\ CL\ 1}(s)}}{1 + \dfrac{Z_{out\ CL\ 2}(s)}{Z_{D\ CL\ 1}(s)}} \tag{4.30}$$

$$Z_{v_{pan1}\ i_{g2}\ SYS}(s) = -\frac{G_{i_{bulk}\ i_g\ CL\ 2}(s)G_{v_{pan}\ v_{bulk}\ CL\ 1}(s)Z_{out\ CL\ 1}(s)}{1 + \dfrac{Z_{out\ CL\ 1}(s)}{Z_{out\ CL\ 2}(s)}} \tag{4.31}$$

$$G_{v_{pan1}\ v_{bulk}\ SYS}(s) = \frac{G_{v_{pan}v_{bulk}\ CL\ 1}(s)Z_{out\ CL\ 1}(s)}{Z_{out\ CL\ 1}(s) + Z_{out\ CL\ 2}(s)} \tag{4.32}$$

Transfer function (4.25) relates perturbations of the control variable of SCPVM 1 (V_{ref1}) to perturbations of its output variable (v_{pan1}). This transfer function is responsible for settling time and overshoot in the step response of SCPVM 1 as part of the string under study and can be used to optimize MPPT parameters (e.g., perturbation interval and amplitude in a P&O algorithm [9, 76]). Transfer function (4.29) relates perturbations of the control variable of SCPVM 2 (V_{ref2}) to perturbations of the output variable of SCPVM 1 and represents the effect of the dynamics of a module on the remaining ones. Transfer functions (4.30), (4.31), and (4.32) relate perturbations of v_{pan1} to external variables; indeed, V_{bulk} is controlled by the inverter or battery charger (it is constant in a battery), whereas i_{g1} and i_{g2} are directly related to the solar irradiation G.

4.5.3 Stability of a String of SCPVMs

In this section local stability analysis of a string of SCPVMs will be presented. Local stability can be studied by deriving the loop transfer function of each SCPVM as part of the string under study [79]. Applying Middlebrook's extra element theorem, the loop transfer function of SCPVM 1 in Figure 4.113 can be written as

$$T_{c1\ SYS}(s) = G_{vpan1\ d_1SYS}(s) K_{sen} G_c(s) \frac{1}{V_m} \tag{4.33}$$

where

$$G_{v_{pan1}\ d_1\ SYS}(s) = G_{v_{pan}\ d\ OL1}(s) \frac{1 + \dfrac{Z_{out\ CL\ 2}(s)}{Z_{N\ OL\ 1}(s)}}{1 + \dfrac{Z_{out\ CL\ 2}(s)}{Z_{D\ OL\ 1}(s)}} \tag{4.34}$$

The terms $Z_{N\ OL\ 1}(s)$ and $Z_{D\ OL\ 1}(s)$ represent the impedances defined in Middlebrook's theorem and calculated in open-loop condition. Their expressions can be obtained as shown for the impedances $Z_{N\ CL1}(s)$ and $Z_{D\ CL1}(s)$, leading to

$$Z_{D\ OL\ 1}(s) = Z_{out\ OL\ 1}(s) \tag{4.35}$$

$$Z_{N\ OL\ 1}(s) = \frac{Z_{out\ OL\ 1}(s)}{1 + \dfrac{G_{v_{pan}\ v_{bulk}\ OL\ 1}(s) G_{i_{bulk}\ d\ OL\ 1}(s) Z_{out\ OL\ 1}(s)}{G_{v_{pan}\ d\ OL\ 1}(s)}} \tag{4.36}$$

Bode or Nyquist criteria can be applied to the transfer function $T_{c1\ SYS}(s)$. In order to predict stability of a string of N SCPVMs, like the one depicted in Figure 4.108, such criteria must be satisfied for all SCPVMs in all possible operating conditions. In principle if it were

$$Z_{N\ OL\ 1}(s) \gg Z_{out\ CL\ 2}(s) \tag{4.37}$$

$$Z_{D\ OL\ 1}(s) \gg Z_{out\ CL\ 2}(s) \tag{4.38}$$

it could be assumed that the loop function of converter 1 is not affected by the operation of converter 2. This condition is not easy to be ensured since the transfer function $Z_{out\ CL\ 2}(s)$ entirely depends on converter 2. Moreover, in the general case of a string composed by N SCPVMs, (4.34) would assume the form

$$G_{v_{pan1}\ d_1\ SYS}(s) = G_{v_{pan}\ d\ OL1}(s)\frac{1+\dfrac{\sum_{i=2}^{N} Z_{out\ CL\ i}(s)}{Z_{N\ OL\ 1}(s)}}{1+\dfrac{\sum_{i=2}^{N} Z_{out\ CL\ i}(s)}{Z_{D\ OL\ 1}(s)}} \tag{4.39}$$

However, analysis of (4.39) leads to the conclusion that if it was $Z_{N\ OL\ 1}(s) \sim Z_{D\ OL\ 1}(s)$, then $G_{v_{pan1}\ d_1\ SYS}(s) \sim G_{v_{pan}\ d\ OL1}(s)$. In this case stability of SCPVM 1 could be predicted a priori, ignoring the other SCPVMs connected in series and their properties, since the effect of their presence would be negligible. In conclusion, if it was $Z_{N\ OL\ 1}(s) \sim Z_{D\ OL\ 1}(s)$, then the design approach would be modular, at least to some extent. Figure 4.114 shows the Bode diagrams of the transfer functions $T_{cL}\ (s)$, $T_{cL\ SYS}(s)$, $T_{cH}\ (s)$, and $T_{cH\ SYS}(s)$ in the following operating conditions (subscript H labels quantities referring to the SCPVMs operating under an irradiance level equal to G_H, while subscript L labels quantities referring to the SCPVMs operating under an irradiance level equal to G_L): $NH = 1$, $NL = 12$, $G_H = 1000\ \text{W/m}^2$, $G_L = 500\ \text{W/m}^2$, $v_{panH} = 19.8\ \text{V}$, $v_{panL} = 15\ \text{V}$, $I_{gH} = 11\ \text{A}$, $I_{gL} = 4\ \text{A}$, $R_{panH} = 2.6\ \Omega$, $R_{panL} = 84\ \Omega$, and converter parameters as for Figure 4.112. Figures 4.115 and 4.116 instead show Bode diagrams of impedances $Z_{N\ OL\ H}(s)$, $Z_{D\ OL\ H}(s)$, $Z_{N\ OL\ L}(s)$, and $Z_{D\ OL\ L}(s)$ used to compute transfer functions $T_{cL\ SYS}(s)$ and $T_{cH\ SYS}(s)$, respectively.

Bode diagrams of Figure 4.114 show that, in the case under study, the effect of the series interconnection is negligible around crossover (10 kHz) and at higher frequencies, whereas it is remarkable at lower frequencies. In fact, as evident from Figures 4.115 and 4.116, at frequencies higher than about 10 kHz it is $Z_{N\ OL\ H}(s) \sim Z_{D\ OL\ H}(s)$ and $Z_{N\ OL\ L}(s) \sim Z_{D\ OL\ L}(s)$. As a consequence,

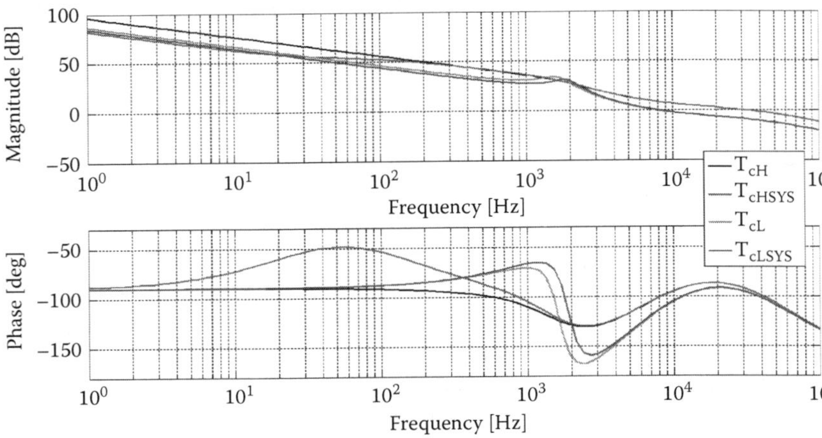

FIGURE 4.114
Loop transfer functions of the SCPVMs as isolated systems ($T_{cL}(s)$, $T_{cH}(s)$) and as parts of the string of N SCPVMs ($T_{cL\ SYS}(s)$, $T_{cH\ SYS}(s)$).

stability of each SCPVM is not affected by the dynamics of the other SCPVMs in the considered series configuration. Thus for each constant voltage reference given by the MPPT block (in the range of applicability of this model), the respective SCPVM will reach a steady-state condition in which $V_{pan} = V_{ref}$. Therefore we can conclude that, in order to prevent stability of an SCPVM from being affected by modules interaction, it is important to locate the crossover frequency in a range where it is $Z_{N\ OL\ 1}(s) \sim Z_{D\ OL\ 1}(s)$. The system under

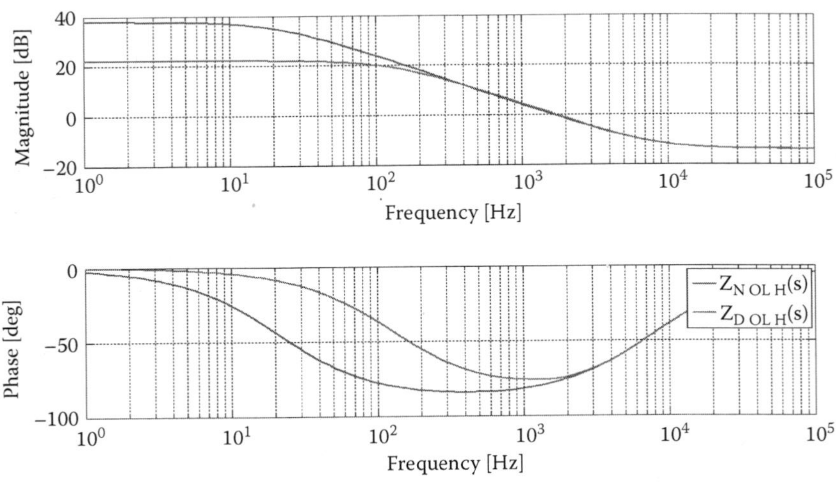

FIGURE 4.115
Bode diagrams of impedances $Z_{N\ OL\ H}(s)$ and $Z_{D\ OL\ H}(s)$.

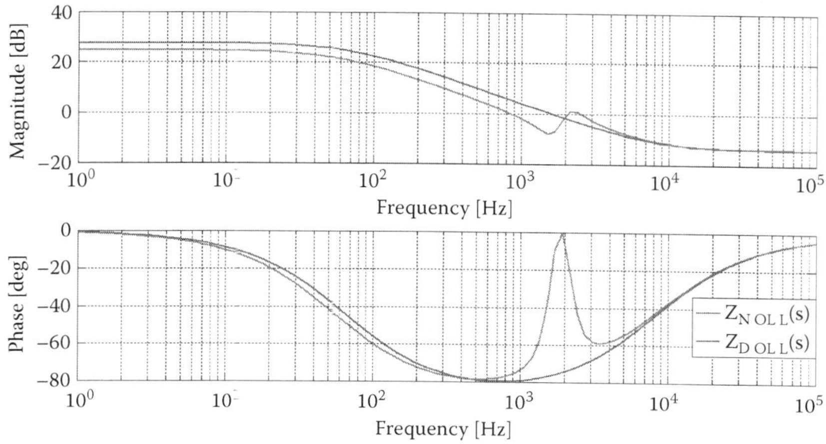

FIGURE 4.116
Bode diagrams of impedances $Z_{N\ OL\ L}(s)$ and $Z_{D\ OL\ L}(s)$.

study has been analyzed in a wide range of shade intensity and a number of shaded PV modules. The condition $Z_{N\ OL\ 1}(s) \sim Z_{D\ OL\ 1}(s)$ is always fulfilled around the crossover frequency, and therefore the stability has never been found to be affected by modules interaction. Under these conditions, if the global MPP belongs to the feasibility map of the system the DMPPT will be able to achieve it. Cases where global MPP does not belong to the feasibility map of the system (e.g., Figure 4.7) deserve further considerations [79]. In this case the P&O algorithm gradually varies the PV module voltage until one or more of the inequalities (4.1g) or (4.1k) take over, and therefore the trajectory of the system in the feasibility map hits the boundary of an unfeasible region. In this condition a net DC difference between the reference voltage and the controlled voltage causes the amplitude of the output voltage of the error amplifier to grow indefinitely (of course until saturation occurs), and any further perturbation of the reference voltage in the same direction would not result in any actual perturbation of the operating point of the SCPVM. Thus when duty cycle limitation intervenes (let it be due to the limited conversion ratio of the chosen DC/DC converter topology or to the duty cycle limitation block), the MPPT controller should reverse perturbation direction since, after the perturbation of the reference voltage, no increase of the PV power is obtained. This way, once a constrained MPP is reached, the MPPT controller starts bouncing back and forth and the trajectory of the system settles on a boundary point corresponding to a local constrained MPP. When duty cycle limitation or saturation (ideally to 0 or 1) takes over, the SCPVM operates in open-loop condition and (4.39) is no longer valid. If the MPPT algorithm is such that in these cases duty cycle limitation is kept active for a whole perturbation interval, the model presented above can be extended by deriving the transfer functions of an SCPVM when some of the remaining ones are operated at a fixed duty cycle [79]. In this condition (4.39) becomes

$$G_{v_{pan1}\,d_1\,SYS}(s) = G_{v_{pan}\,d\,OL1}(s)\frac{1+\dfrac{\sum_{\substack{i\neq 2\\ i\in S_{CL}}} Z_{out\,CL\,i}(s)+\sum_{\substack{i\neq 2\\ i\in S_{OL}}} Z_{out\,OL\,i}(s)}{Z_{N\,OL\,1}(s)}}{1+\dfrac{\sum_{\substack{i\neq 2\\ i\in S_{CL}}} Z_{out\,CL\,i}(s)+\sum_{\substack{i\neq 2\\ i\in S_{OL}}} Z_{out\,OL\,i}(s)}{Z_{D\,OL\,1}(s)}} \tag{4.40}$$

where S_{CL} is the set collecting the indices of those SCPVMs operating in closed-loop conditions, while S_{OL} is the set collecting the indices of those SCPVMs operating in open-loop conditions.

Simulation results of the system under study are presented in the following and show that the assumption of considering the duty cycle constant under limitation or saturation conditions is correct. The values of the parameters adopted for the numerical simulations follow: L = 100 μH, r_L = 0.082 Ω, $r_{ds\,hs}$ = 0.021 Ω, $r_{ds\,ls}$ = 0.013 Ω, C_{out} = 99 μF, r_{Cout} = 0.12 Ω, C_{in} = 94 μF, r_{Cin} = 0.18 Ω, fs = 160 kHz, V_f = 0.6 V, R_{s1} = 1 kΩ, R_{s2} = 0.39 kΩ, R_{i1} = 100 Ω, C_{i1} = 47 nF, C_{f1} = 2.2 nF, R_{f2} = 1 kΩ, C_{f2} = 47 nF, K_{sen} = 0.454, V_m = 1.8 V, $V_{ds\text{-}max}$ = 90 V, V_{bulk} = 400 V, and R_{bulk} = 0; PV module: Kyocera 175W, ΔV_{ref} = 0.08 V (P&O amplitude of perturbation), $T_{sampling}$ = 0.002 s (P&O perturbation interval), and N = 13.

In particular, the system of Figure 4.4 has been simulated for G_H = 1000 W/m², NH = [1, 3, 5, 7, 9, 11], G_L = [200, 500, 800] W/m². Time domain data are mapped on feasibility maps and are shown in Figures 4.117a–4.134a. The trajectories of the system on feasibility maps are drawn in green, when the duty cycle limitation is not active in any of the DC/DC MPPT converters. Points corresponding to duty cycle limitation in the NH fully lit SCPVMs are drawn in magenta, the duty cycle limitation of the shaded SCPVMs is indicated with yellow, and red points indicate duty cycle limitation in all of the SCPVMs.

System trajectories are not fully contained in the feasibility regions since these are representative of steady-state operation, whereas trajectories refer to the instantaneous behavior of the system. The peculiar shape of the trajectories is due to the combined effects of MPPT and duty cycle limitation. In Figures 4.117b–4.134b the PV voltages from the same simulations are plotted vs. simulation time, while in Figures 4.117c–4.134c the duty cycles are plotted vs. simulation time.

Figure 4.117 shows a case in which global MPP is a feasible point of the system. Also, Figures 4.120, 4.123, 4.124, 4.126, 4.127, 4.129, 4.130, 4.132, and 4.133 refer to cases in which the global MPP belongs to the feasible region

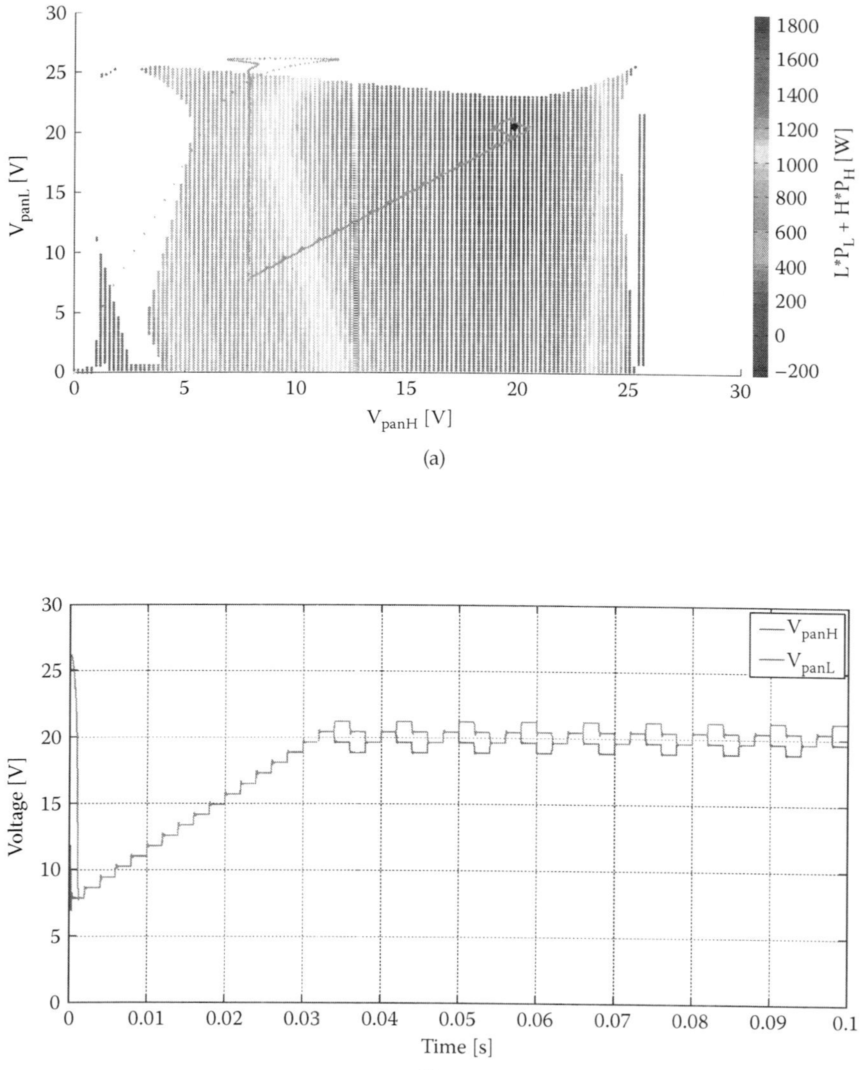

FIGURE 4.117 (See color insert)
Simulation of the system in Figure 4.4 for $NH = 11$ and $G_L = 800$ W/m^2. (a) Time domain data plotted on the corresponding feasibility map (the black-filled triangle indicates global MPP, the black-filled circle indicates average steady-state operation point of the system). (b) PV panel voltage. (c) DC/DC MPPT converter duty cycle.

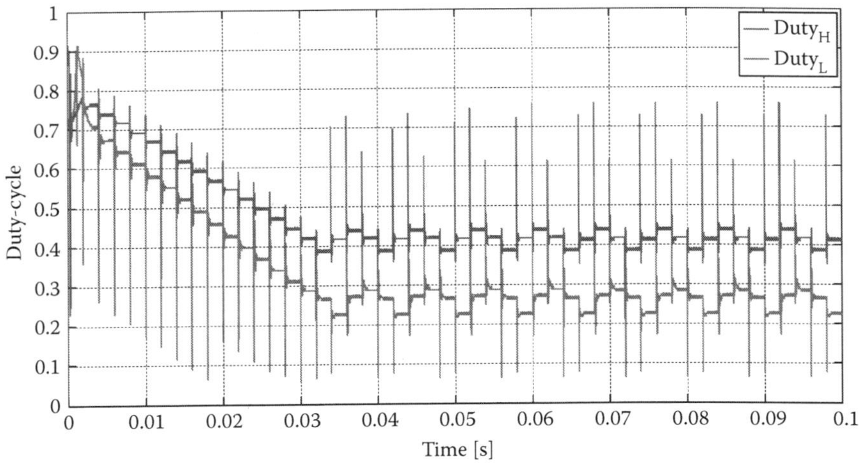

(c)

FIGURE 4.117 ***(continued)***
Simulation of the system in Figure 4.4 for $NH = 11$ and $G_L = 800$ W/m^2. (a) Time domain data plotted on the corresponding feasibility map (the black-filled triangle indicates global MPP, the black-filled circle indicates average steady-state operation point of the system). (b) PV panel voltage. (c) DC/DC MPPT converter duty cycle.

of the system. In all these cases, the discussed DMPPT technique is able to achieve the global MPP and stability can be predicted by applying Nyquist or Bode criteria to Equation (4.33), in which $G_{vpan1\ d_1SYS}$ is calculated by using Equation (4.39).

Figure 4.118 refers instead to a case in which global MPP does not belong to the set of feasible points of the system. In these cases the P&O algorithm gradually changes panel voltages v_{panH} and v_{panL} until duty cycle $duty_L$ saturates to its minimum allowed value. The same situation takes place in the cases of Figures 4.119 and 4.121. In these, stability can be predicted by applying Nyquist or Bode criteria to Equation (4.33), in which $G_{vpan1\ d_1SYS}$ is calculated by using Equation (4.40).

Figures 4.122, 4.125, and 4.128 refer to cases in which a constrained MPP is achieved because the shaded modules are simply bypassed. In these cases the output bypass diode (shown in Figure 4.107) of the NL low-irradiated SCPVMs starts conducting and neither v_{panL} nor D_L exhibits any variation.

Figure 4.131 shows a case in which, at the same time, duty cycle upper limitation takes place in the NH highly irradiated SCPVMs and duty cycle lower saturation takes place in the remaining NL SCPVMs.

Figure 4.134 refers to a case in which duty cycle upper limitation takes place in the NH highly irradiated SCPVMs. Indeed, the fact that the orbit of the system on the feasibility map is drawn in magenta (Figure 4.134a) indicates that duty cycle limitation is active. Therefore the NH fully lit SCPVMs

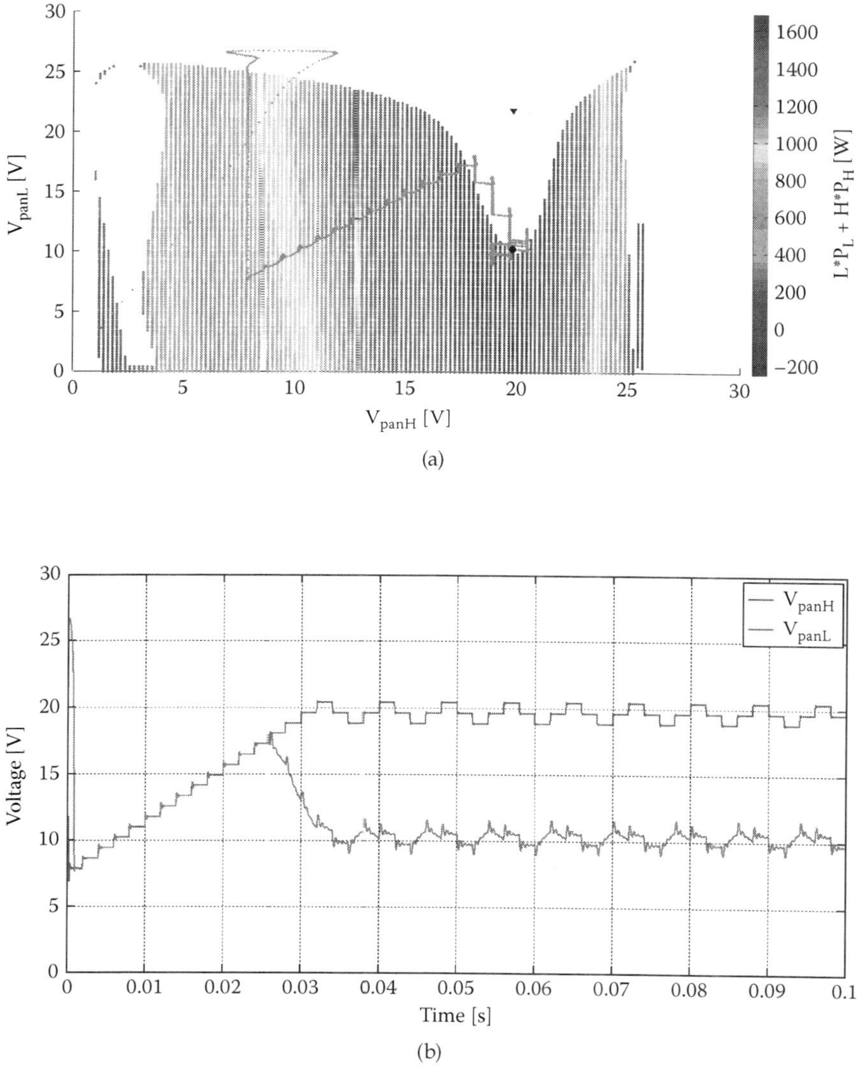

FIGURE 4.118 (See color insert for part a)
Simulation of the system in Figure 4.4 for $NH = 11$ and $G_L = 500$ W/m^2. (a) Time domain data plotted on the corresponding feasibility map (the black-filled triangle indicates global MPP, the black-filled circle indicates average steady-state operation point of the system). (b) PV panel voltage. (c) DC/DC MPPT converter duty cycle.

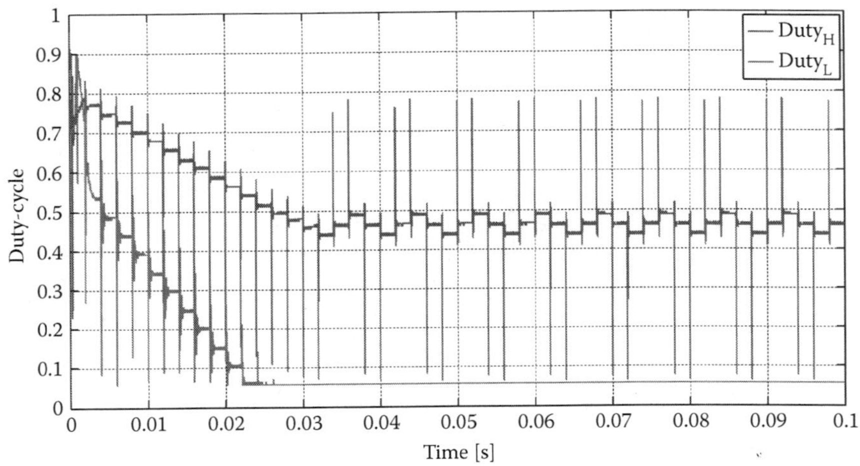

FIGURE 4.118 *(continued)*
Simulation of the system in Figure 4.4 for $NH = 11$ and $G_L = 500$ W/m². (a) Time domain data plotted on the corresponding feasibility map (the black-filled triangle indicates global MPP, the black-filled circle indicates average steady-state operation point of the system). (b) PV panel voltage. (c) DC/DC MPPT converter duty cycle.

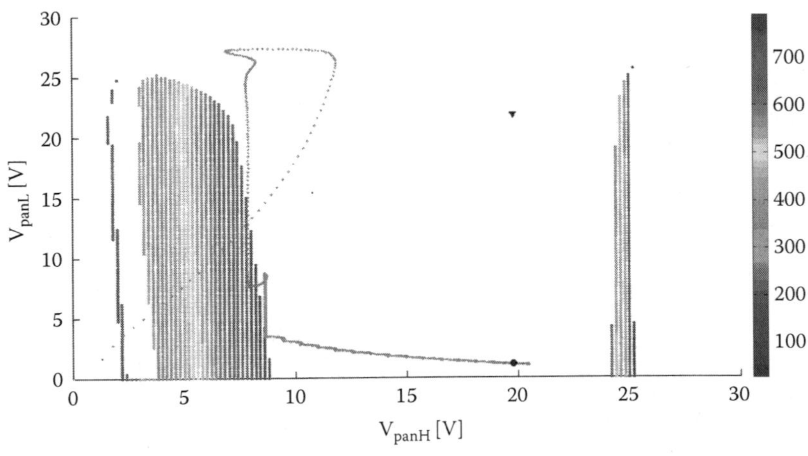

FIGURE 4.119 (See color insert for part a)
Simulation of the system in Figure 4.4 for $NH = 11$ and $G_L = 200$ W/m². (a) Time domain data plotted on the corresponding feasibility map (the black-filled triangle indicates global MPP, the black-filled circle indicates average steady-state operation point of the system). (b) PV panel voltage. (c) DC/DC MPPT converter duty cycle.

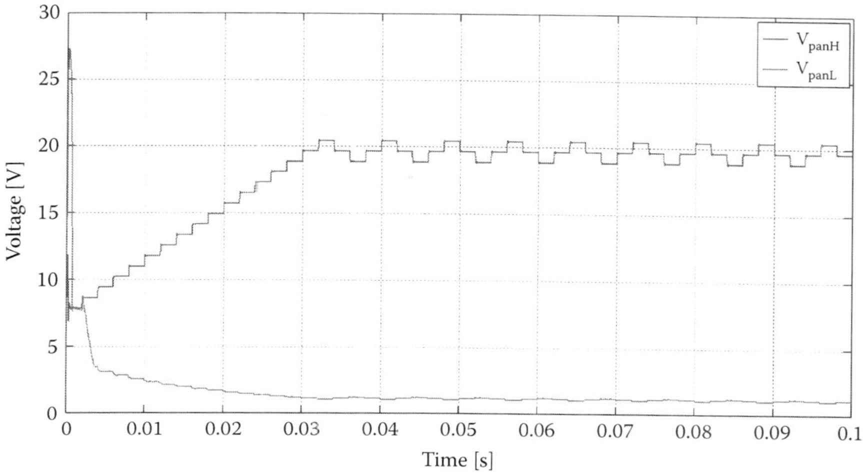

(b)

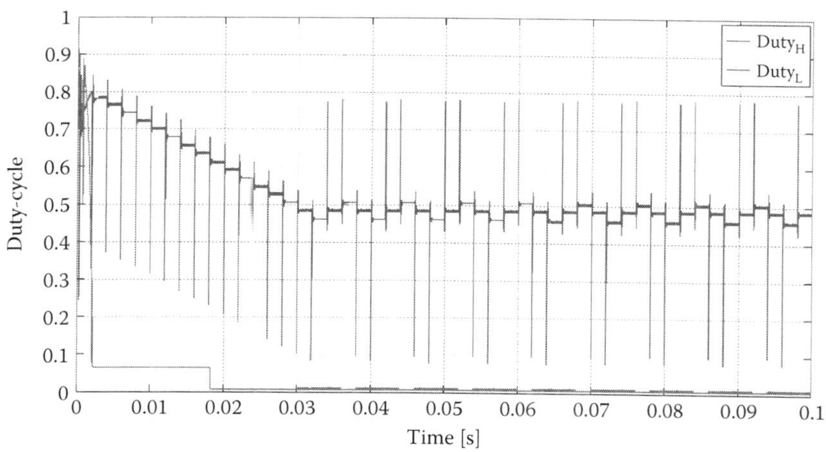

(c)

FIGURE 4.119 (***continued***)
Simulation of the system in Figure 4.4 for $NH = 11$ and $G_L = 200$ W/m². (a) Time domain data plotted on the corresponding feasibility map (the black-filled triangle indicates global MPP, the black-filled circle indicates average steady-state operation point of the system). (b) PV panel voltage. (c) DC/DC MPPT converter duty cycle.

operate in open loop. As a consequence, perturbations of the reference voltage of the NL low-irradiated SCPVMs in the string affect panel voltage of the NH SCPVMs operating in open-loop condition.

Also in the cases of Figures 4.122, 4.125, 4.128, and 4.131, stability can be predicted by applying Nyquist or Bode criteria to Equation (4.33) in which $G_{vpan1\ d_1SYS}$ is calculated by using Equation (4.40).

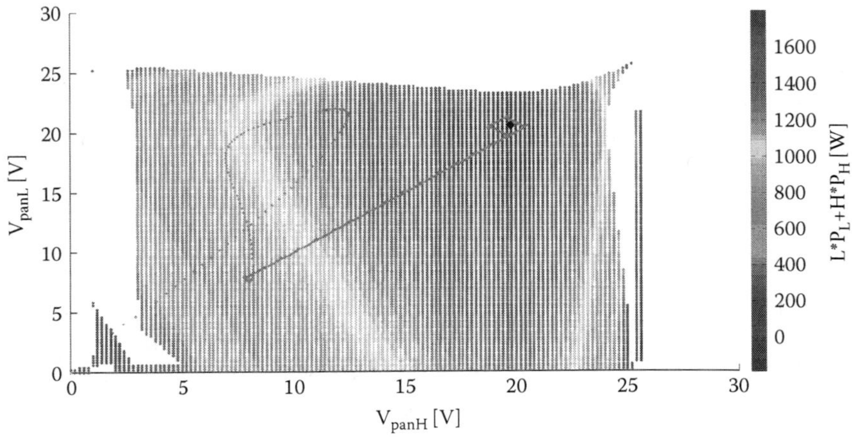

(a)

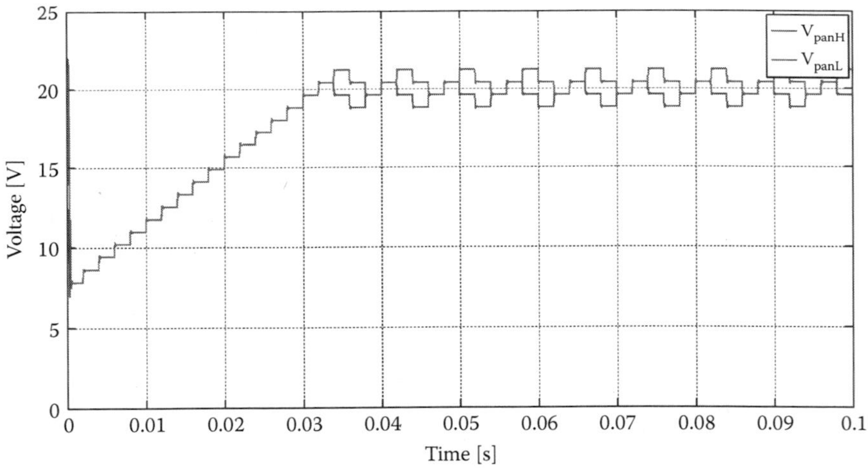

(b)

FIGURE 4.120 (See color insert for part a)
Simulation of the system in Figure 4.4 for $NH = 9$ and $G_L = 800$ W/m^2. (a) Time domain data plotted on the corresponding feasibility map (the black-filled triangle indicates global MPP, the black-filled circle indicates average steady-state operation point of the system). (b) PV panel voltage. (c) DC/DC MPPT converter duty cycle.

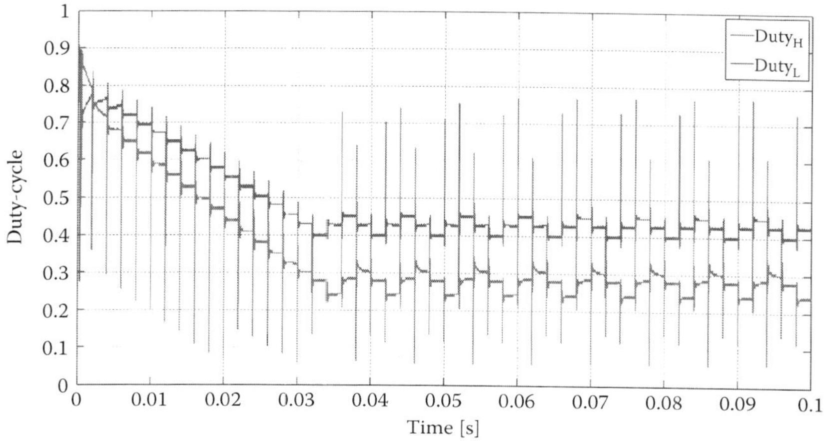

(c)

FIGURE 4.120 (*continued*)
Simulation of the system in Figure 4.4 for $NH = 9$ and $G_L = 800$ W/m^2. (a) Time domain data plotted on the corresponding feasibility map (the black-filled triangle indicates global MPP, the black-filled circle indicates average steady-state operation point of the system). (b) PV panel voltage. (c) DC/DC MPPT converter duty cycle.

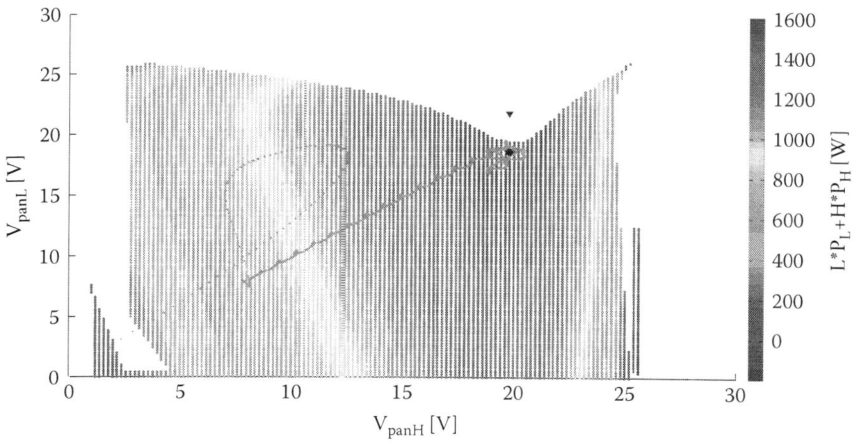

(a)

FIGURE 4.121 (See color insert for part a)
Simulation of the system in Figure 4.4 for $NH = 9$ and $G_L = 500$ W/m^2. (a) Time domain data plotted on the corresponding feasibility map (the black-filled triangle indicates global MPP, the black-filled circle indicates average steady-state operation point of the system). (b) PV panel voltage. (c) DC/DC MPPT converter duty cycle.

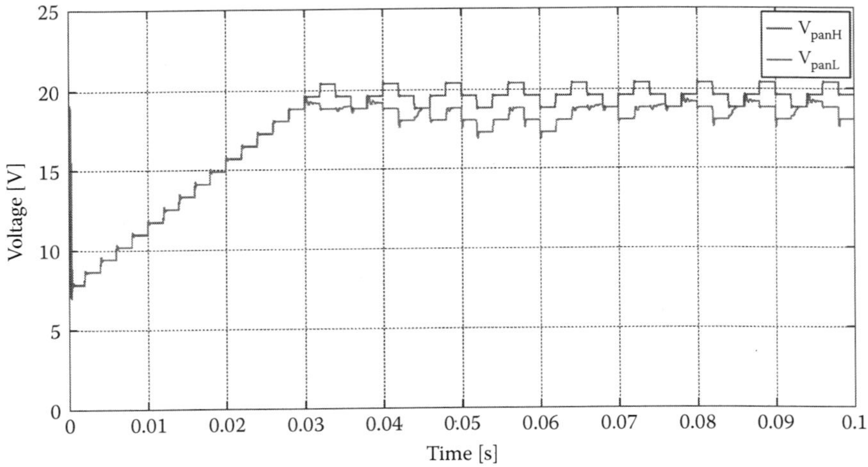

(b)

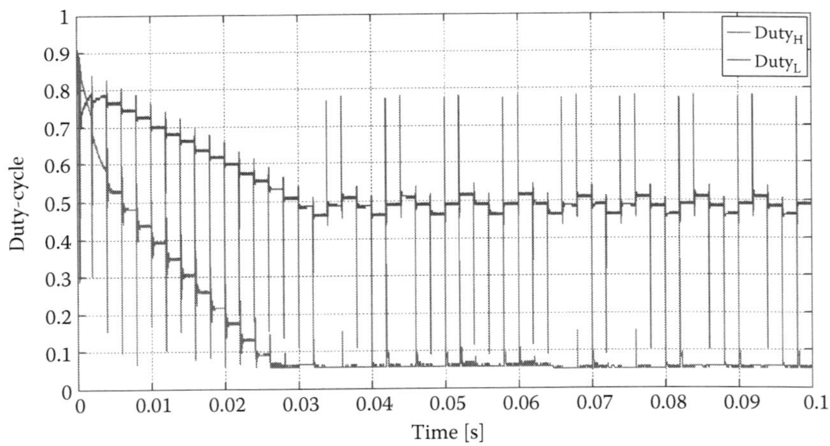

(c)

FIGURE 4.121 ***(continued)***
Simulation of the system in Figure 4.4 for $NH = 9$ and $G_L = 500$ W/m^2. (a) Time domain data plotted on the corresponding feasibility map (the black-filled triangle indicates global MPP, the black-filled circle indicates average steady-state operation point of the system). (b) PV panel voltage. (c) DC/DC MPPT converter duty cycle.

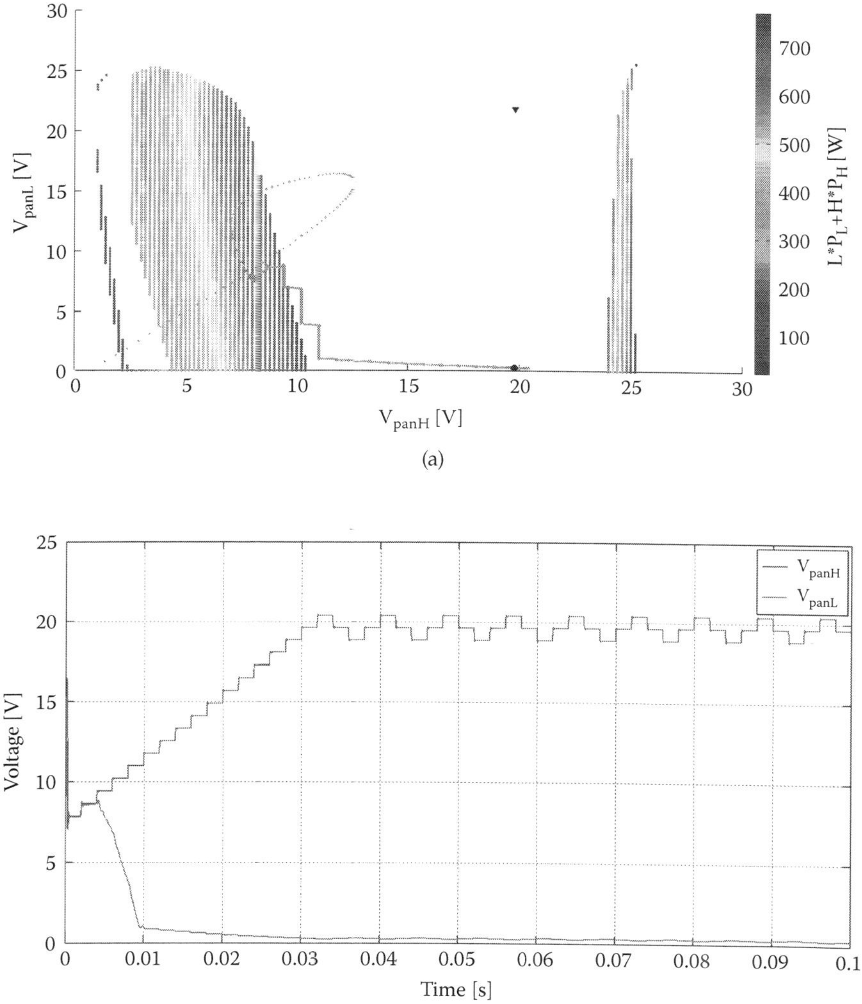

FIGURE 4.122 (See color insert for part a)
Simulation of the system in Figure 4.4 for $NH = 9$ and $G_L = 200$ W/m^2. (a) Time domain data plotted on the corresponding feasibility map (the black-filled triangle indicates global MPP, the black-filled circle indicates average steady-state operation point of the system). (b) PV panel voltage. (c) DC/DC MPPT converter duty cycle.

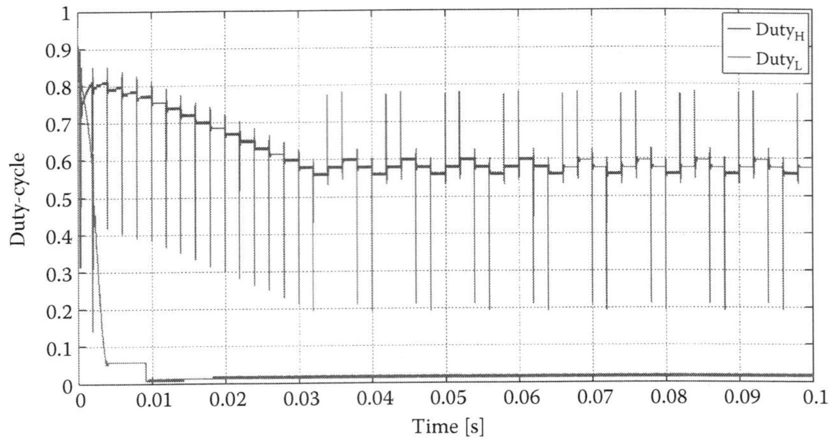

(c)

FIGURE 4.122 ***(continued)***
Simulation of the system in Figure 4.4 for $NH = 9$ and $G_L = 200$ W/m^2. (a) Time domain data plotted on the corresponding feasibility map (the black-filled triangle indicates global MPP, the black-filled circle indicates average steady-state operation point of the system). (b) PV panel voltage. (c) DC/DC MPPT converter duty cycle.

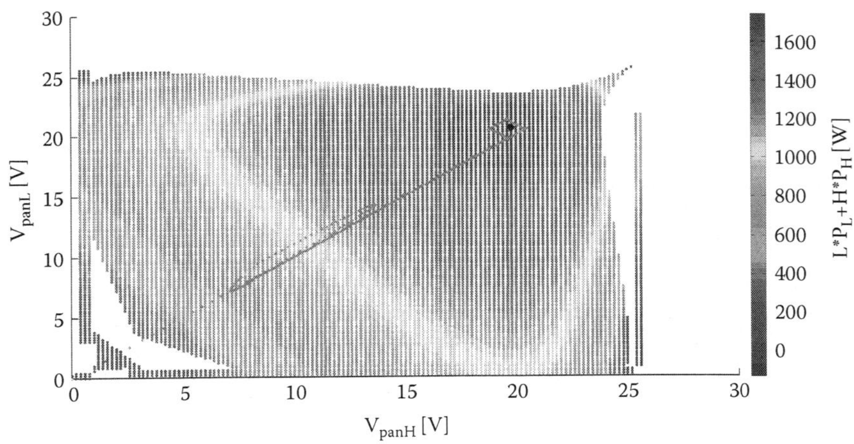

(a)

FIGURE 4.123 (See color insert for part a)
Simulation of the system in Figure 4.4 for $NH = 7$ and $G_L = 800$ W/m^2. (a) Time domain data plotted on the corresponding feasibility map (the black-filled triangle indicates global MPP, the black-filled circle indicates average steady-state operation point of the system). (b) PV panel voltage. (c) DC/DC MPPT converter duty cycle.

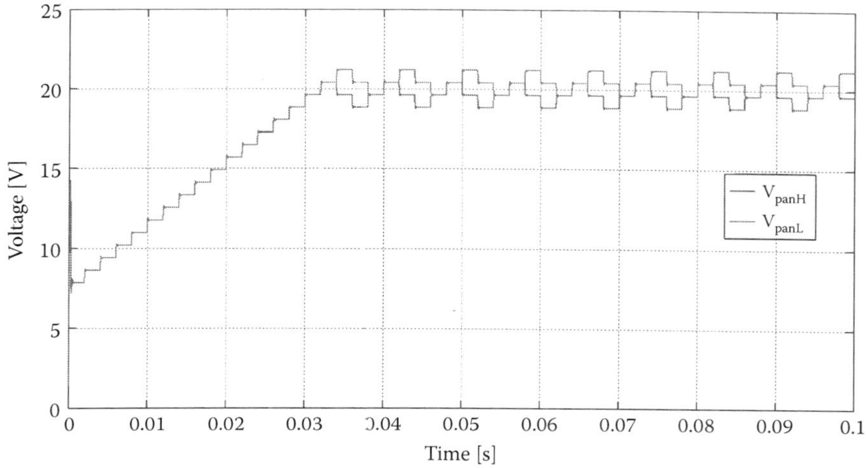

(b)

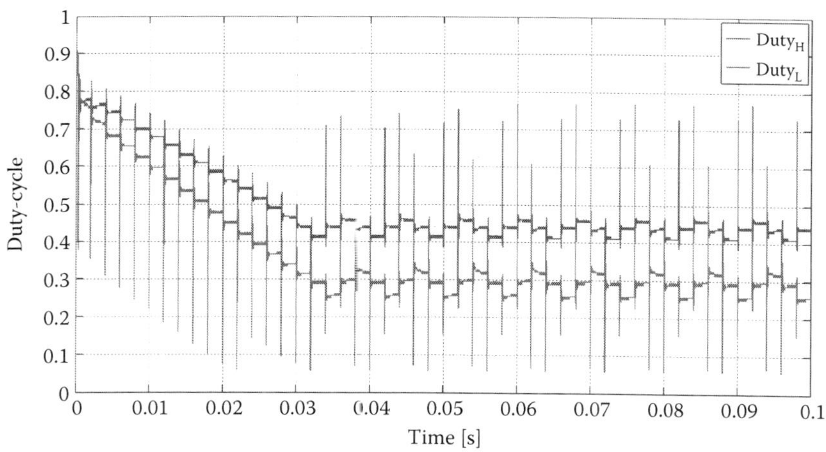

(c)

FIGURE 4.123 ***(continued)***
Simulation of the system in Figure 4.4 for $NH = 7$ and $G_L = 800$ W/m². (a) Time domain data plotted on the corresponding feasibility map (the black-filled triangle indicates global MPP, the black-filled circle indicates average steady-state operation point of the system). (b) PV panel voltage. (c) DC/DC MPPT converter duty cycle.

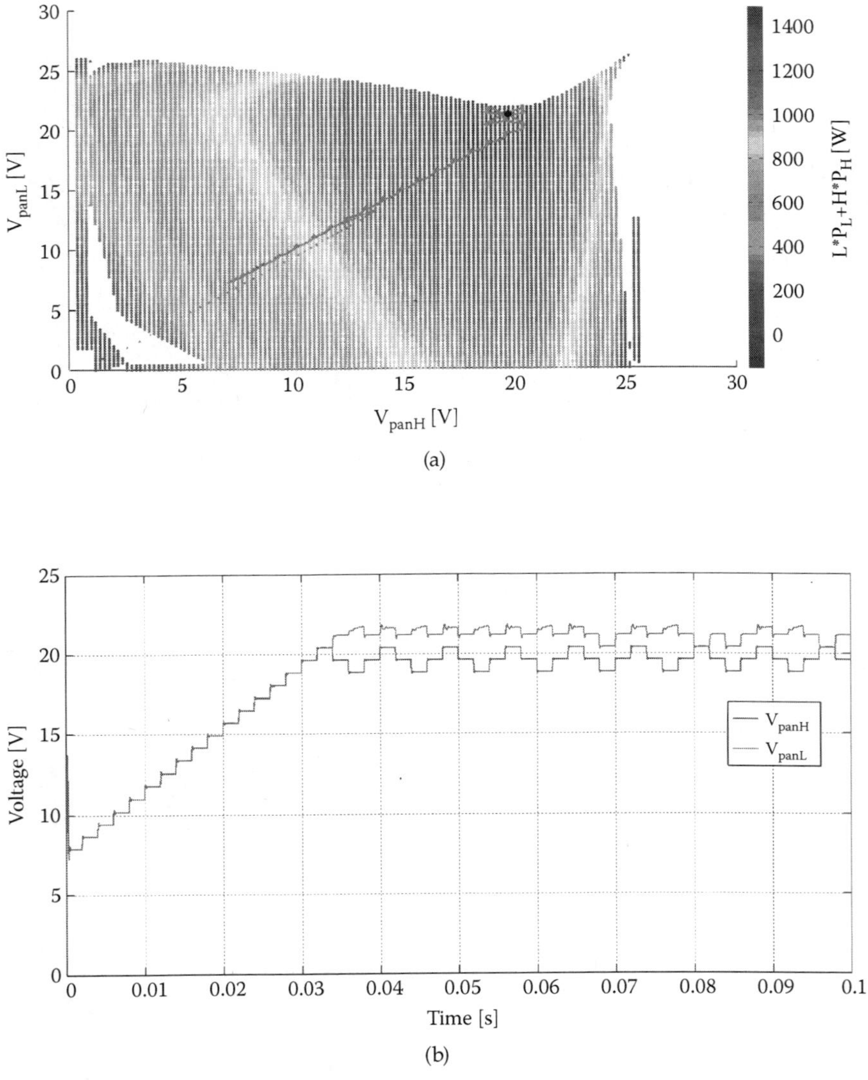

FIGURE 4.124 (See color insert for part a)
Simulation of the system in Figure 4.4 for $NH = 7$ and $G_L = 500$ W/m^2. (a) Time domain data plotted on the corresponding feasibility map (the black-filled triangle indicates global MPP, the black-filled circle indicates average steady-state operation point of the system). (b) PV panel voltage. (c) DC/DC MPPT converter duty cycle.

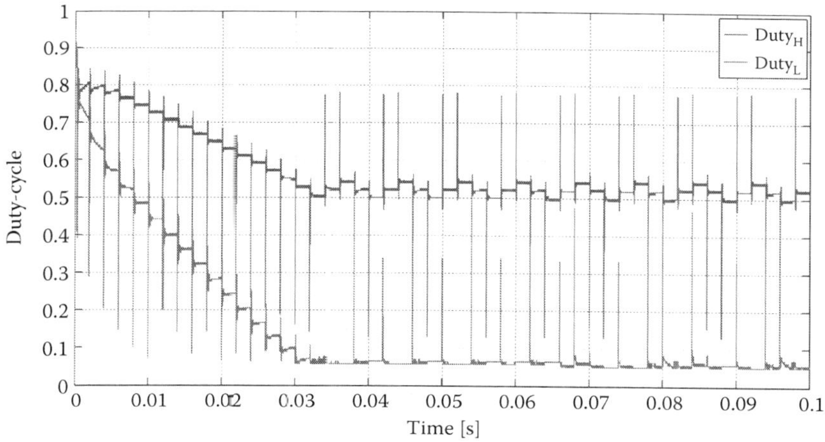

(c)

FIGURE 4.124 (*continued*)
Simulation of the system in Figure 4.4 for $NH = 7$ and $G_L = 500$ W/m^2. (a) Time domain data plotted on the corresponding feasibility map (the black-filled triangle indicates global MPP, the black-filled circle indicates average steady-state operation point of the system). (b) PV panel voltage. (c) DC/DC MPPT converter duty cycle.

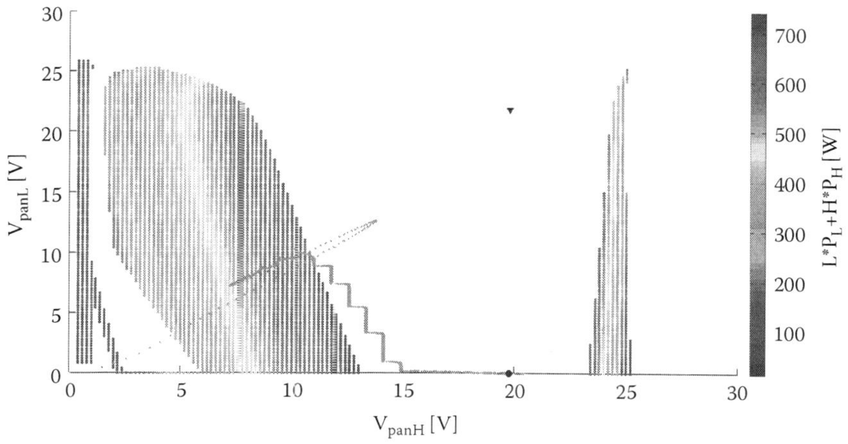

(a)

FIGURE 4.125 (See color insert for part a)
Simulation of the system in Figure 4.4 for $NH = 7$ and $G_L = 200$ W/m^2. (a) Time domain data plotted on the corresponding feasibility map (the black-filled triangle indicates global MPP, the black-filled circle indicates average steady-state operation point of the system). (b) PV panel voltage. (c) DC/DC MPPT converter duty cycle.

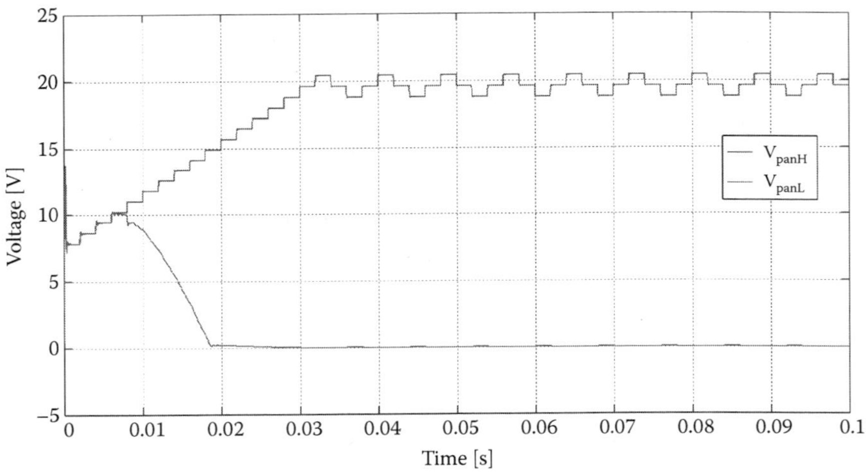

(b)

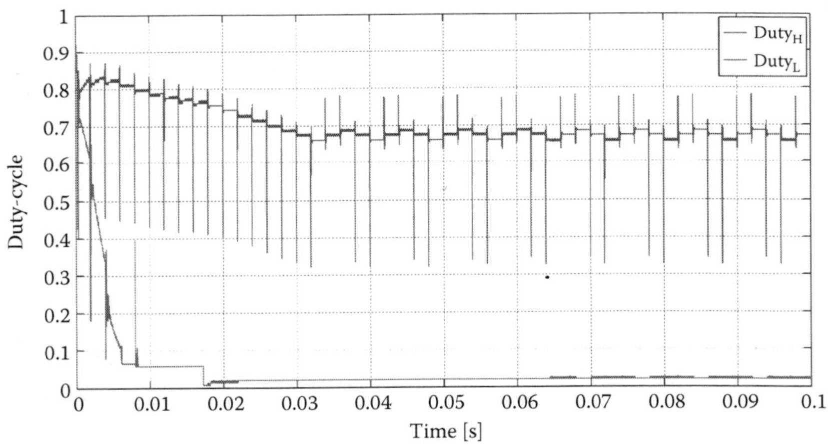

(c)

FIGURE 4.125 (*continued*)
Simulation of the system in Figure 4.4 for $NH = 7$ and $G_L = 200$ W/m^2. (a) Time domain data plotted on the corresponding feasibility map (the black-filled triangle indicates global MPP, the black-filled circle indicates average steady-state operation point of the system). (b) PV panel voltage. (c) DC/DC MPPT converter duty cycle.

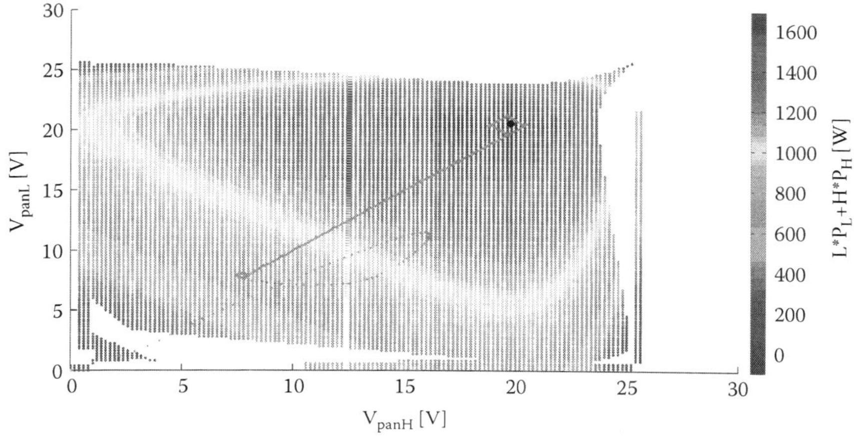

(a)

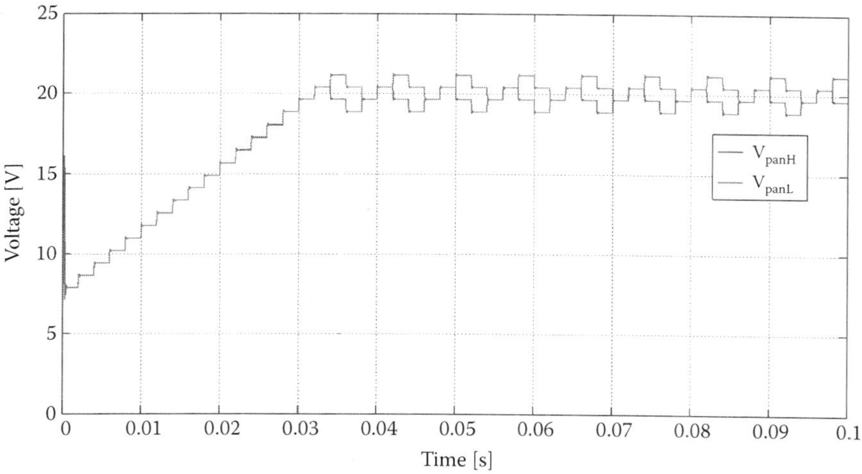

(b)

FIGURE 4.126 (See color insert for part a)
Simulation of the system in Figure 4.4 for $NH = 5$ and $G_L = 800$ W/m^2. (a) Time domain data plotted on the corresponding feasibility map (the black-filled triangle indicates global MPP, the black-filled circle indicates average steady-state operation point of the system). (b) PV panel voltage. (c) DC/DC MPPT converter duty cycle.

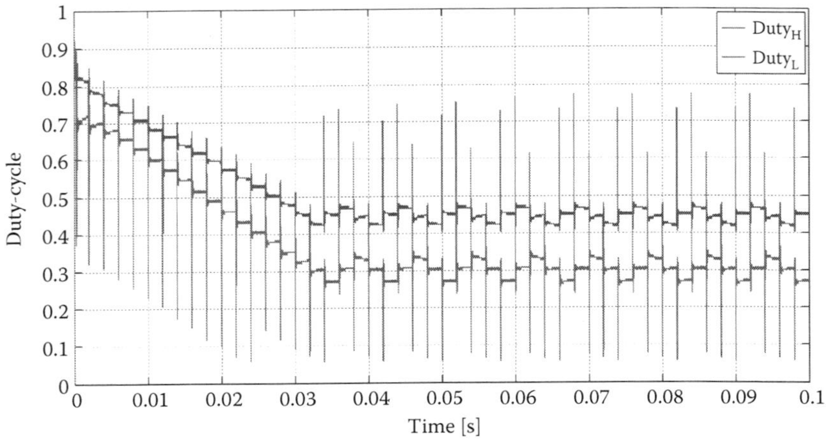

(c)

FIGURE 4.126 (*continued*)
Simulation of the system in Figure 4.4 for $NH = 5$ and $G_L = 800$ W/m². (a) Time domain data plotted on the corresponding feasibility map (the black-filled triangle indicates global MPP, the black-filled circle indicates average steady-state operation point of the system). (b) PV panel voltage. (c) DC/DC MPPT converter duty cycle.

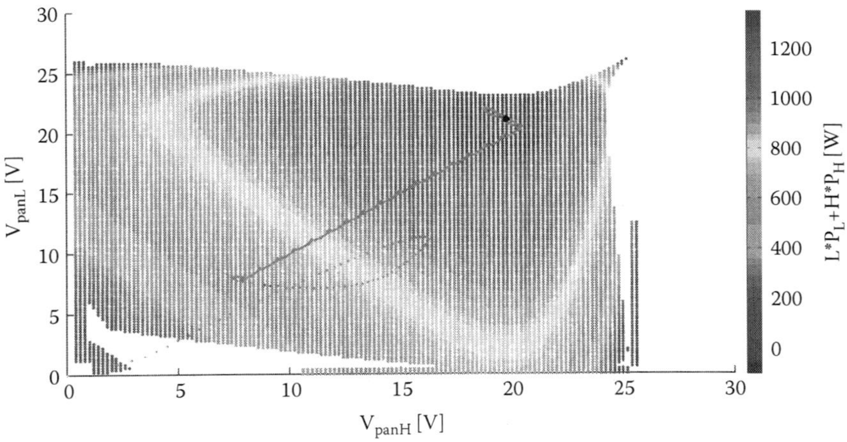

(a)

FIGURE 4.127 (See color insert for part a)
Simulation of the system in Figure 4.4 for $NH = 5$ and $G_L = 500$ W/m². (a) Time domain data plotted on the corresponding feasibility map (the black-filled triangle indicates global MPP, the black-filled circle indicates average steady-state operation point of the system). (b) PV panel voltage. (c) DC/DC MPPT converter duty cycle.

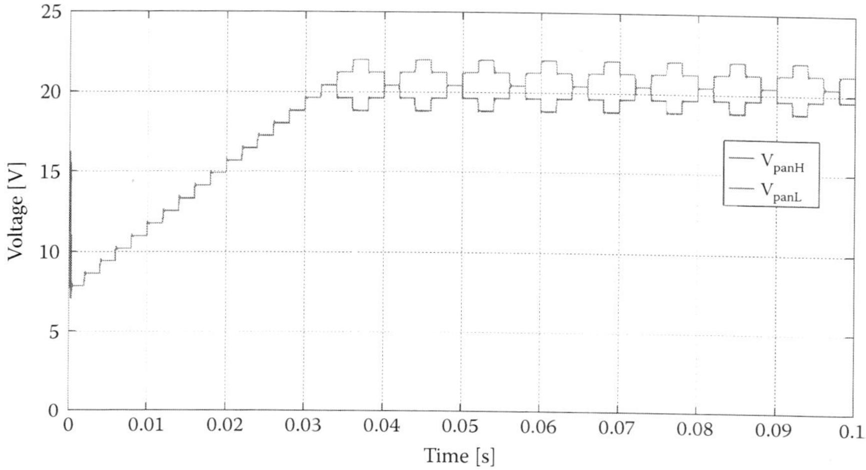

(b)

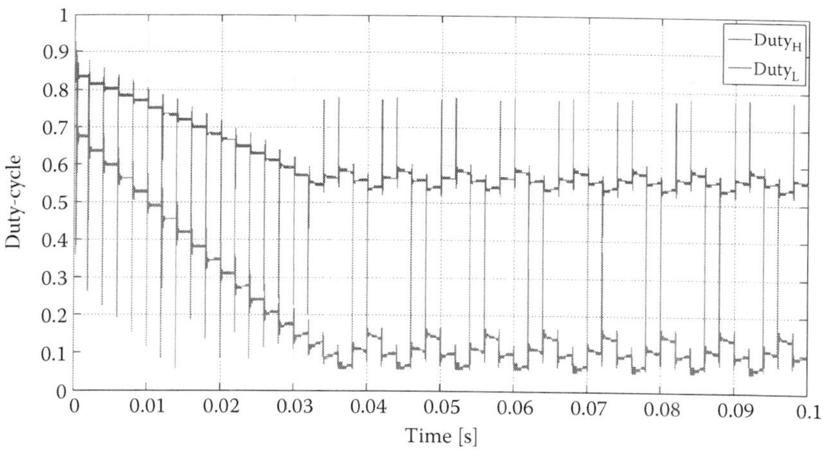

(c)

FIGURE 4.127 (***continued***)
Simulation of the system in Figure 4.4 for $NH = 5$ and $G_L = 500$ W/m^2. (a) Time domain data plotted on the corresponding feasibility map (the black-filled triangle indicates global MPP, the black-filled circle indicates average steady-state operation point of the system). (b) PV panel voltage. (c) DC/DC MPPT converter duty cycle.

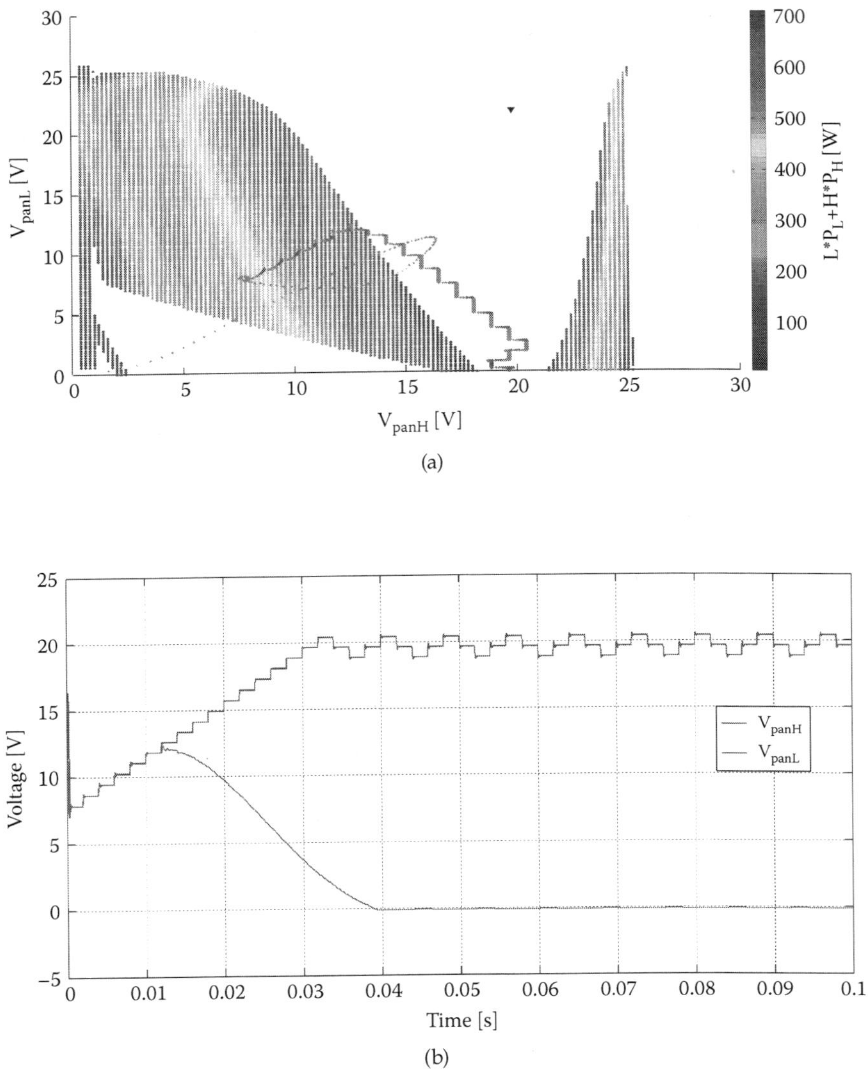

FIGURE 4.128 (See color insert for part a)
Simulation of the system in Figure 4.4 for $NH = 5$ and $G_L = 200$ W/m^2. (a) Time domain data plotted on the corresponding feasibility map (the black-filled triangle indicates global MPP, the black-filled circle indicates average steady-state operation point of the system). (b) PV panel voltage. (c) DC/DC MPPT converter duty cycle.

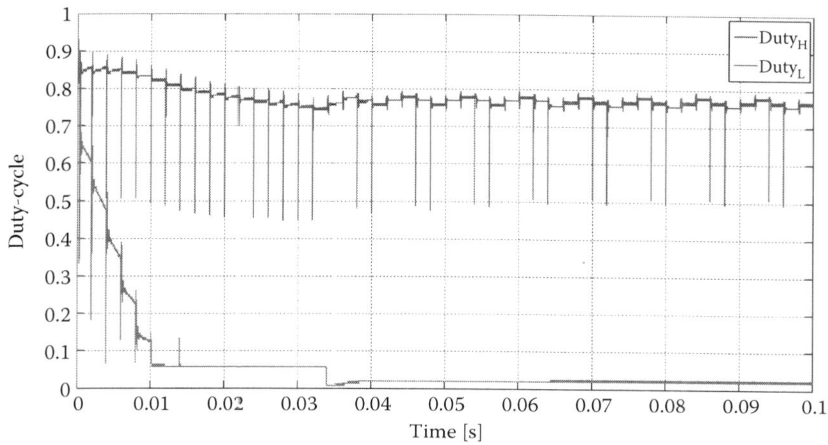

(c)

FIGURE 4.128 ***(continued)***
Simulation of the system in Figure 4.4 for $NH = 5$ and $G_L = 200\ W/m^2$. (a) Time domain data plotted on the corresponding feasibility map (the black-filled triangle indicates global MPP, the black-filled circle indicates average steady-state operation point of the system). (b) PV panel voltage. (c) DC/DC MPPT converter duty cycle.

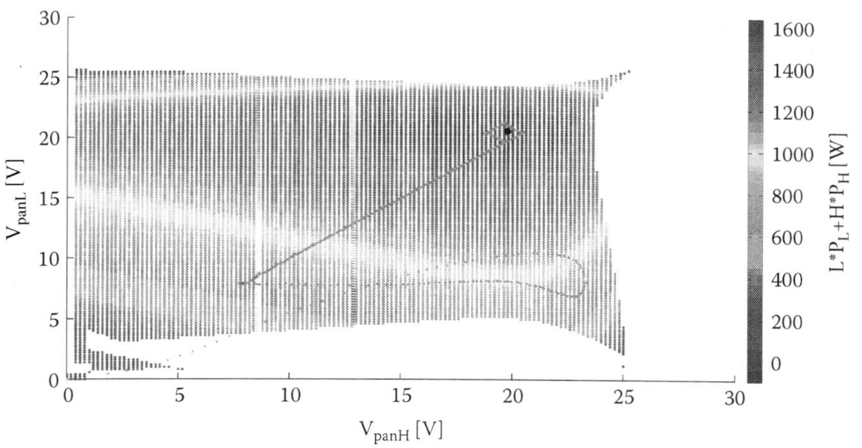

(a)

FIGURE 4.129 (See color insert for part a)
Simulation of the system in Figure 4.4 for $NH = 3$ and $G_L = 800\ W/m^2$. (a) Time domain data plotted on the corresponding feasibility map (the black-filled triangle indicates global MPP, the black-filled circle indicates average steady-state operation point of the system). (b) PV panel voltage. (c) DC/DC MPPT converter duty cycle.

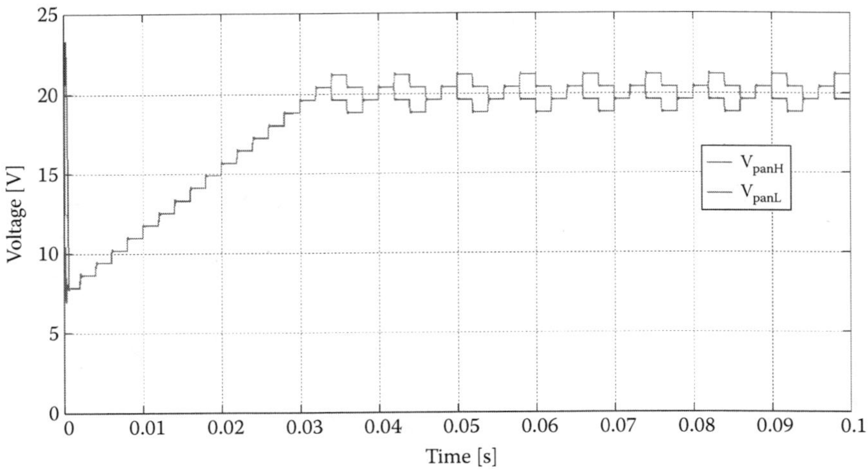

(b)

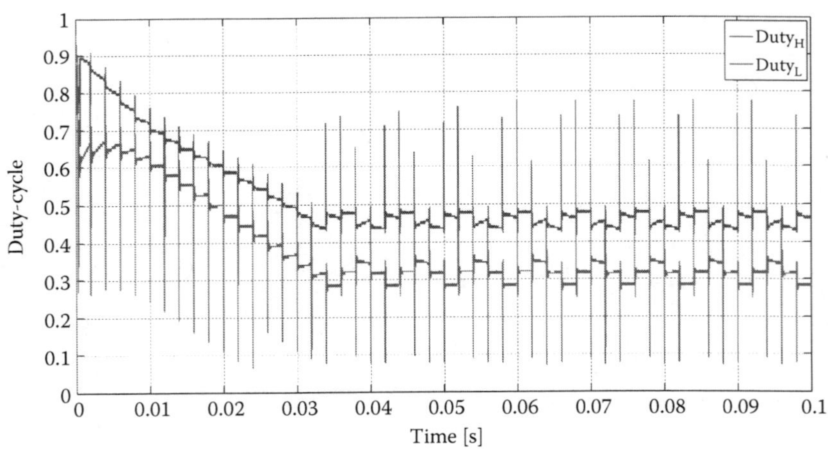

(c)

FIGURE 4.129 ***(continued)***
Simulation of the system in Figure 4.4 for $NH = 3$ and $G_L = 800$ W/m². (a) Time domain data plotted on the corresponding feasibility map (the black-filled triangle indicates global MPP, the black-filled circle indicates average steady-state operation point of the system). (b) PV panel voltage. (c) DC/DC MPPT converter duty cycle.

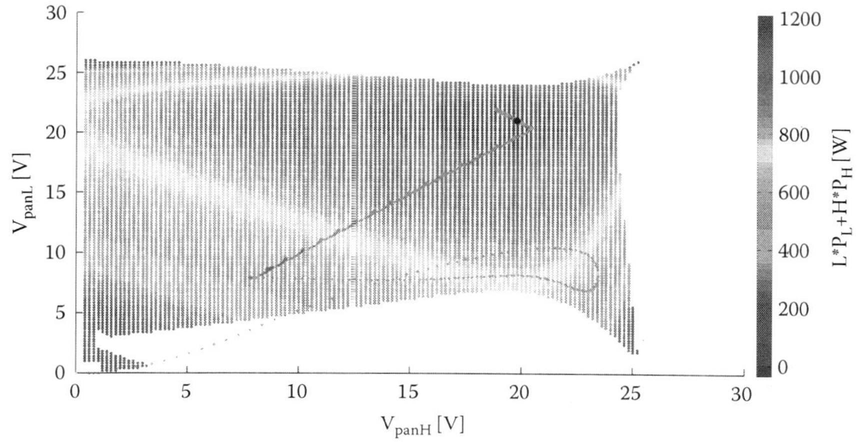

(a)

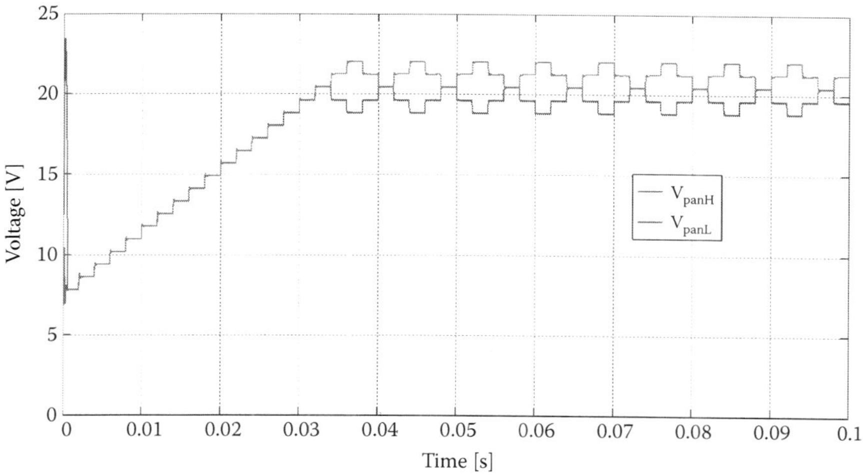

(b)

FIGURE 4.130 (See color insert for part a)
Simulation of the system in Figure 4.4 for $NH = 3$ and $G_L = 500$ W/m². (a) Time domain data plotted on the corresponding feasibility map (the black-filled triangle indicates global MPP, the black-filled circle indicates average steady-state operation point of the system). (b) PV panel voltage. (c) DC/DC MPPT converter duty cycle.

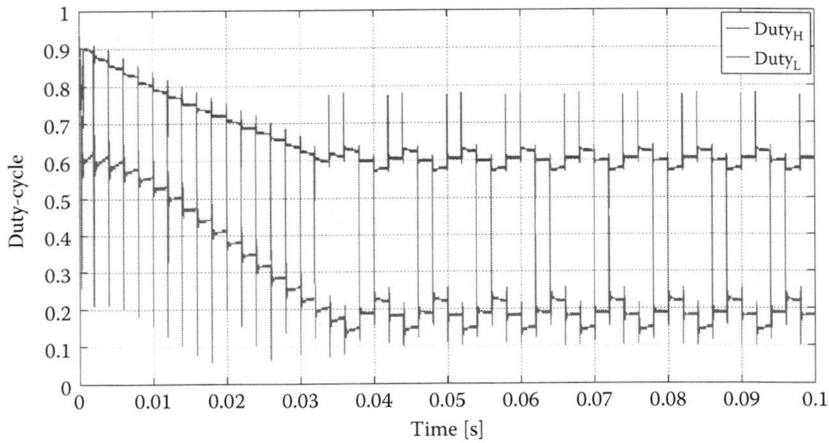

(c)

FIGURE 4.130 ***(continued)***
Simulation of the system in Figure 4.4 for $NH = 3$ and $G_L = 500$ W/m². (a) Time domain data plotted on the corresponding feasibility map (the black-filled triangle indicates global MPP, the black-filled circle indicates average steady-state operation point of the system). (b) PV panel voltage. (c) DC/DC MPPT converter duty cycle.

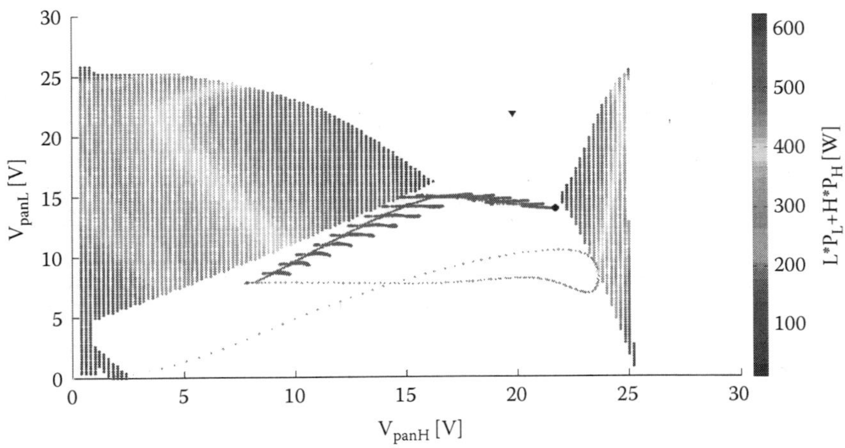

(a)

FIGURE 4.131 (See color insert for part a)
Simulation of the system in Figure 4.4 for $NH = 3$ and $G_L = 200$ W/m². (a) Time domain data plotted on the corresponding feasibility map (the black-filled triangle indicates global MPP, the black-filled circle indicates average steady-state operation point of the system). (b) PV panel voltage. (c) DC/DC MPPT converter duty cycle.

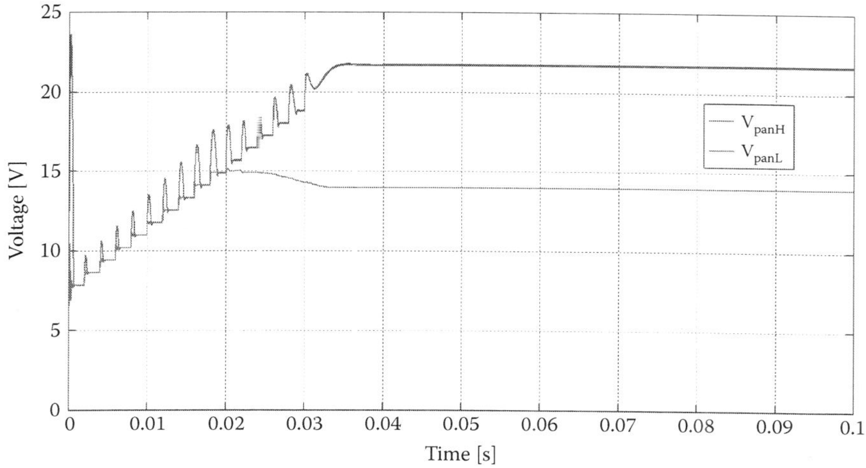

(b)

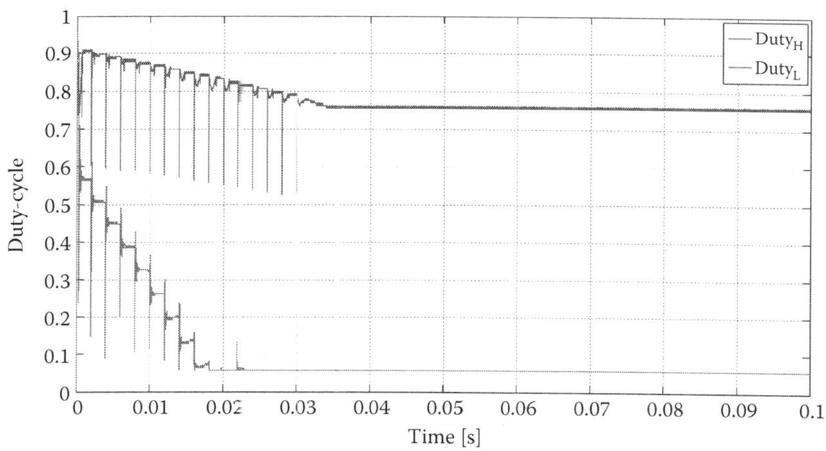

(c)

FIGURE 4.131 *(continued)*
Simulation of the system in Figure 4.4 for $NH = 3$ and $G_L = 200$ W/m^2. (a) Time domain data plotted on the corresponding feasibility map (the black-filled triangle indicates global MPP, the black-filled circle indicates average steady-state operation point of the system). (b) PV panel voltage. (c) DC/DC MPPT converter duty cycle.

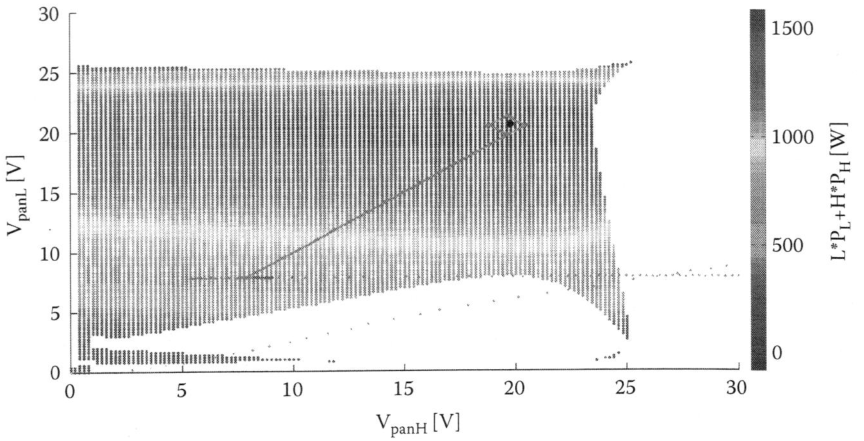

(a)

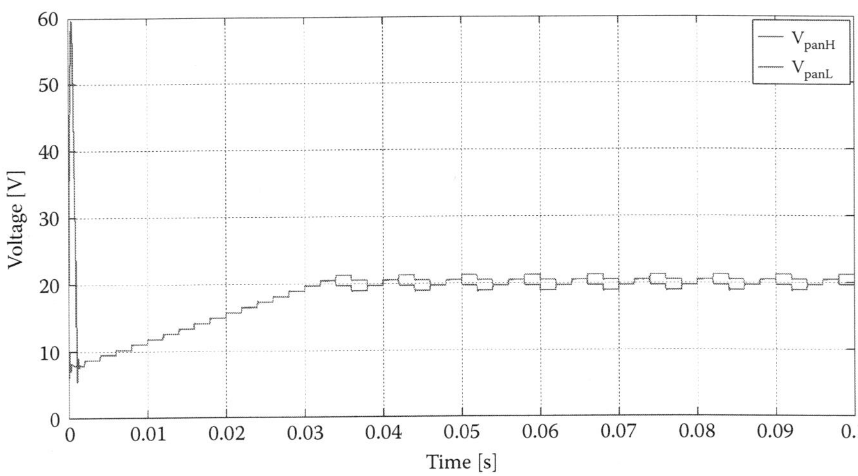

(b)

FIGURE 4.132 (See color insert for part a)
Simulation of the system in Figure 4.4 for $NH = 1$ and $G_L = 800$ W/m^2. (a) Time domain data plotted on the corresponding feasibility map (the black-filled triangle indicates global MPP, the black-filled circle indicates average steady-state operation point of the system). (b) PV panel voltage. (c) DC/DC MPPT converter duty cycle.

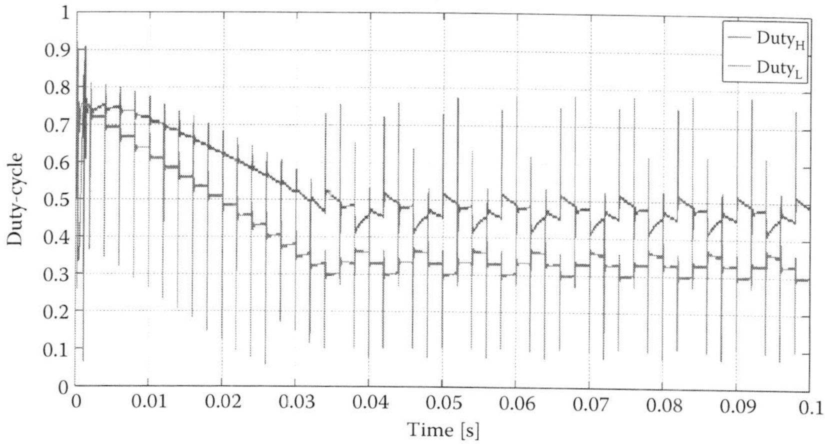

(c)

FIGURE 4.132 (*continued*)
Simulation of the system in Figure 4.4 for $NH = 1$ and $G_L = 800$ W/m². (a) Time domain data plotted on the corresponding feasibility map (the black-filled triangle indicates global MPP, the black-filled circle indicates average steady-state operation point of the system). (b) PV panel voltage. (c) DC/DC MPPT converter duty cycle.

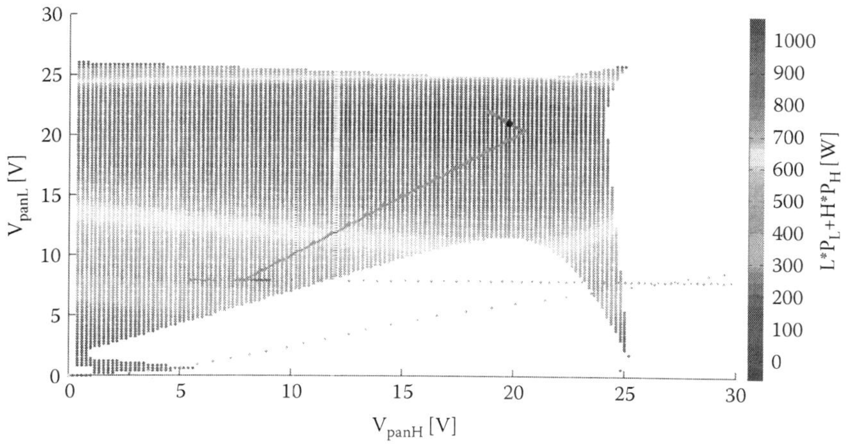

(a)

FIGURE 4.133 (See color insert for part a)
Simulation of the system in Figure 4.4 for $NH = 1$ and $G_L = 500$ W/m². (a) Time domain data plotted on the corresponding feasibility map (the black-filled triangle indicates global MPP, the black-filled circle indicates average steady-state operation point of the system). (b) PV panel voltage. (c) DC/DC MPPT converter duty cycle.

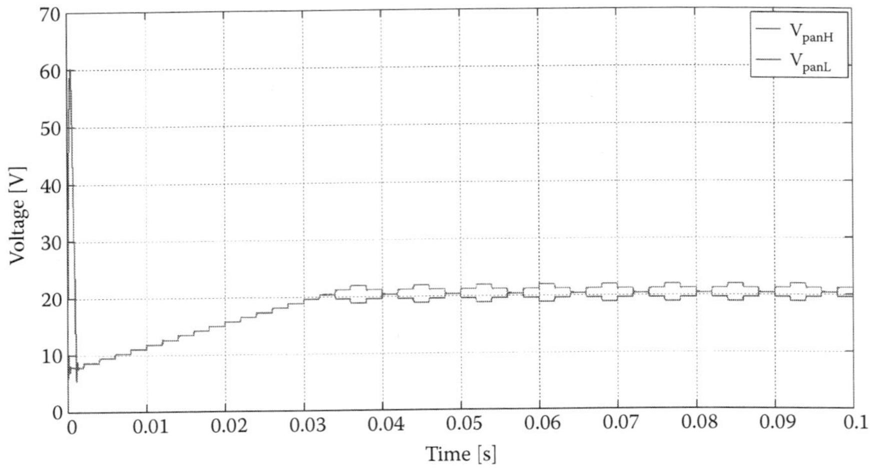

(b)

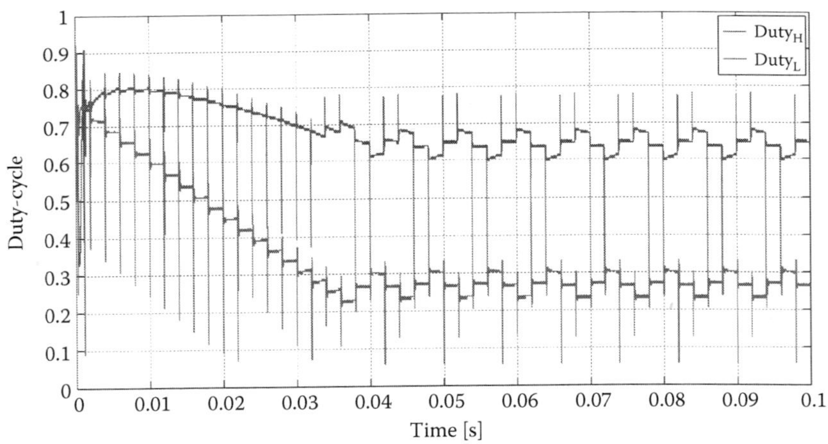

(c)

FIGURE 4.133 ***(continued)***
Simulation of the system in Figure 4.4 for $NH = 1$ and $G_L = 500$ W/m^2. (a) Time domain data plotted on the corresponding feasibility map (the black-filled triangle indicates global MPP, the black-filled circle indicates average steady-state operation point of the system). (b) PV panel voltage. (c) DC/DC MPPT converter duty cycle.

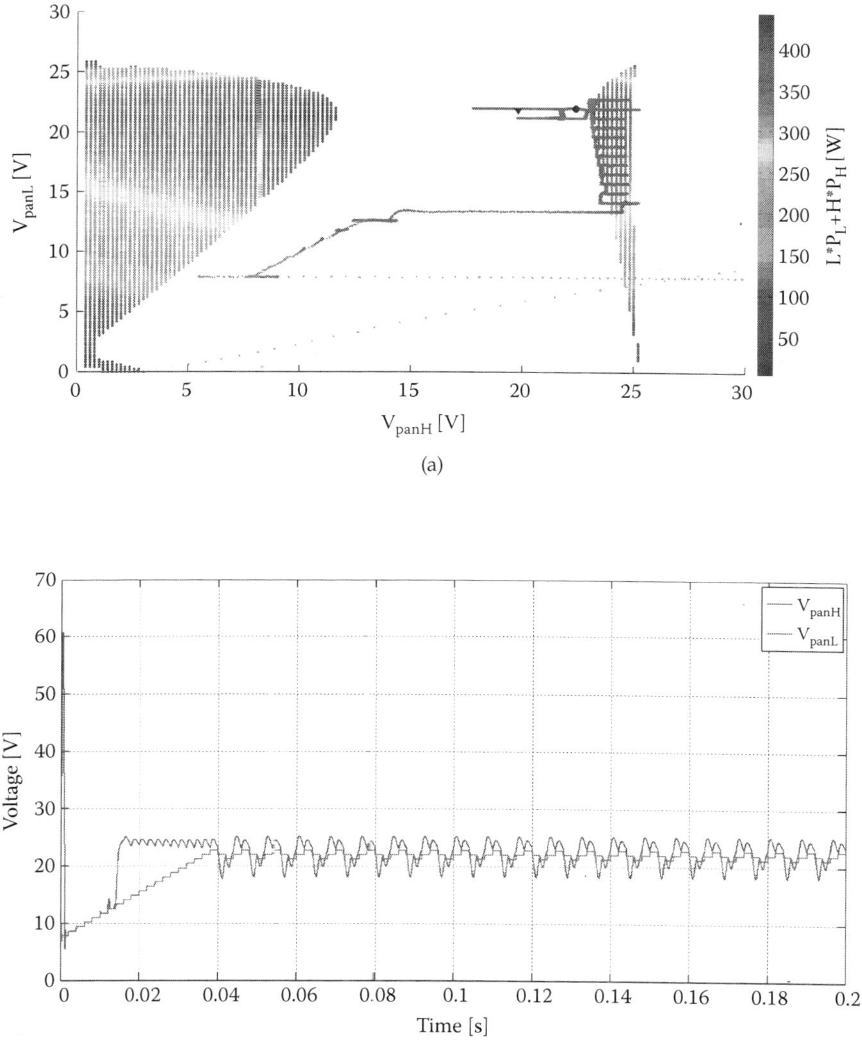

(b)

FIGURE 4.134 (See color insert for part a)
Simulation of the system in Figure 4.4 for $NH = 1$ and $G_L = 200$ W/m^2. (a) Time domain data plotted on the corresponding feasibility map (the black-filled triangle indicates global MPP, the black-filled circle indicates average steady-state operation point of the system). (b) PV panel voltage. (c) DC/DC MPPT converter duty cycle.

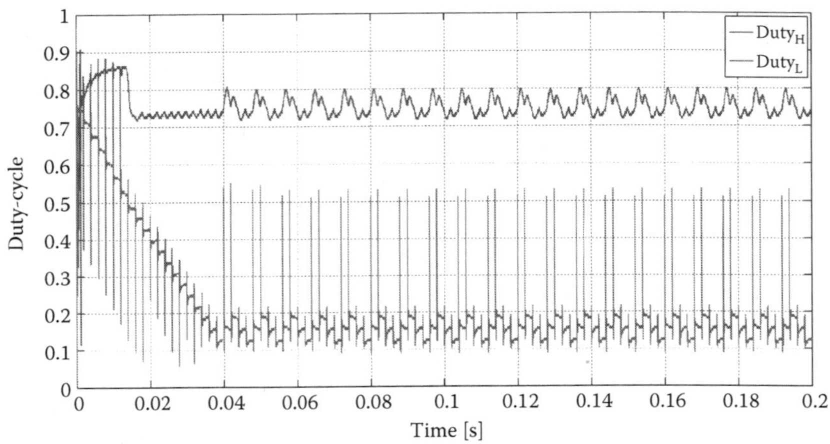

(c)

FIGURE 4.134 ***(continued)***
Simulation of the system in Figure 4.4 for $NH = 1$ and $G_L = 200$ W/m^2. (a) Time domain data plotted on the corresponding feasibility map (the black-filled triangle indicates global MPP, the black-filled circle indicates average steady-state operation point of the system). (b) PV panel voltage. (c) DC/DC MPPT converter duty cycle.

The approach described above allows us to effectively analyze modules interaction and stability of a string of SCPVMs provided that the dynamics of the MPPT block in each SCPVM is slow enough for the closed-loop DC/DC converter to reach steady-state operation after each perturbation. Note, however, that the MPPT DC/DC converter is controlled by two feedback loops; the first one is the input voltage control loop and the second one is the MPPT loop. The approach presented so far holds for systems where quasi-stationary MPPT techniques are employed (e.g., perturb and observe, incremental conductance) [79]. Particular care must be taken in cases in which MPPT techniques involving faster dynamics are used. Moreover, the MPPT block can involve nonlinear functions. Nonlinear systems stability criteria or small-signal double-control-loop design techniques should be adopted in these cases.

References

1. S. Liu and R.A. Dougal. Dynamic multiphysics model for solar array. *IEEE Transactions on Energy Conversion*, 17(2):285–294, 2002.
2. G. Petrone, G. Spagnuolo, and M. Vitelli. Analytical model of mismatched photovoltaic fields by means of Lambert W-function. *Solar Energy Materials and Solar Cells*, 91(18):1652–1657, 2007.

3. D. Roche, H. Outhred, and R.J. Kaye. Analysis and control of mismatch power loss in photovoltaic arrays. *Progress in Photovoltaics: Research and Applications*, 3(2):115–127, 1995.
4. M. Garcıa, J.M. Maruri, L. Marroyo, E. Lorenzo, and M. Perez. Partial shadowing, MPPT performance and inverter configurations: Observations at tracking PV plants. *Progress in Photovoltaics: Research and Applications*, 16(6):529–536, 2008.
5. P. Sanchis, J. Lopez, A. Ursua, E. Gubıa, and L. Marroyo. On the testing, characterization, and evaluation of PV inverters and dynamic MPPT performance under real varying operating conditions. *Progress in Photovoltaics: Research and Applications*, 15(6):541–556, 2007.
6. E.L. Meyer and E. Ernest van Dyk. The effect of reduced shunt resistance and shading on photovoltaic module performance. In *Thirty-First IEEE Photovoltaic Specialists Conference*, January 3–7, 2005, pp. 1331–1334.
7. R. Bruendlinger, B. Bletterie, M. Milde, and H. Oldenkamp. Maximum power point tracking performance under partially shaded PV array conditions. In *Twenty-First European Photovoltaic Solar Energy Conference*, Dresden, Germany, September 4–8, 2006, pp. 2157–2160.
8. T. Oozeki, K. Otani, and K. Kurokawa. Accuracy of estimated shading loss ratio by means of SV method—An extraction algorithm of maximum pattern. In *Thirty-First IEEE Photovoltaic Specialists Conference*, January 3–7, 2005, pp. 1804–1807.
9. N. Femia, G. Petrone, G. Spagnuolo, and M. Vitelli. Optimization of perturb and observe maximum power point tracking method. *IEEE Transactions on Power Electronics*, 20(4):963–973, 2005.
10. M. Veerachary, T. Senjyu, and K. Uezato. Voltage-based maximum power point tracking control of PV system. *IEEE Transactions on Aerospace and Electronic Systems*, 38(1):262–270, 2002.
11. K.K. Tse, M.T. Ho, H.S.-H. Chung, and S.Y. Hui. A novel maximum power point tracker for PV panels using switching frequency modulation. *IEEE Transactions on Power Electronics*, 17(6):980–989, 2002.
12. K.H. Hussein, I. Muta, T. Hshino, and M. Osakada. Maximum photovoltaic power tracking: An algorithm for rapidly changing atmospheric conditions. *Proceedings of the Institute of Electrical Engineering*, 142(1):59–64, 1995.
13. P. Midya, P. Krein, R. Turnbull, R. Reppa, and J. Kimball. Dynamic maximum power point tracker for photovoltaic application. In *Proceedings of 27th Annual IEEE Power Electronics Specialists Conference*, June 1996, vol. 2, pp. 1710–1716.
14. E. Koutroulis, K. Kalaitzakis, and N.C. Voulgaris. Development of a microcontroller-based, photovoltaic maximum power point tracking control system. *IEEE Transactions on Power Electronics*, 16(1):46–54, 2001.
15. T. Wu, C. Chang, and Y. Chen. A fuzzy-logic-controlled single-stage converter for PV-powered lighting system applications. *IEEE Transactions on Industrial Electronics*, 47(2):287–296, 2000.
16. I. Batarseh, T. Kasparis, K. Rustom, W. Qiu, N. Pongratananukul, and W. Wu. DSP-based multiple peak power tracking for expandable power system. In *Proceedings of 18th Annual IEEE Applied Power Electronics Conference Expo*, February 2003, vol. 1, pp. 525–530.

17. D.P. Hohm and M.E. Ropp. Comparative study of maximum power point tracking algorithms. *Progress in Photovoltaics: Research and Applications*, 11(1):47–62, 2003.
18. C. Hua and J. Linb. A modified tracking algorithm for maximum power tracking of solar array. *Energy Conversion and Management*, 45(6):911–925, 2004.
19. R. Leyva, C. Alonso, I. Queinnec, A. Cid-Pastor, D. Lagrange, and L. Martinez-Salamero. MPPT of photovoltaic systems using extremum-seeking control. *IEEE Transactions on Aerospace and Electronic Systems*, 42(1):249–258, 2006.
20. N. Femia, G. Petrone, G. Spagnuolo, and M. Vitelli. A new analog MPPT technique: TEODI. In *Progress in photovoltaics: Research and applications*. Vol. 18. Wiley-Blackwell, 2010, pp. 28–41.
21. M. Veerachary. Power tracking for nonlinear PV sources with coupled inductor SEPIC converter. *IEEE Transactions on Aerospace and Electronic Systems*, 41(3):1019–1029, 2005.
22. D. Shmilovitz. On the control of photovoltaic maximum power point tracker via output parameters. *IEE Proceedings—Electric Power Applications*, 152(2):239–248, 2005.
23. W. Xiao, M.G.J. Lind, W.G. Dunford, and A. Capel. Real-time identification of optimal operating points in photovoltaic power systems. *IEEE Transactions on Industrial Electronics*, 53(4):1017–1026, 2006.
24. I.-S. Kim, M.-B. Kim, and M.-J. Youn. New maximum power point tracker using sliding-mode observer for estimation of solar array current in the grid-connected photovoltaic system. *IEEE Transactions on Industrial Electronics*, 53(4):1027–1035, 2006.
25. J.-M. Kwon, K.-H. Nam, and B.-H. Kwon. Photovoltaic power conditioning system with line connection. *IEEE Transactions on Industrial Electronics*, 53(4):1048–1054, 2006.
26. N. Mutoh, M. Ohno, and T. Inoue. A method for MPPT control while searching for parameters corresponding to weather conditions for PV generation systems. *IEEE Transactions on Industrial Electronics*, 53(4):1055–1065, 2006.
27. N. Mutoh and T. Inoue. A control method to charge series-connected ultra-electric double-layer capacitors suitable for photovoltaic generation systems combining MPPT control method. *IEEE Transactions on Industrial Electronics*, 54(1):374–383, 2007.
28. J.-H. Park, J.-Y. Ahn, B.-H. Cho, and G.-J. Yu. Dual-module-based maximum power point tracking control of photovoltaic systems. *IEEE Transactions on Industrial Electronics*, 53(4):1036–1047, 2006.
29. W. Xiao, W.G. Dunford, P.R. Palmer, and A. Capel. Regulation of photovoltaic voltage. *IEEE Transactions on Industrial Electronics*, 54(3):1365–1374, 2007.
30. T. Noguchi, S. Togashi, and R. Nakamoto. Short-current pulse-based maximum-power-point tracking method for multiple photovoltaic-and-converter module system. *IEEE Transactions on Industrial Electronics*, 49(1):217–223, 2002.
31. K. Irisawa, T. Saito, I. Takano, and Y. Sawada. Maximum power point tracking control of photovoltaic generation system under nonuniform insolation by means of monitoring cells. In *Proceedings of 28th IEEE Photovoltaic Specialists Conference*, September 2000, pp. 1707–1710.
32. R. Gules, P.J. De Pellegrin, H.L. Hey, and J. Imhoff. A maximum power point tracking system with parallel connection for PV stand-alone applications. *IEEE Transactions on Industrial Electronics*, 55:2674–2683, 2008.

33. I.-S. Kim and M.-J. Youn. Variable-structure observer for solar-array current estimation in a photovoltaic power-generation system. *IEE Proceedings—Electric Power Applications*, 152(4):953–959, 2005.
34. G. Petrone, G. Spagnuolo, R. Teodorescu, M. Veerachary, and M. Vitelli. Reliability issues in photovoltaic power processing systems. *IEEE Transactions on Industrial Electronics*, 55(7):2569–2580, 2008.
35. B. Burger, B. Goeldi, S. Rogalla, and H. Schmidt. Module integrated electronics—An overview. Presented at 25th European Photovoltaic Solar Energy Conference and Exhibition, Valencia, Spain, September 2010.
36. Q. Li and P. Wolfs. A review of the single phase photovoltaic module integrated converter topologies with three different DC link configurations. *IEEE Transactions on Power Electronics*, 23(3):1320–1333, 2008.
37. S.B. Kjaer, J.K. Pedersen, and F. Blaabjerg. A review of single phase grid-connected inverters for photovoltaic modules. *IEEE Transactions on Industry Applications*, 41(5):1292–1306, 2005.
38. www.enphaseenergy.com.
39. http://www.greenraysolar.com.
40. www.petrasolar.com.
41. www.enecsys.com.
42. http://www.exeltech.com.
43. http://solarbridgetech.com.
44. http://www.involar.com.
45. G.R. Walker and P.C. Sernia. Cascaded DC/DC converter connection of photovoltaic modules. *IEEE Transactions on Power Electronics*, 19(4):1130–1139, 2004.
46. E. Roman, R. Alonso, P. Ibanez, S. Elorduizapatarietxe, and D. Goitia. Intelligent PV module for grid-connected PV systems. *IEEE Transactions on Industrial Electronics*, 53(4):1066–1073, 2006.
47. N. Femia, G. Lisi, G. Petrone, G. Spagnuolo, and M. Vitelli. Distributed maximum power point tracking of photovoltaic arrays: Novel approach and system analysis. *IEEE Transactions on Industrial Electronics*, 55(7), 2610–2621, 2008.
48. E. Roman, V. Martinez, J.C. Jimeno, R. Alonso, P. Ibanez, and S. Elorduizapatarietxe. Experimental results of controlled PV module for building integrated PV systems. *Solar Energy*, 82:471–480, 2008.
49. W. Yao, M. Gao, Z. Ren, M. Chen, and Z. Qian. Improvement of performance and flexibility for photovoltaic module using individual DC/DC converter. In *6th IEEE International Power Electronics and Motion Control Conference (IPEMC)*, May 17–20, 2009, pp. 441–444.
50. T. Zhao, H. Wang, X. Li, Z. Ju, X. Wei, and S. Zhang. The distributed maximum power point tracking method and application in the PV grid-connected generation. In *2010 International Conference on Intelligent System Design and Engineering Application*, 2010, pp. 639–642.
51. A. Elasser, M. Agamy, J. Sabate, R. Steigerwald, R. Fisher, and M. Harfman-Todorovic. A comparative study of central and distributed MPPT architectures for megawatt utility and large scale commercial photovoltaic plants. In *IECON 2010*, 2010, pp. 2753–2758.
52. Q. Li and P. Wolfs. A preliminary study of the distributed maximum power point tracker designs for different types of solar cells in solar and electric vehicle arrays. In *AUPEC 2007*, pp. 1–6.

53. R. Shao, M. Kaye, and L. Chang. Advanced building blocks of power converters for renewable energy based distributed generators. In *8th International Conference on Power Electronics—ECCE Asia*, Shilla Jeju, Korea, May 2011, pp. 2168–2174.
54. S. Poshtkouhi, J. Varley, R. Popuri, and O. Trescases. Analysis of distributed peak power tracking in photovoltaic systems. In *2010 International Power Electronics Conference*, 2010, pp. 942–947.
55. R. Alonso, P. Ibáñez, V. Martínez1, E. Román, and A. Sanz. Analysis of performance of new distributed MPPT architectures. In *ISIE 2010*, 2010, pp. 3450–3455.
56. D. Ahmadi, S.A. Mansouri, and J. Wang. Circuit topology study for distributed MPPT in very large scale PV power plants. In *Applied Power Electronics Conference and Exposition (APEC)*, 2010, pp. 786–791.
57. U. Kamnarn, S. Yousawat, S. Sreeta, W. Muangjai, and T. Somsak. Design and implementation of a distributed solar controller using modular buck converter with maximum power point tracking. In *UPEC 2010*, 2010, pp. 1–6.
58. Y.-M. Chen, C.-W. Chen, and Y.-L. Chen. Development of an autonomous distributed maximum power point tracking PV system. In *Distributed Energy Conversion Congress and Exposition (ECCE)*, 2011, pp. 3614–3619.
59. S. Poshtkouhi, V. Palaniappan, M. Fard, and O. Trescases. A general approach for quantifying the benefit of distributed power electronics for fine grained MPPT in photovoltaic applications using 3D modeling. *IEEE Transactions on Power Electronics*, vol. 27, 11, November 2012, 4656–4666.
60. G. Petrone, G. Spagnuolo, and M. Vitelli. An analog technique for Distributed MPPT PV applications. *IEEE Transactions on Industrial Electronics*, vol. 59, 12 December 2012, 4713–4722.
61. P. Tsao, S. Sarhan, and I. Jorio. Distributed maximum power point tracking for photovoltaic arrays. In *PVSC PV Photovoltaic Specialists Conference*, 2009, pp. 2293–2298.
62. S. Poshtkouhi and O. Trescases. Multi-input single-inductor DC-DC converter for MPPT in parallel-connected photovoltaic applications. In *Applied Power Electronics Conference and Exposition (APEC)*, 2011, pp. 41–47.
63. R. Giral, C.E. Carrejo, M. Vermeersh, A.J. Saavedra-Montes, and C.A. Ramos-Paja. PV field distributed maximum power point tracking by means of an active bypass converter. In *ICCEP 2011*, 2011, pp. 94–98.
64. Z. Cheng, H. Zhou, and H. Yang. Research on MPPT control of PV system based on PSO algorithm. In *2010 Chinese Control and Decision Conference*, pp. 887–892.
65. http://solarmagic.com/en/index.html.
66. http://www.solaredge.com.
67. http://www.tigoenergy.com.
68. http://www.xandex.com.
69. http://eiqenergy.com.
70. G. Adinolfi, N. Femia, G. Petrone, G. Spagnuolo, and M. Vitelli. Design of DC/DC converters for DMPPT PV applications based on the concept of energetic efficiency. *Journal of Solar Energy Engineering*, 132:021005-1–021005-10, 2010.
71. M. Balato, N. Femia, G. Petrone, G. Spagnuolo, and M. Vitelli. Factors limiting the efficiency of DMPPT in PV applications. In *International Conference on Clean Electrical Power*, Ischia, June 14–16, 2011, pp. 604–608.

72. R.W. Erikson and D. Maksimovic. *Fundamentals of power electronics*. 2nd ed. Kluwer Academic Publishers, Norwell, MA.
73. J.M. Carrasco, L.G. Franquelo, J.T. Bialasiewicz, E. Galvan, R.C.P. Guisado, Ma.A.M. Prats, J.I. Leon, and N. Moreno-Alfonso. Power-electronic systems for the grid integration of renewable energy sources: A survey. *IEEE Transactions on Industrial Electronics*, 53(4):1002–1016, 2006.
74. S.B. Kjaer, J.K. Pedersen, and F. Blaabjerg. A review of single-phase grid-connected inverters for photovoltaic modules. *IEEE Transactions on Industry Applications*, 41(5):1292–1306, 2005.
75. F. Blaabjerg, R. Teodorescu, M. Liserre, and A. Timbus. Overview of control and grid synchronization for distributed power generation systems. *IEEE Transactions on Industrial Electronics*, 53(5):1398–1409, 2006.
76. N. Femia, G. Petrone, G. Spagnuolo, and M. Vitelli. A technique for improving P&O MPPT performances of double stage grid-connected photovoltaic systems. *IEEE Transactions on Industrial Electronics*, 56(11):4473–4482, 2009.
77. R.D. Middlebrook. Null double injection and the extra element theorem. *IEEE Transactions on Education*, 32:167–180, 1989.
78. R.D. Middlebrook and V. Vorperian. The extra element theorem. *IEEE Transactions on Circuits and Systems I: Fundamental Theory and Applications*, 45(9):919–935, 1998.
79. G. Lisi. Distributed maximum power point tracking of photovoltaic arrays. PhD dissertation, University of Salerno, February 2008.

5

Design of High-Energy-Efficiency Power Converters for PV MPPT Applications

5.1 Introduction

In the previous chapters, issues related to topologies and control techniques in the achievement of maximum power point tracking of PV sources have been discussed. We are now ready to focus on the ultimate objective of the methodologies discussed in this book: achieving maximum energy harvesting in PV systems. Such achievement is the fruit of appropriate combinations of power converter topology, power devices, and control techniques. On the topology side, we have seen in Chapter 4 that buckboost DC/DC converters allow a potentially wider DMPPT operating range, thanks to step-up/down capabilities. However, this result can be compromised if silicon devices of the power converter are characterized by a too low breakdown voltage, imposing restrictive voltage limitations that can paradoxically cause a cutoff of peak power deliverable by unshaded panels in PV strings subjected to mismatched irradiation conditions. Other topologies, like the boost or the buck one, though potentially less flexible, due to their intrinsic monomodal voltage conversion ratio (either step up or step down), might ensure overall DMPPT effectiveness comparable with step-up/down topologies thanks to the lower voltage stress, which reduces the undesired action of voltage limiting in mismatched conditions and allows selection of devices with lower channel resistance. On the control side, we have seen in Chapter 3 that the effectiveness of MPPT in variable sun irradiation conditions is much conditioned by the proper setup of P&O parameters. These require careful tuning, based on the intrinsic dynamic behavior of the power converter, which depends on passive components whose size is conditioned by the switching frequency and then by the silicon device losses. Again, silicon devices have a great influence on the achievement of high energy efficiency. As the goal of MPPT in PV applications is to help in harvesting as much energy as possible, the global impact of the silicon devices on the energy efficiency must be necessarily considered, and their correct selection becomes one of the key steps in DC/DC converter design for PV applications. In this chapter, after a brief preliminary overview of energy efficiency issues, the analysis of voltage and current stresses on silicon devices is discussed in correlation with the power converter topologies for a PV application. Models allowing loss calculation for commercial silicon devices starting from datasheets are then overviewed and discussed. Some hints for selection of devices allowing trade-off of switching frequency vs. energy efficiency are presented. The case study considered for the discussion is the same one investigated in Chapter 4, regarding a PV DMPPT application. Indeed, this is the case where a deep analysis of the global combined impact of power converter topology, silicon power devices, and relevant DMPPT control active operating range is needed for the true achievement of maximum energy harvesting.

5.2 Power, Energy, Efficiency

During the last decades many energy saving programs have been promoted by governments and public agencies worldwide to encourage high-efficiency Switch-Mode Power Supplies (SMPS) manufacturing. Some pro-actions connected to these initiatives are aimed at:

- Development of energy saving consumers' awareness
- Product grading (energy efficiency qualifying labels)
- Incentives (including cash rebate)
- Adoption of government standards and specifications
- Definition of test specifications

In addition to efficiency specifications, restrictions on absolute power loss in no-load and active-load operating conditions are specified by the greater part of energy efficiency standards to achieve energy qualifications, such as UE Energy Star. Developing new SMPS characterized by high energy efficiency is a good way to increase competitiveness and profits. In the PV context, energy efficiency is a major issue. PV cells have intrinsically a very low efficiency, and the energy processing devices in the power chain from the PV source to the final user must have very high efficiency in order to harvest as much as possible of the available sun energy. The conversion efficiency in PV applications is typically expected to be higher than 96%, both for high-power and for low-power devices. In distributed MPPT applications, where a DC/DC converter is used to extract the maximum power from each PV module, the energy efficiency of the converter is expected to be even higher than 98%, as the use of additional power electronics to increase the energy productivity of PV plants subjected to mismatching conditions makes sense if its efficiency is so high that in absence of mismatch the additional power losses are negligible. Achieving 98% and even higher efficiency with a DC/DC converter is not difficult if power devices are sufficiently oversized; however, this likely routes to expensive solutions that are out of market, considering the cost breakdown of PV modules. This makes loss analysis a very crucial task to identify feasible commercial solutions complying with tight energy efficiency and cost constraints. The efficiency of a switch-mode power supply can be defined and analyzed in different ways, depending on the application and context. The basic definition of efficiency is based on the difference between the input and the output power, according to Equation (5.1):

$$\eta_P = \frac{P_{load}}{P_{source}} = \frac{V_{out}I_{out}}{V_{in}I_{in}} = \frac{P_{load}}{P_{load} + P_{loss}} = \frac{V_{out}I_{out}}{V_{out}I_{out} + P_{loss}(V_{in}, V_{out}, I_{in}, I_{out}, \underline{p}, f_s, T_a)} \quad (5.1)$$

where the voltages, currents, and powers are referred to the block scheme of Figure 5.1.

The SMPS power losses P_{loss} depend on many influence factors, like the input and output ports' *termination conditions* (V_{in}, I_{in}, V_{out}, I_{out}), the *switching frequency* (f_s), the *physical parameters* of the devices where the losses take place (p), and the *ambient temperature* (T_a). The exact dependence of P_{loss} on the influence factors is quite involved. The achievement of an SMPS design complying with given efficiency requirements is possible only if adequate loss models are used. Efficiency requirements greatly depend on the application. If both source and load are time invariant, the analysis of efficiency is limited to a fixed set of port termination conditions. This is the simplest case, as the efficiency reduces to a function—though involved—of switching frequency and temperature. Unfortunately, most applications involve some variation of load, source, or both. Consequently, the efficiency also has to be analyzed with respect

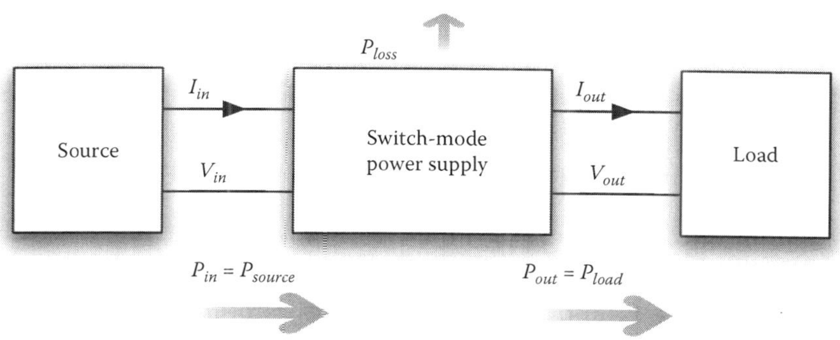

FIGURE 5.1
SMPS intput, output and loss power.

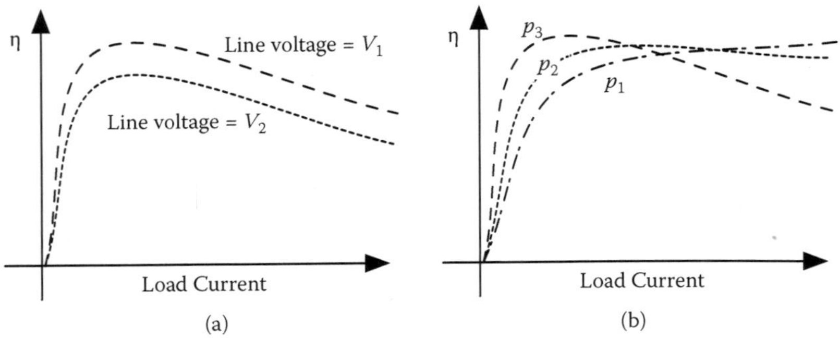

FIGURE 5.2
Efficiency curves of a switch-mode power supply. (a) Fixed parameters, different line voltage values. (b) Fixed line voltage, different FET parameter values.

to the termination conditions to identify the worst case. Typical plots of SMPS power efficiency with respect to load current and line voltage variations are shown in Figure 5.2. The plots show two fundamental elements characterizing the efficiency.

On one side, as shown in Figure 5.2a, we have a different dependence of efficiency on load current and line voltage. Depending on the topology of the converter, the mode of operation, the specific power devices, and the operating ranges, the dependence of the efficiency on line voltage and load current may exhibit different behaviors. A dual characterization can be done with respect to load voltage and line current. The first one is suitable, for example, to analyze the efficiency of SMPS dedicated to feed voltage-controlled variable loads from unregulated sources (e.g., *point-of-load* applications). The second one fits, for example, with the needs of analyzing the efficiency of SMPS dedicated to extract power from variable sources to be delivered to sink loads (e.g., photovoltaic applications, belonging to the *point-of-source* applications category).

On the other side, the shape of the efficiency curve vs. load greatly changes with the parameters of the power devices used to realize the SMPS. The curves of Figure 5.2b show that the peak power efficiency is achieved at different load rates, depending on the set of parameters $\{p_1, p_2, p_3\}$ characterizing the power devices.

The main SPMS design issues related to efficiency are:

1. The analysis of the application efficiency targets
2. The analysis of the efficiency as a function of the SMPS topology, operation mode, termination conditions, switching frequency, thermal conditions, and power device parameters
3. The selection of power devices and switching frequency allowing a reliable achievement of the application efficiency targets

Point 1 mainly depends on design context and goals. If SPMS design specifications are fixed by a customer, then typically a minimum guaranteed efficiency at nominal load current and line voltage is required. If the design goal is the development of a new product for a given field of application, then regulations, recommended practices, and qualification standards related to the specific application context provide the due references and constraints. Qualification of photovoltaic inverters is based on metrics like the European efficiency h_{EU} and the California Energy Commission efficiency h_{CA}, given by (5.2) and (5.3):

$$\eta_{EU} = 0.03\eta_{5\%} + 0.06\eta_{10\%} + 0.13\eta_{20\%} + 0.10\eta_{30\%} + 0.48\eta_{50\%} + 0.20\eta_{100\%} \quad (5.2)$$

$$\eta_{CA} = 0.04\eta_{10\%} + 0.05\eta_{20\%} + 0.12\eta_{30\%} + 0.21\eta_{50\%} + 0.53\eta_{75\%} + 0.05\eta_{100\%} \quad (5.3)$$

which are weighted combinations of the inverter efficiency calculated at given load levels. As a consequence of the two different metrics adopted by the two definitions (5.2) and (5.3), given the inverter efficiency curves, different values of the efficiency are obtained. As an example, the application of the EU and California efficiency formulas (5.2) and (5.3) to the efficiency curves of the PowerOne Corporation PVI-4.2-OUTD inverter yields the results summarized in Table 5.1.

The difference between the two efficiency evaluations is around 0.5%, which is not negligible for PV inverters. So, if the inverter has to be designed for the U.S. market, it would be better to select power devices so higher California efficiency is achieved, and vice versa for the EU market target. The concept underlying the EU and California efficiency definitions goes beyond the mere peak power efficiency or maximum load power efficiency.

TABLE 5.1

EA and CA Efficiency of a Commercial PV Inverter

Efficiency Samples	EU Weights	CA Weights	215Vdc Efficiency	345Vdc Efficiency	480Vdc Efficiency
0,05	0,03		0,900	0,910	0,880
0,1	0,06	0.04	0,920	0,930	0,910
0,2	0,13	0.05	0,945	0,955	0,945
0,3	0,1	0.12	0,953	0,965	0,955
0,5	0,48	0.21	0,960	0,968	0,963
0,75		0.53	0,960	0,968	0,965
1	0,2	0.05	0,958	0,962	0,967
		EU eff	**0,953**	**0,961**	**0,955**
		CA eff	**0,957**	**0,965**	**0,960**

Weighting the efficiency over the 0 to 100% SMPS rated power takes into account variable power operation. This naturally brings us to the *energy efficiency* (EE) concept, given by

$$\eta_P = \frac{U_{load}}{U_{source}} = \frac{U_{out}}{U_{in}} = \frac{\int_{t_s}^{t_e} v_{out}(t)i_{out}(t)dt}{\int_{t_s}^{t_e} v_{in}(t)i_{in}(t)dt} \tag{5.4}$$

where t_s and t_e are respectively the start time and end time of a typical interval of operation over which either the source or the load (or both) operates its typical *work cycle*, $U_{load} = U_{out}$ is the energy exiting the power supply and delivered to the load, and $U_{source} = U_{in}$ is the energy exiting the PV source and delivered to the power supply. The *work cycle* characterizes the specific SMPS application and then the consequent EE constraints. Some examples of work cycles are given in Figure 5.3.

The SMPS of Figure 5.3a is a PV maximum power point tracker. It works with a variable source, the PV generator, whose current goes from zero up to a maximum peak value and down back to zero over an interval of time of several hours. The MPP voltage V_{MPP} varies in the range from about 50% to about 80% of the open-circuit voltage V_{oc} over the range of sun irradiance G from 10% to 100% of maximum irradiance G_{max} and over the range of ambient temperature T_a from 10% to 100% of maximum temperature T_{max}. The MPP current I_{MPP}, instead, varies proportionally to sun irradiance. The load works as an adaptive impedance, which sinks the maximum power deliverable by the PV source. Normally its voltage is regulated, so that the output current of the SMPS varies according to the input one. The SMPS of Figure 5.3b is a point-of-load regulator feeding a low-voltage/high-current variable load, like a µP or a Field Programmable Gate Array (FPGA). Such loads need a well-regulated voltage, and their current work cycles depend on the application. Also in this case there is a peak current value and the swing from zero to the peak current can pass through some intermediate levels very quickly (e.g., tens to hundreds of nanoseconds for µP and Digital Signal processors (DSPs), generating step-up/down staircase waveshapes whose envelope shape varies very much over hours (e.g., work cycles of FPGAs in telecom systems processing daily data traffic). The source in this case may provide regulated as well as unregulated voltage (depending on SMPS power architecture), and the current it delivers follows the load current. The evaluation of the SMPS energy efficiency for these two applications is based on the calculation of input and output energies, given respectively by (5.5) and (5.6):

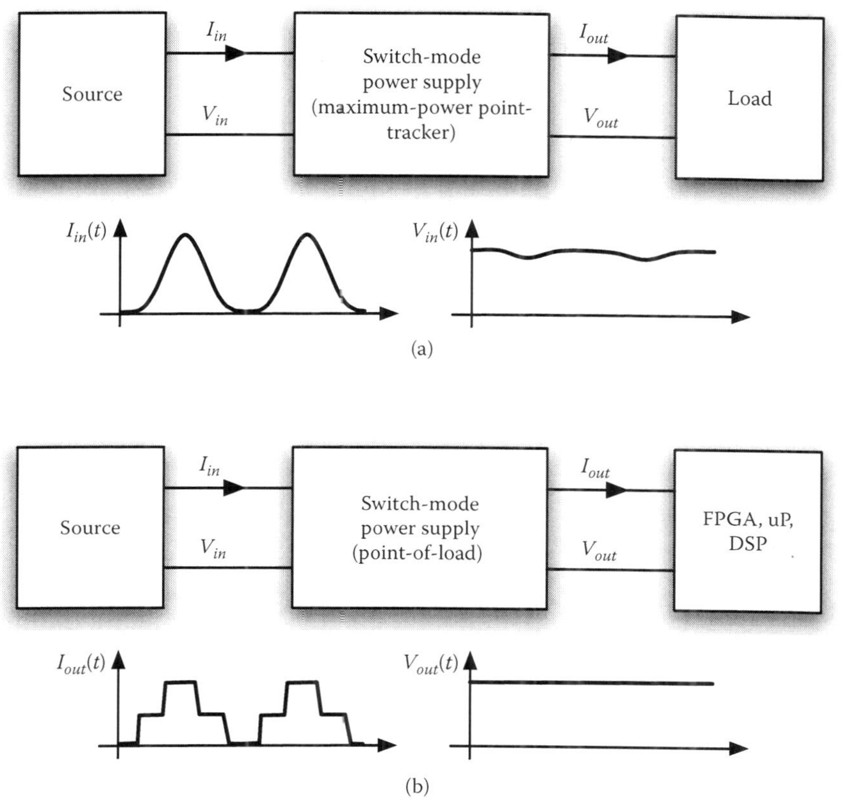

FIGURE 5.3
(a) PV-MPPT SMPS. (b) Point-of-load SMPS.

$$\text{PV: } U_{in} = \int_{t_s}^{t_e} v_{in}(t) i_{in}(t) dt \quad U_{out} = \int_{t_s}^{t_e} v_{in}(t) i_{in}(t) \eta(\underline{tc}, \underline{p}, f_s, T_a) dt \tag{5.5}$$

$$\text{POL: } U_{in} = \int_{t_s}^{t_e} \frac{v_{out}(t) i_{out}(t)}{\eta(\underline{tc}, \underline{p}, f_s, T_a)} dt \quad U_{out} = \int_{t_s}^{t_e} v_{out}(t) i_{out}(t) dt \tag{5.6}$$

where $\underline{tc} = (v_{in}, i_{in}, v_{out}, i_{out})$ is the vector of SMPS port termination conditions. Thus the achievement of a given energy efficiency design target is subordinated to the knowledge of the efficiency function $\eta(\underline{t_c}, \underline{p}, f_s, T_a)$, for whatever application we are dealing with. In the next paragraph we focus on the energy efficiency analysis of DMPPT PV applications.

5.3 Energy Harvesting in PV Plant Using DMPPT Power Converters

Let us consider the case discussed in Chapter 4, where the PV source is made of a couple of Sunmodule SW225 PV panels (V_{oc} = 36.8 V, I_{sc} = 8.17 A, V_{MPP} = 29.5 V, I_{MPP} = 7.63 A, $NOCT$ = 46°C), each one equipped with a DMPPT DC/DC power converter of boost or buckboost topology. In Chapter 4 the analysis of the mismatched operation was done with reference to a generic condition characterized by sun irradiances of 1000 W/m^2 for one panel and 500 W/m^2 for the other one. Here we consider in more detail a realistic condition referred to real daily sun irradiance and temperature data and to the presence of a body shading the two panels in nonuniform way. Plots of Figure 5.4 show the daily evolution of sun irradiance and ambient temperature recorded on September 15, 2001, in Napoli, Italy (40° 50′ 23.940″ N, 14° 15′ 9.152″ E). During that day, the sun irradiance reached a quite high peak value, near 1200 W/m^2, due to a very clean sky. The temperature held mostly in the range from 20°C to 23°C. The day was close to autumn equinox, so that there were almost 12 light hours. Table 5.2 summarizes the azimuth and incidence angles of sun irradiance with respect to the PV panel surface.

The panels are assumed to be installed over a flat surface, oriented south and tilted 40°, equal to the city of Napoli latitude. Right side plots of Figure 5.4 show the corresponding path of MPP of one unshaded 225 W PV panel in the I vs. V and P vs. V planes, with a 10 min sampling time interval.

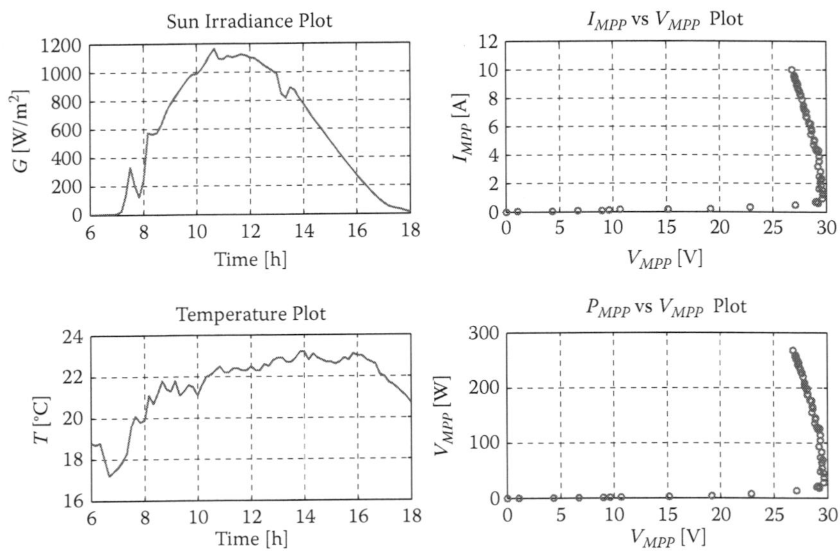

FIGURE 5.4
Sun irradiance, temperature data, and related MPP voltage, current, and power data of 225 Wp PV panel.

TABLE 5.2

Daily Sun Position over September 15, 2001, in Napoli, Italy

day hour	6	7	8	9	10	11	12	13	14	15	16	17	18
sun elevation angle [°]	2,29	13,60	24,62	34,89	43,69	49,88	52,06	49,57	43,16	34,23	23,88	12,81	14,80
sun azimuth angle [°]	–91,99	–82,05	–71,29	–58,68	–42,94	–22,91	0,73	24,23	43,96	59,44	71,87	82,54	92,44
sun incidence angle [°]	42,29	53,60	64,62	74,89	83,69	89,88	92,06	89,57	83,16	74,23	63,88	52,81	54,80
tilt factor	0,67	0,80	0,90	0,97	0,99	1,00	1,00	1,00	0,99	0,96	0,90	0,80	0,82

Source: http://www.sunearthtools.com.

The I_{MPP} vs. V_{MPP} path of Figure 5.4 provides the input for the calculation of nominal current and voltage stresses on the silicon devices of a DC/DC power converter for DMPPT in matched irradiance conditions. Due to irradiance mismatch determined by the shadow, the real voltage stress can be much different with respect to nominal ones, as already illustrated in the example in Chapter 4, referring to a generic condition characterized by sun irradiances of 1000 W/m^2 for one panel and 500 W/m^2 for the other. Herein we consider in detail a realistic mismatch condition determined by the shadow of a vertical body, like a chimney or a pile, sweeping the surface of the two panels during the day, as sketched in Figure 5.5. In the following, the

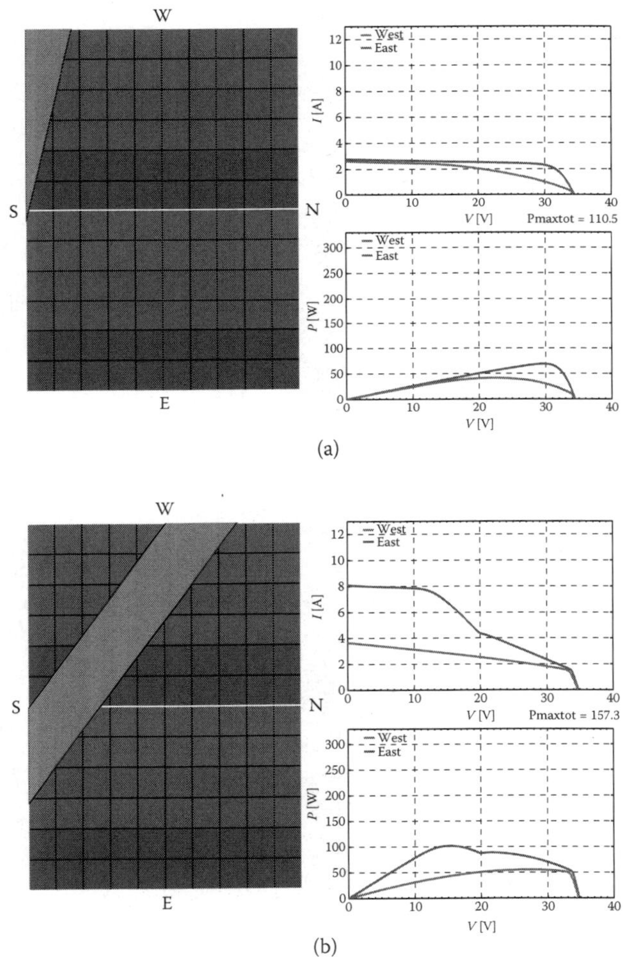

FIGURE 5.5 (See color insert)
Positions of the shadow in six times of the day, with related P vs. V plots of PV panels. (a) 7:30 a.m., S = 333 W/m^2. (b) 9:30 a.m., S = 908 W/m^2.

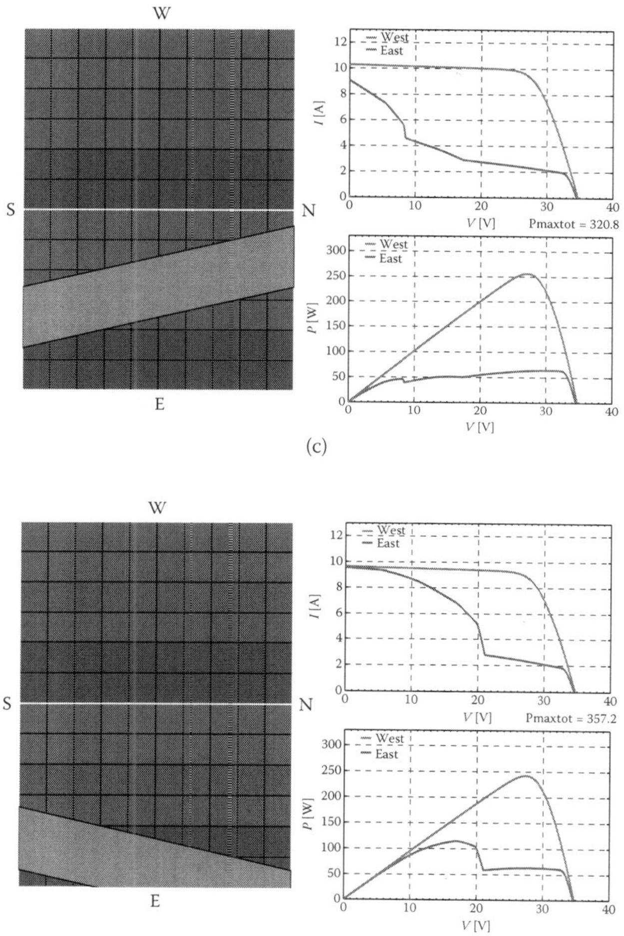

FIGURE 5.5 ***(continued)***
Positions of the shadow in six times of the day, with related P vs. V plots of PV panels. (c) 11:30 a.m., S = 1112 W/m^2. (d) 12:30 a.m., S = 1050 W/m^2.

two panels are identified as west side panel and east side panel according to their position. For each time instant, the plots of I-V and P-V curves of the two panels are also shown. These plots have been obtained by using two different sun irradiance levels for unshaded and shaded PV cells. For unshaded cells, the equivalent sun irradiance values shown in Figure 5.5 have been determined by correcting the nominal instant sun irradiance values of Figure 5.4 with the tilt factor of Table 5.1. Instead, for shaded cells, the equivalent sun irradiance has been assumed to be one-fourth of the value used for

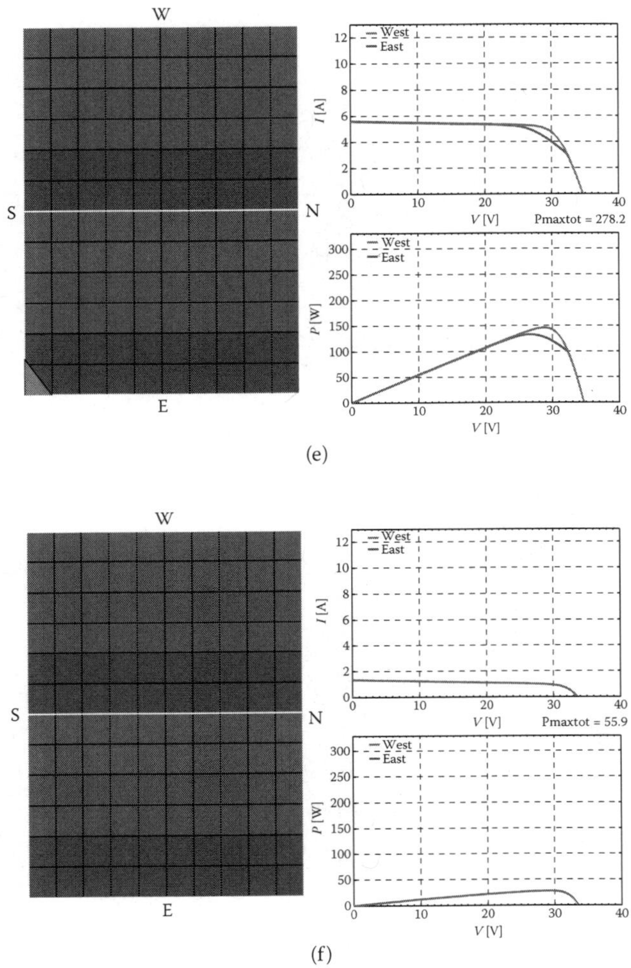

FIGURE 5.5 (*continued*)
Positions of the shadow in six times of the day, with related P vs. V plots of PV panels. (e) 2:30 p.m., S = 641 W/m^2. (f) 4:30 p.m., S = 163 W/m^2.

unshaded cells. The curves have been determined by modeling separately for each PV cell its unshaded portion and its shaded portion, as discussed in [17]. The two portions have been considered in parallel. The bypass diode action has also been considered for each 20-cell string. The system of 2 × 60 equations representing each panel has been numerically solved by means

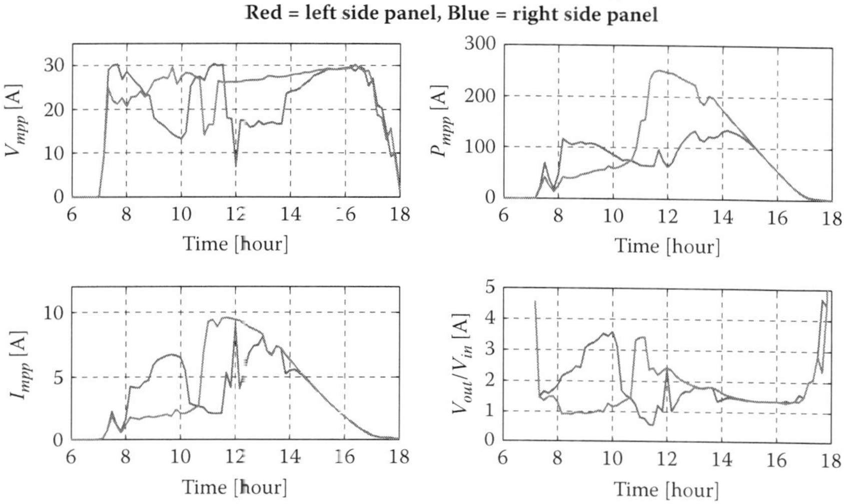

FIGURE 5.6 (See color insert)
MPP voltage, current, and power and input-output voltage conversion ratio: red = west side panel, blue = east side panel.

of a Newton-Raphson algorithm, and the I-V and P-V characteristics of both panels have been determined over 97 time instants from 6:00 a.m. to 6:00 p.m. using sun irradiance and temperature data of Figure 5.4. In Chapter 4, nonisolated boost and buckboost topologies have been considered to discuss the effects of mismatch on the DMPPT efficiency in the case study of a PV system composed of two PV panels, each one served by a DC/DC MPPT converter of one of the two topologies, whose outputs are put in series and terminated on a 80 V DC link.

In this chapter the same case study is analyzed. The plots of Figure 5.6 show the daily evolution of the MPP voltage, current, and power for each panel and the expected voltage conversion ratio of each of the two DMPPT DC/DC converters. The plots show the effect of the shadow on the maximum power of the two panels in the different times of the day due to the effect of the sweeping shadow. The voltage conversion ratio plots show that both DMPPT converters should be able to work in step-up/down mode. Let us analyze the way the topology selected for the DMPPT converter influences the voltage and current ratings of the silicon devices. The daily evolution of the voltage and current stresses V_{off} and I_{on} on the silicon devices of the two power converter topologies are plotted in Figures 5.7 and 5.8 for boost-based and for buckboost-based DMPPT converters, respectively. The plots

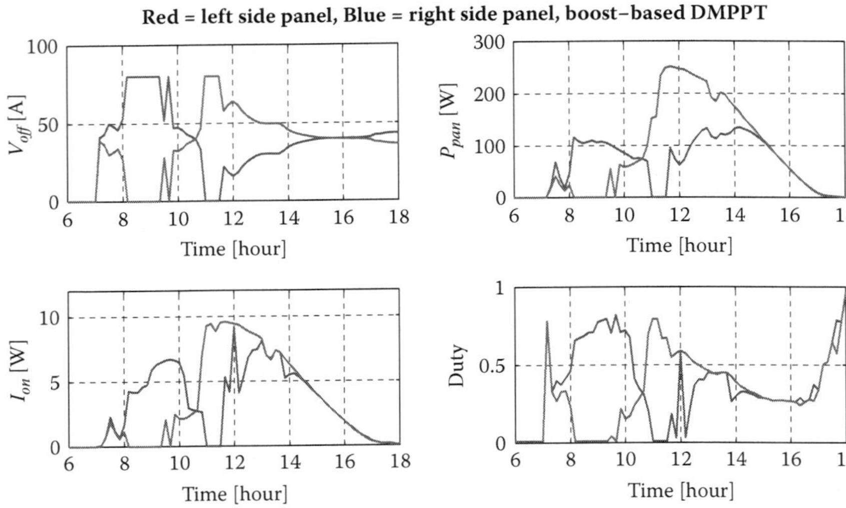

FIGURE 5.7 (See color insert)
Boost DMPPT converter voltage and current stresses: red = west side panel, blue = east side panel.

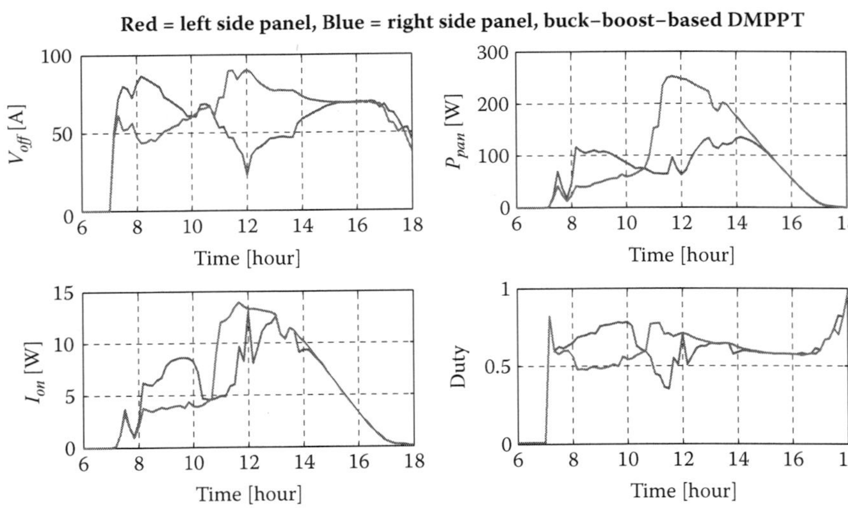

FIGURE 5.8 (See color insert)
Buckboost DMPPT converter voltage and current stresses: red = west side panel, blue = east side panel.

of Figure 5.7 show that, due to the step-up nature of boost topology, during some hours of the day, depending on the position of the sweeping shadow, only one of the two DMPPT converters will be operating and its output voltage saturates at the 80 V bulk voltage.

Let us assume that 80 V is lower than the breakdown voltage of the two MOSFETs, and let us calculate the energy loss due to the missing power extraction in the two intervals of Figure 5.7b, where one of the two converters is not in operation. The energy of west side panel lost in the first interval is about 79 Wh, over a total available daily energy of 1107 Wh. The percent loss is 7%. The energy of east side panel lost in the second interval is about 43 Wh, over a total available daily energy of 846 Wh. The percent loss is 5%. The overall average percent energy loss for both panels is 6.2%. Looking at plots of Figure 5.8 shows that there is no cutoff due to unsustainable voltage conversion ratio, thanks to the twofold step-up/down operation of buckboost topology. Accordingly, buckboost topology looks better performing than boost topology in terms of energy harvesting. This is true only if the conversion efficiencies of boost and buckboost are comparable. Buckboost devices have, indeed, to sustain higher voltage and currents stresses than boost ones. Comparing the plots of Figure 5.8a and 5.8c with the plots of Figure 5.7a and 5.7c shows that while buckboost topology imposes about 90 V/14 A maximum stress, the boost topology imposes 80 V/10 A maximum stress.

In addition, it must be considered that the worst-case voltage stress for the buckboost solution must be calculated in open-circuit operation rather than at MPP, the open-circuit voltage being about 25% bigger than the MPP voltage. In the case under study, this means about 6 V additional stress. Adequate devices have to be adopted for each topology to sustain the operating stresses and guarantee the highest energy efficiency. To this purpose let us first analyze the ratings of commercial MOSFETs. Figures 5.9 and 5.10 show the channel resistance R_{ds} and the total gate charge Q_g as functions of the maximum forward current I_{DS} of some 100 and 150 V *N*-channel power MOSFETs from International Rectifier and Vishay. The plots highlight that choosing a 100 V rather than a 150 V MOSFET, for the same current rating, makes a difference. Depending on the manufacturer, 100 V and 150 V FETs may exhibit much different channel resistance; thus achieving sufficiently low channel resistance without increasing very much the device current rating might be quite difficult at the 150 V rating.

As the MPP varies during the day, a proper design strategy based on accurate loss analysis is required to find the most convenient combination of MOSFETs for DMPPT applications. The next section provides the fundamentals for switching power converters' loss analysis.

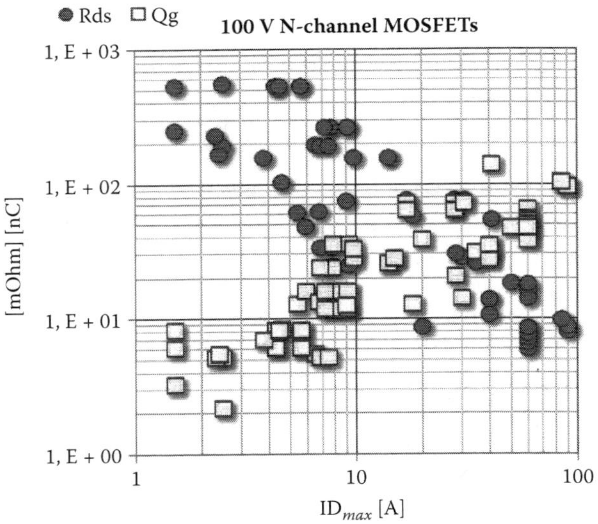

(a)

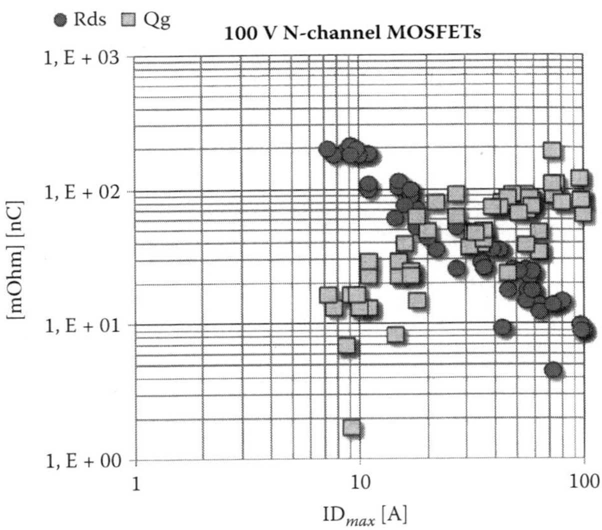

(b)

FIGURE 5.9
100 V *N*-channel power MOSFETs. (a) Vishay. (b) International Rectifier.

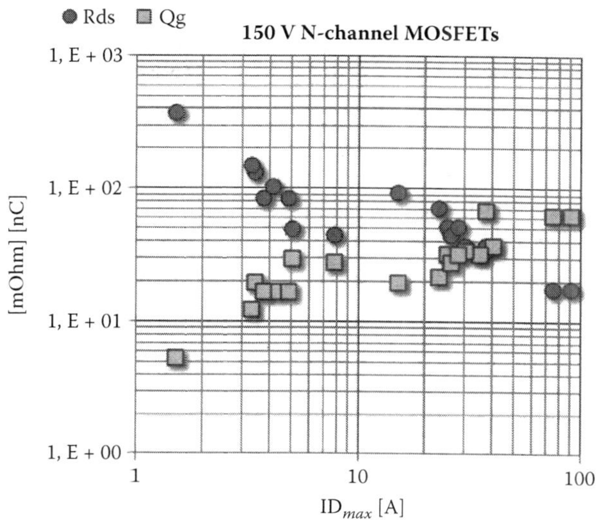

(a)

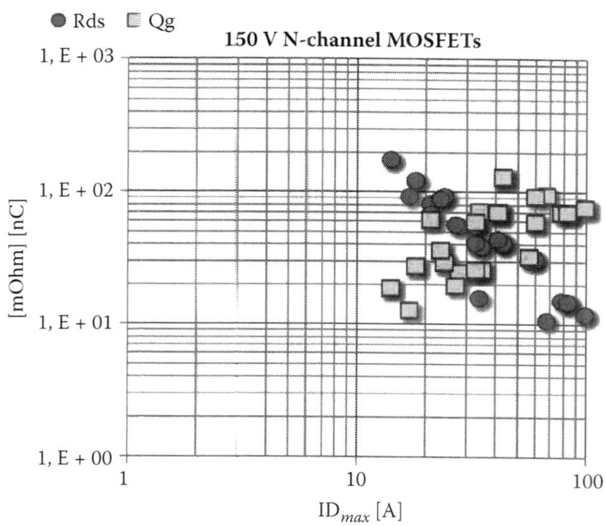

(b)

FIGURE 5.10
150 V *N*-channel power MOSFETs. (a) Vishay. (b) International Rectifier.

5.4 Losses in Power Converters

The selection of power devices and switching frequency allowing a reliable achievement of the application efficiency targets can be achieved only through the analysis of the efficiency as a function of the SMPS topology, operation mode, termination conditions, switching frequency, thermal conditions, and power device parameters. This is possible only if adequate loss models at device level and circuit level are adopted. The models must be sufficiently detailed and accurate to provide reliable results; however, they cannot be too detailed because some elements influencing the power losses, like stray inductances of Printed Circuit Board (PCB) traces, are not known until the SMPS design is complete, including power devices and housekeeping circuitry, and the PCB routing is done. A second important issue is related to how the parameters of power devices used in the loss models are extracted from datasheets, where some numerical data are provided for certain reference operating conditions and more complete data are available in graphic format. Using different combinations of loss model approximations and parameter approximations can lead to much different power loss predictions: This may have a big impact on power device selection and the achievement of energy efficiency design goals.

The silicon and magnetic devices are responsible for the majority of power losses of SMPS. The loss mechanisms of these devices are not easy to model, as they are connected to various nonlinear physical phenomena and are heavily conditioned by frequency and temperature. For silicon devices, switching losses are much more difficult to calculate than conduction losses. They are influenced by many factors:

- *Converter topology*: Different topologies involve different voltage/current stresses.
- *Operation mode*: SMPS may operate as hard switching, soft switching, resonant, etc.
- *Load/line range*: Some applications require operation over wide line/load ranges, which involve wide changes of device stress and related losses and possible lack of soft switching.
- *Device technology*: Materials, physical structures, operation principles, and parameter sensitivity of devices are much diversified, and thus so are their loss mechanisms.
- *Drivers*: Switching times depend on drive maximum current and adaptive dead time management capabilities.

An important element characterizing silicon device behavior is that both conduction and switching losses occur in the same physical body of the components. Actually, they are time interleaved, as switching losses are concentrated during very short time intervals that are alternated with long intervals where the conduction losses take place. By a thermal point of view this leads to the definition of a univocal die temperature that is used to determine the values of parameters upon which losses depend. In other words, there are no interdevice heat transfer mechanisms to be modeled in order to obtain loss prediction. Moreover, thermal resistance of silicon device packages is known, and this makes the thermal analysis easier.

In magnetic devices, instead, there are windings and core losses, occurring in two separate material bodies. Conduction losses occur in the windings and are influenced by skin and proximity effects. Eddy currents cause conduction losses in the core, too, where hysteresis losses also occur. The main factors influencing magnetic device losses in SMPS are:

- *Converter topology, mode of operation, line/load ranges*: These factors determine the voltage and current stresses.
- *Core material*: The power loss density and its dependence on frequency and temperature change very much depending on which ferrite or powdered iron core is used.
- *Windings*: Windings losses can dramatically increase due to skin and proximity effects.

Heuristic formulas are provided by magnetic materials and core manufacturers to determine total core losses. These formulas approximate the heavily nonlinear dependence of losses on frequency, magnetic flux density, and temperature and allow us to obtain plausible loss prediction. Thus losses of magnetic devices realized with a custom design can be predicted, even though only with a certain degree of approximation. The situation is much worse for off-the-shelf inductors and transformers, for which the manufacturers do not provide details for loss calculation. Mostly, only losses at nominal voltage/currents and ratings and at a reference frequency are specified. Finally, thermal analysis of magnetic components [19] is much more difficult than for silicon devices because of:

Intradevice heat flows and transfers: Heat flows in a quite complicated way through the core and also passes from the core to the windings or vice versa, depending on which part is hotter.

Lack of thermal coefficients: The physical structure of most inductors and transformers makes it very difficult to determine intradevice and device-to-air convection coefficients needed to model the heat transfer; most device manufacturers do not provide any thermal data.

Nonlinearity: The complete magneto-electro-thermal model of a magnetic device is quite puzzling, due to heavy nonlinearity involving interdependence among physical and geometrical parameters.

All the facts mentioned above make the calculation of magnetic device losses a challenging task, especially for commercial parts. Nevertheless, loss models are needed for reliable energy efficiency design, which has to take into account all the above elements.

In the next paragraphs, some models and methods for loss calculation and selection of power MOSFETs for high-energy-efficiency SMPS design are discussed. The application context considered is restricted to DC/DC converters for PV applications, specifically for MPPT at the module or submodule level. Nevertheless, most fundamental aspects and considerations can be extended to other SMPS applications.

5.5 Losses in the Synchronous FET Switching Cells

A wide variety of topologies and switching techniques are adopted for power converter design and realization. Nevertheless, the loss mechanisms in switching devices can be described by referring to the basic configuration made of a couple of synchronous FETs shown in Figure 5.11a. The switching cell of Figure 5.11a is the brick of most PWM hard-switching inductive power converters, like the boost and buckboost of Figure 5.11b and 5.11c.

The following discussion will be focused on this specific and elementary case, as the goal is to illustrate in depth a loss-controlled design methodology for high-efficiency applications, aimed at supporting a systematic and reliable real power components selection using the parameter values available in component datasheets and application notes. Similarly, other elementary cells for soft-switching converters, as well as for switched capacitor converters, could be identified, providing the fundamentals for the derivation of pertinent loss models that can be used with the same design approach.

The switching cell of Figure 5.11, in manifold variants, has been widely discussed and modeled in literature to describe the behavior of switching converters for both loss analysis and static and dynamic behavioral analysis. In the present discussion, the cell will be used to highlight the basic voltage/current stress correlations, which provide the reference operating parameters for loss calculations and device selection. The two switches operate under the action of a driver controller, ensuring that when a FET is ON, the

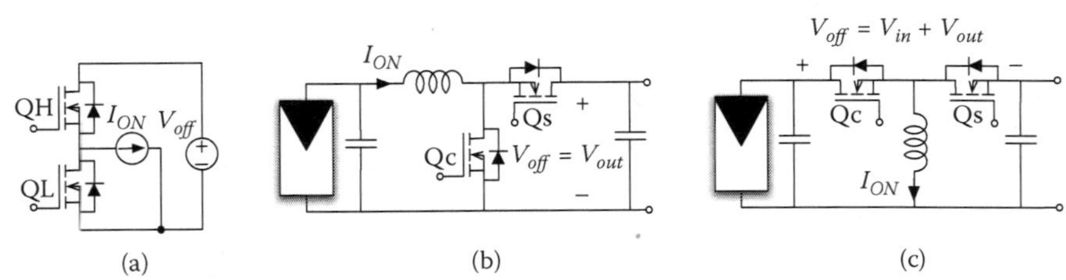

FIGURE 5.11
(a) Synchronous switching cell. (b) Boost topology. (c) Buckboost topology.

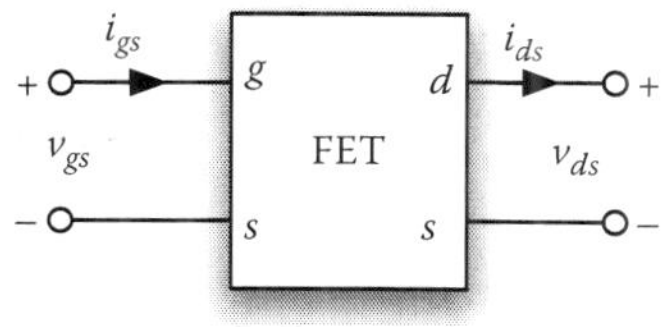

FIGURE 5.12
Two-port model of a MOSFET.

other one is OFF. The fundamental function implemented by the switching cell of Figure 5.11a, consists in allowing dual-energy flow paths between the voltage source and the current source. Depending on the specific converter, these archetypal sources may represent either real sources (e.g., battery) or passive energy storage devices (inductors, capacitors). Based on the topological connection and the complementary state operation of the two devices, under ideal conditions they would be subjected to the same voltage and current stresses, V_{off} when OFF and I_{on} when ON, respectively. Moreover, ideal switches would operate with no losses due to zero resistance when ON, infinite resistance when OFF, and instantaneous commutation between ON and OFF states and vice versa. In literature the two switches are commonly identified as control (CTR) or high-side (HS) switch and synchronous (SYN) or low-side (LS) switch. In high-current or high-efficiency applications FETs or other active devices are used to physically realize both switches, whereas in low-current or low-efficiency applications a diode is mostly used in place of the SYN switch. In the following, the CTR FET will be labeled C-FET, while the SYN FET will be labeled S-FET. The characteristics and behavior of real switches and real devices represented by the voltage and current sources, in addition to real drive circuitry and PCB ones, define the real voltage and current stresses acting on the two switches, which can change not negligibly with respect to the ideal value, and determine the real losses in ON and OFF states and during the commutations. In this paragraph, the behavior of the two switches will be analyzed in detail, assuming that both FETs are *N*-channel type. Each FET will be modeled as a 2-port, as shown in Figure 5.12. Accordingly, the average losses caused by the FET over an entire switching period T_s will be expressed by Equation (5.7):

$$P_d = f_s \int_0^{T_s} \left[v_{gs}(t) i_{gs}(t) + v_{ds}(t) i_{ds}(t) \right] dt = P_c - P_{sw,on} + P_{sw,off}$$

$$= f_s \left[\int_{T_c} v_{ds}(t) i_{ds}(t) dt + \sum_{i=1}^{N_{on}} \int_{T_{on,i}} \left(v_{gs}(t) i_{gs}(t) + v_{ds}(t) i_{ds}(t) \right) dt + \sum_{j=1}^{N_{off}} \int_{T_{off,j}} \left(v_{gs}(t) i_{gs}(t) + v_{ds}(t) i_{ds}(t) \right) dt \right] \tag{5.7}$$

Both conduction and switching losses are included in Equation (5.7). The switching period $T_s = 1/f_s$ is divided in $1 + N_{on} + N_{off}$ subintervals, each one corresponding to a certain phase of operation. *Conduction phase* corresponds to the subinterval $T_c = [t_{cs}, t_{ce}]$, wherein the switch is permanently in the ON state, whereas *commutation phases* correspond to the subintervals $T_{on,i} = [t_{on,i-1}, t_{on,i}]$, $i = 1, \ldots, N_{on}$, and $T_{off,j} = [t_{off,j-1}, t_{off,j}]$, $j = 1, \ldots, N_{off}$, during which the transitions from the ON to the OFF state, and vice versa, occur. References [2–8] can be of help to get insight in detailed FET loss analysis. Simplified formulas for loss calculation of the switches of the elementary switching cell of Figure 5.11a are provided in many technical papers and FET manufacturer application notes (e.g., [11, 14]). Several different assumptions and simplifications are adopted in these documents, so that much different results are obtained depending on the model adopted and the way the FET parameters are extracted from their datasheets. The two contributions of conduction and switching losses can be calculated starting from FET parameters provided in the datasheets, as illustrated below.

5.6 Conduction Losses

The first integral term in square brackets in Equation (5.7) provides the *conduction losses*. The integral argument of this term is limited to the power product $v_{ds}(t)i_{ds}(t)$ as $i_{gs}(t) = 0$ during the ON state conduction subinterval T_c. If $v_{gs}(t)$ is sufficiently high, the FET will be operating in the ohmic region, and the voltage $v_{ds}(t)$ can be approximately expressed as $v_{ds}(t) = R_{ds,on}i_{ds}(t)$, where $R_{ds,on}$ is the FET channel resistance. For some devices, if $v_{gs}(t)$ is sufficiently high and if the magnitude swing of current $i_o(t)$ during the subinterval T_c is sufficiently limited, then $R_{ds,on}$ can be considered constant and then conduction losses are expressed by the simplified Equation (5.8):

$$P_c = f_s \int_{t_{cs}}^{t_{ce}} v_{ds}(t)i_{ds}(t)\,dt = R_{ds,on} f_s \int_{t_{cs}}^{t_{ce}} i_{ds}^2(t)\,dt = R_{ds,on} I_{ds,rms}^2 \tag{5.8}$$

Equation (5.8) is used in application notes of FET manufacturers and power supplies design books, handbooks, and cookbooks. Figure 5.13 shows some families of curves describing the dependence of $R_{ds,on}$ as a function of i_{ds} parameterized with respect to gate-source (i.e., gate drive) voltage v_{gs}. From high to low v_{gs}, the function $R_{ds,on}(i_{ds,on}, v_{gs})$ of some devices is linear with respect to i_{ds} (Figure 5.13a), whereas for other devices it is always nonlinear (Figure 5.13b), and for others, it goes from linear to nonlinear (Figure 5.13c)

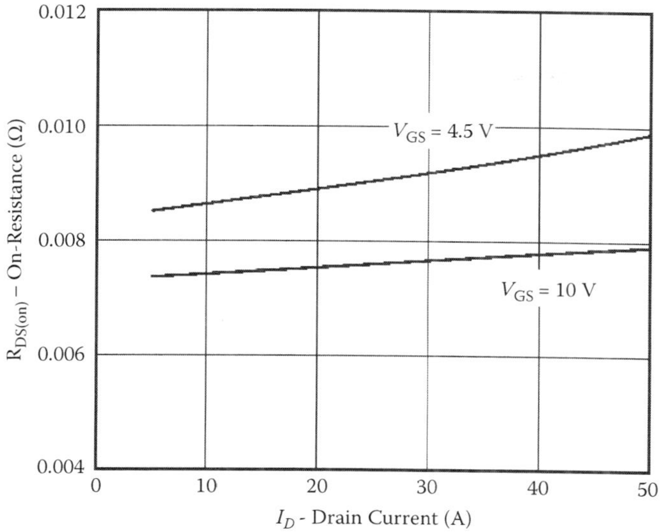

On-Resistance vs. Drain Current and Gate Voltage

(a)

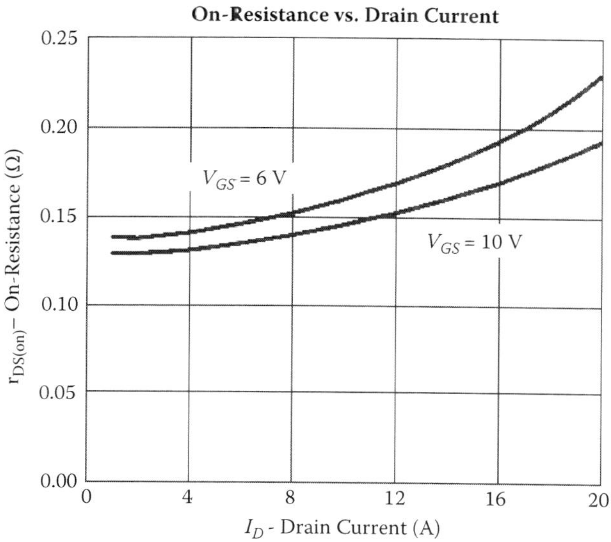

(b)

FIGURE 5.13
Dependency of channel resistance with respect to drain current. (a) Si4840BDY. (b) Si4982DY. (c) Si4892DY.

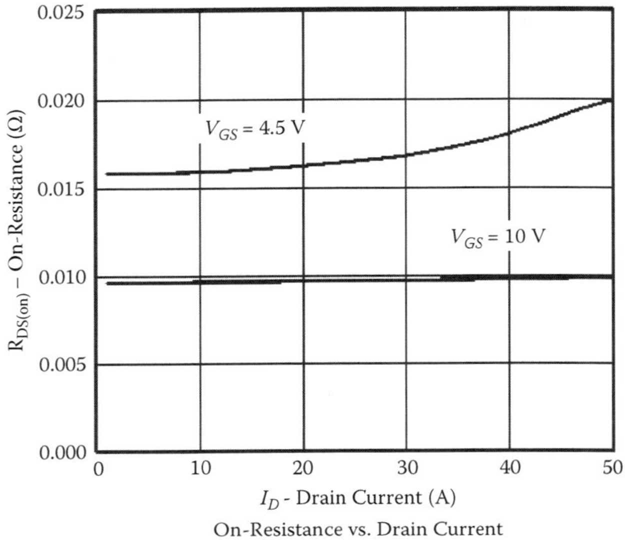

(c)

FIGURE 5.13 *(continued)*
Dependency of channel resistance with respect to drain current. (a) Si4840BDY. (b) Si4982DY. (c) Si4892DY.

while gate voltage decreases. For some devices a sufficiently high v_{gs} flattens the $R_{ds,on}$ with respect to i_{ds}, thus making Equation (5.8) valid. The simplified Equation (5.8) cannot be used for conduction loss calculation, either when v_{gs} is low or when the swing of current $i_o(t)$ during the subinterval T_c is wide. It should be noted that there are several situations where big current ripples are accepted to keep inductors small. Indeed, when current is high the switching frequency must be kept limited because of losses, and ensuring small current ripples would require big inductances.

It is quite difficult to find inductors with high inductance and high-current rating, unless big size (and cost) is accepted. Then the dependence of $R_{ds,on}$ with respect to i_{ds} must be considered for correct conduction loss evaluation. Using simple curve fitting tools, it is possible to obtain different analytical equations approximating the function $R_{ds,on}(i_{ds,on}, v_{gs})$. If N-th-order polynomials are adopted to fit the plots of Figure 5.13b, Equation (5.9) should be used for conduction loss calculation:

$$P_c = f_s \int_{t_{cs}}^{t_{ce}} R_{ds,on}(i_{ds}) i_{ds}^2(t)\,dt = f_s \int_{t_{cs}}^{t_{ce}} \left[\sum_{k=0}^{N} a_k i_{ds}^k(t) \right] i_{ds}^2(t)\,dt = f_s \int_{t_{cs}}^{t_{ce}} \left[\sum_{k=0}^{N} a_k i_{ds}^{k+2}(t) \right] dt \quad (5.9)$$

Normally, a low-order polynomial ($N = 2 \div 3$) is sufficient for good approximation. Exact calculation of conduction losses requires the knowledge of $R_{ds,on}$ polynomial coefficients a_k and of the waveform $i_{ds}(t)$. If the switching

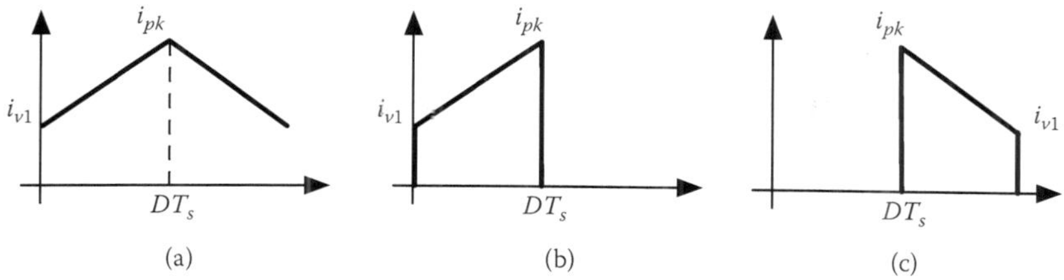

FIGURE 5.14
(a) $i_o(t)$. (b) $i_{Qc}(t)$. (c) $i_{Qs}(t)$.

cell of Figure 5.11a is working in a hard-switching PWM DC/DC converter, $I_{on}(t)$ looks like the typical triangular waveform of Figure 5.14a (e.g., inductor current in DC/DC converters), and the control and synchronous switch currents, $i_{Qc}(t)$ and $i_{Qs}(t)$, respectively, look like the waveforms of Figure 5.14b and Figure 5.14c, respectively.

Assuming $i_{Qc}(t) = i_{vl} + \alpha\, t$ (Figure 5.14b), where $\alpha = \Delta i_{opp}/t_{on}$, yields

$$P_c = f_s \int_{t_{cs}}^{t_{ce}} \left[\sum_{k=0}^{N} a_k (i_{vl} + \alpha t)^{k+2} \right] dt = f_s \left[\sum_{k=0}^{N} \frac{a_k \left((i_{vl} + \alpha t_{ce})^{k+3} - (i_{vl} + \alpha t_{cs})^{k+3} \right)}{(k+3)\alpha} \right] \tag{5.10}$$

A simplified form of Equation (5.10) can be obtained by using the average value of $R_{ds,on}$:

$$\begin{aligned} < R_{ds,on} > &= \frac{1}{\Delta i_{opp}} \int_{i_{ds,\min}}^{i_{ds,\max}} R_{ds,on}(i_{ds})\, di_{ds} = \frac{1}{\Delta i_{opp}} \int_{i_{ds,\min}}^{i_{ds,\max}} \left[\sum_{k=0}^{N} a_k i_{ds}^k \right] di_{ds} \\ &= \frac{1}{\Delta i_{opp}} \left[\sum_{k=0}^{N} \frac{a_k}{k+1} \left(i_{pk}^{k+1} - i_{vl}^{k+1} \right) \right] \end{aligned} \tag{5.11}$$

According to (5.11), an estimation of conduction losses is given by (5.12):

$$P_c = < R_{ds,on} > I_{ds,rms}^2 = \left\{ \begin{array}{ll} < R_{ds,onQc} > D I_{on} \left[1 + \dfrac{1}{12} \left(\dfrac{\Delta I_{on,pp}}{I_{on}} \right)^2 \right] & Q_c \\ < R_{ds,onQs} > (1-D) I_{on} \left[1 + \dfrac{1}{12} \left(\dfrac{\Delta I_{on,pp}}{I_{on}} \right)^2 \right] & Q_s \end{array} \right\} = K_c(\underline{tc}, \underline{p}, f_s, T_a) \tag{5.12}$$

which is sufficiently accurate for practical purposes. $\Delta I_{on,pp}$ represents the peak-to-peak ripple magnitude of the source current, whose average value is I_{on}. This current indeed represents the inductor current in DC/DC converters of Figure 5.14.

5.7 Switching Losses

Calculation is much more complicated for switching losses than for conduction losses. The reason is that the waveforms of voltages and currents $v_{ds}(t)$, $i_{ds}(t)$, $v_{gs}(t)$, and $i_{gs}(t)$ during the commutations must be determined taking into account the FET physical parameters influencing the transition from the interdiction region to the linear region passing through the saturation region. A detailed loss analysis is allowed by the knowledge of complete equations describing:

- The nonlinear dependence of the FET channel current on gate-to-source and drain-to-source voltages
- The nonlinear/parametric dependence of FET inter-electrode capacitances on drain-to-source voltage
- The nonlinear dependence of FET body diode current on drain-to-source voltage

Capacitances are fundamental FET parameters for switching loss calculation. The three physical capacitances C_{gs}, C_{gd}, and C_{ds} depend nonlinearly on the absolute value of the drain-to-source voltage $|V_{ds}|$. Their values can be reconstructed from the values of the inter-electrode input capacitance $C_{iss} = C_{gs} + C_{gd}$, reverse capacitance $C_{rss} = C_{gd}$, and output capacitance $C_{oss} = C_{ds} + C_{gd}$. The FET manufacturers provide the curves of the capacitances C_{iss}, C_{rss}, and C_{oss} as functions of $|V_{ds}|$. An example is shown in Figure 5.15, for the same FETs of Figure 5.13.

The three capacitances C_{gs}, C_{gd}, and C_{ds} will be subjected to currents whenever the terminal voltages V_{gs}, V_{gd}, and V_{ds} are subjected to changes. The main difference consists in the fact that C_{gs} and C_{gd} have to be considered as parametric linear capacitances, as their value does not depend directly on their own terminal voltages V_{gs} and V_{gd}, while C_{ds} has to be considered as nonlinear, as its value depends on its own terminal voltage V_{ds}. Accordingly, we have:

$$i_{Cgs} = \frac{dQ_{gs}}{dt} = v_{gs}\frac{dC_{gs}(|v_{ds}|)}{d|v_{ds}|}\frac{d|v_{ds}|}{dt}\mathrm{sign}(V_{ds}) + C_{gs}(|v_{ds}|)\frac{dv_{gs}}{dt} \tag{5.13}$$

$$i_{Cgd} = \frac{dQ_{gd}}{dt} = v_{gd}\frac{dC_{gd}(|v_{ds}|)}{d|v_{ds}|}\frac{d|v_{ds}|}{dt}\mathrm{sign}(V_{ds}) + C_{gd}(|v_{ds}|)\frac{dv_{gd}}{dt} \tag{5.14}$$

$$i_{Cds} = \frac{dQ_{ds}}{dt} = C_{ds}(|V_{ds}|)\frac{dv_{ds}}{dt} \tag{5.15}$$

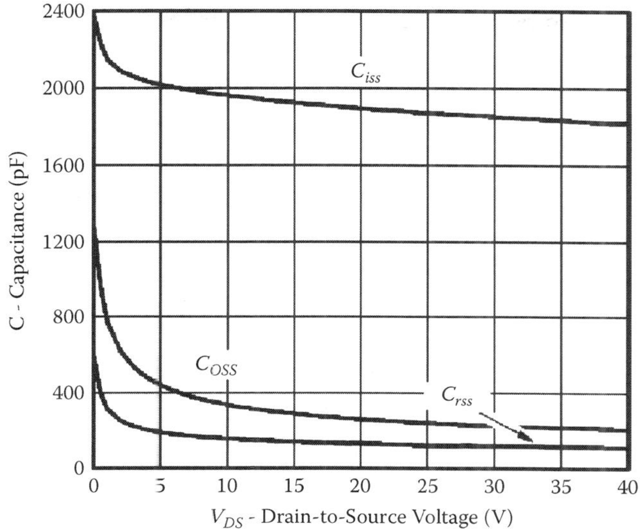

(a)

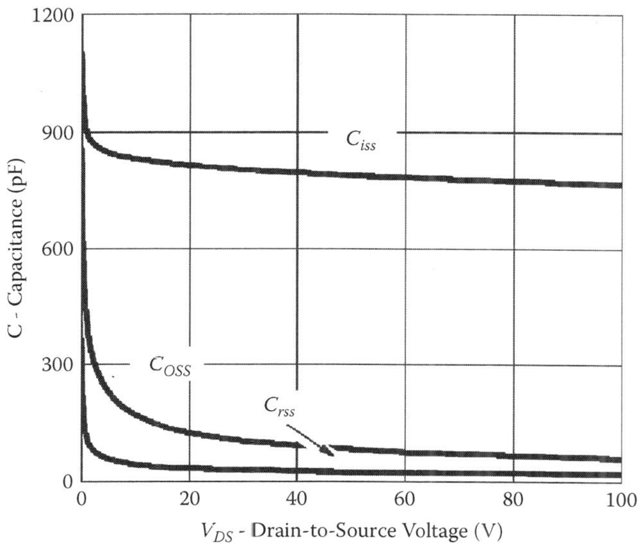

(b)

FIGURE 5.15
Inter-electrode FET capacitances. (a) Si4340BDY. (b) Si4982DY. (c) Si4892DY.

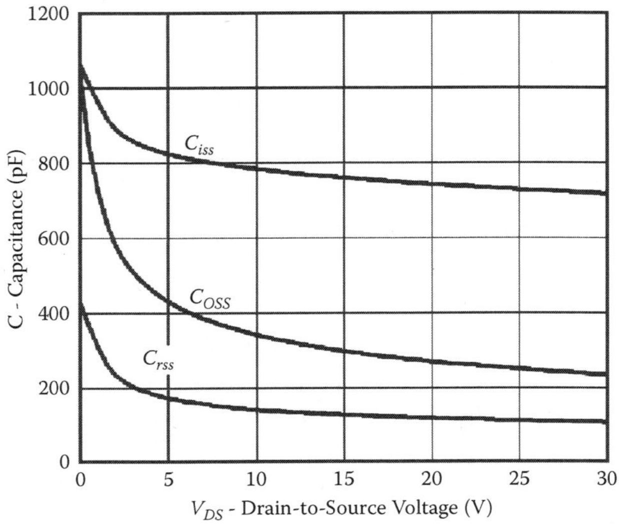

(c)

FIGURE 5.15 (*continued*)
Inter-electrode FET capacitances. (a) Si4840BDY. (b) Si4982DY. (c) Si4892DY.

Equations (5.13) to (5.15) highlight that currents into capacitances C_{gs} and C_{gd} present an additional term with respect to the current into capacitance C_{ds}, due to their parametric nature. This impacts the circulation of currents into the FET channel during the commutations. Accurate loss calculation using Equations (5.13) to (5.15) requires dedicated numerical simulation codes [20]. But correct switching loss prediction is very important in high-efficiency power supplies design, as they limit the maximum operating switching frequency and then the size of the power converter [1]. A reliable and quick prediction of switching losses is then necessary, based on physical parameters of commercial devices available in datasheets. FET modeling and switching loss calculations are treated in depth in the literature [2–5]. Many studies have been presented highlighting the key importance of proper modeling of nonlinear FET capacitance to get correct loss prediction, especially in hard-switching converters. Various approaches to FETs modeling and related parameter extraction have also been proposed [6–8]. Libraries with FET models at different levels of detail are used in circuit simulators [9]. However, when the problem is to design a power supply using existing commercial FETs, parameters provided in manufacturer datasheets only are available to characterize the devices. Consequently, switching power supplies designers use simplified models for switching loss calculation presented in the technical literature, including textbooks, papers, and application notes [10–15]. In this book we are mainly interested in models and methods for evaluation of power losses in high-efficiency converters for PV applications, using FET parameters provided in the component datasheets.

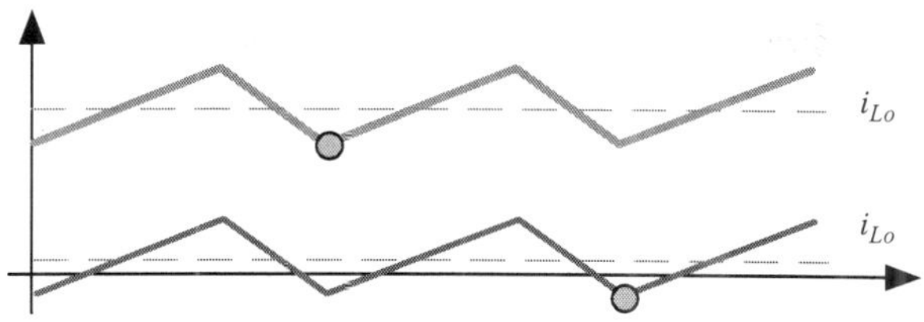

FIGURE 5.16
Top curve = high-current operation. Bottom curve = low-current operation.

The model discussed in this chapter is aimed at providing a practical methodology for quick evaluation of global energy productivity allowed by different topologies in PV applications.

Any commutation of the switching cell of Figure 5.11a involves a split of current (I_{on}) and voltage (V_{off}) between a FET channel and a body diode, and then between that body diode and the other FET channel. The use of synchronous rectification allows the operation of DC/DC converters always in continuous conduction mode (CCM) in PV applications, at whatever PV irradiation level. As a consequence, the body diode of either the C-FET or the S-FET is involved in the commutation, depending on the sign of the current I_{on} at the moment of the commutation. Figure 5.16 shows the typical waveform of the current I_{on} in DC/DC PWM converters, at high and low current, respectively.

The top one, corresponding to heavy-load/heavy-source operation, involves a positive current value at both turn ON and turn OFF of both C-FET and S-FET. In this case, both FET channels exchange the current during the commutations with the body diode of the S-FET. The bottom waveshape involves a negative value at turn ON of C-FET (corresponding to turn OFF of S-FET) and a positive value at turn OFF of C-FET (corresponding to turn ON of S-FET). In this case, each FET channel exchanges the current with its own body diode during the turn ON and with the body diode of the other FET during the turn OFF. This causes a different distribution of losses between C-FET and S-FET compared to heavy-source operation. In fact, whenever a FET exchanges the current during the commutation with its own body diode, it undergoes a soft commutation, whereas when it exchanges the current with the body diode of the other FET, it undergoes a hard commutation. Therefore when the converter operates at low-source current, switching losses will occur in both FETs. This must be taken into account for correct prediction of efficiency at low-source current. Indeed, commercial parts for S-FET functions are normally selected to limit conduction losses (thanks to lower channel resistance), and they will likely have higher switching losses than C-FET parts when operating in hard switching (due to higher capacitances). It should be noted that such change in switching losses distribution starts occurring at not very low-source current. For example, for a boost converter, if the inductor is designed with a typical 40% pk-pk inductor current ripple at maximum standard test conditions (STC), when the current is I_{max},

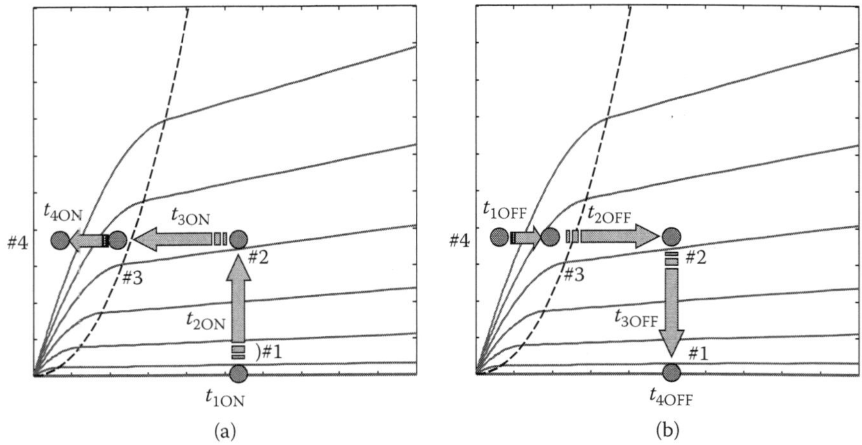

FIGURE 5.17
Paths of FET operating point during (a) turn ON and (b) turn OFF transitions.

the transition to the crossed hard-switching mode occurs below 20% of I_{max}. During a sunny day, a very small fraction of the energy is delivered by PV panels below 20% of I_{max} (about 2 ÷ 3%). However, during cloudy days, the equivalent irradiance conditions are prevalently around 20% of I_{max}.

In the following, the analysis will be focused on heavy-source operation, namely, on the calculation of switching losses for C-FET turn ON and turn OFF hard switching. The case of light-source operation, requiring the calculation of C-FET turn OFF and S-FET turn ON hard switching, can be analyzed using the same formulas basically replacing the C-FET parameters with the S-FET parameters.

Almost all models for switching loss analysis assume that the transition of the FETs from the interdiction, through the saturation to the linear region, and vice versa, is represented as a sequence of four different phases, each one taking a certain time. Figure 5.17 shows the path of the C-FET operating point during turn ON and turn OFF according to such a model.

The waveforms of drain-to-source current and voltage during these four intervals are often approximated as piece-wise linear functions, as shown in Figure 5.18.

Different assumptions characterize the various simplified approaches proposed in technical literature in describing the current rise and voltage fall times (t2,on and t3,on) in the turn ON transition and the voltage rise and current fall times (t2,off and t3,off) in the turn OFF transition [11–15], which also depend on the way the parameters are extracted from datasheets [10, 12–15], where they are provided for given reference conditions. Moreover, while some authors simply neglect switching losses of LS FET, others provide hints for their evaluation [11, 14], and while some authors ascribe output capacitance power loss to both HS and LS FETs, others take them into account just

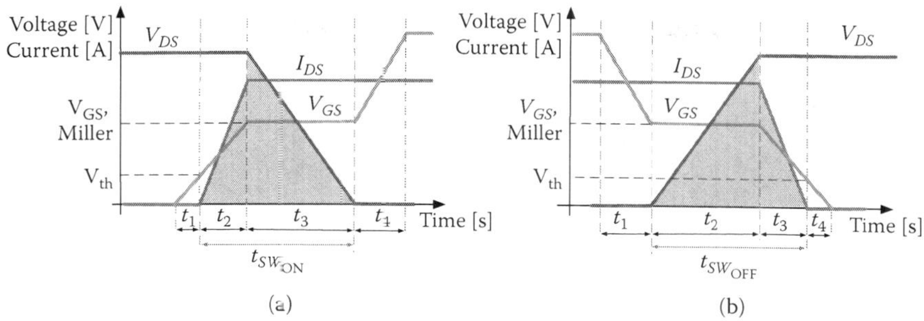

FIGURE 5.18
(a) C-FET turn ON. (b) C-FET turn OFF.

for HS FET [11, 14]. Simplified models normally analyze separately the losses of two FETs of the synchronous couple, whereas during the commutations the two devices strongly interact, because of the crossed circulation of the currents through the body diodes, the channels, and the capacitances [20]. Using the different simplified models and parameter approximations may lead to much different switching time intervals and loss estimation. The four intervals are briefly described in the next paragraphs.

5.7.1 Turn ON

5.7.1.1 Interval t_{1ON}, $0 < v_{gs} < V_{th}$

The gate-to-source voltage exponentially rises from 0 to V_{th}. The FET channel current i_{DS} keeps null, as the current $I_{on} - \Delta I_{on}$ keeps flowing through the diode. The voltage v_{ds} is then clamped at $V_{off} + V_{F,LS}$, where $V_{F,LS}$ is the forward voltage of S-FET body diode. In this interval the gate current i_g mostly charges the input capacitance C_{iss} of the FET, dominated by the gate-to-source capacitance C_{gs}. The operating point on the output characteristic of the FET stands at point 1 of Figure 5.17(a). The duration of this interval is given by Equation (5.16):

$$t_{1,on} = R_{g,on} C_{iss@V_{off}} \log\left[\frac{V_{dr}}{V_{dr} - v_{th}}\right] = R_{g,on} C_{iss@V_{off}} \Lambda_{1,on} \tag{5.16}$$

where V_{dr} is the gate driver voltage and $R_{g,on}$ is the total gate resistance at turn ON, including the pull-up driver resistance, the FET internal gate resistance, and eventual additional external resistance for gate current limiting.

5.7.1.2 Interval t_{2ON}, $0 < i_{DS} < I_{on}$

While the gate-to-source voltage continues rising exponentially from V_{th} to the Miller plateau voltage $V_{sp,on} = V_{th} + (I_{on} - \Delta I_{on})/g_{FS}$, the FET channel current rises from 0 to $I_{on} - \Delta I_{on}$ and the diode current falls from $I_{on} - \Delta I_{on}$ to 0. The voltage v_{ds} keeps clamped at $V_{off} + v_d$. The gate current i_g still mostly charges the input capacitance C_{iss}. The operating point on the output characteristic of the FET lifts up from point 1 to point 2 of Figure 5.17a. The duration of this interval is given by Equation (5.17):

$$t_{2,on} = R_{g,on} C_{iss@V_{off}} \log\left[\frac{V_{dr} - V_{th}}{V_{dr} - V_{sp,on}}\right] = R_{g,on} C_{iss@V_{off}} \Lambda_{2,on} \tag{5.17}$$

5.7.1.3 Interval t_{3ON}, $v_{gs} - V_{th} < v_{ds} < V_{off}$

After the diode turns OFF the FET drain-to-source voltage is unclamped and drops from V_{off} to $v_{gs} - V_{th}$ through the saturation region, Accordingly, the gate current i_g mostly flows through the reverse capacitance C_{rss}, due to its big voltage swing, which also causes an increase of capacitance (FET capacitances increase at low drain-to-source voltage). The gate-to-source voltage v_{gs} varies only slightly around the Miller plateau value $V_{sp,on}$. The FET channel current can peak much above $I_{on} - \Delta I_{on}$, due to the circulation of its own capacitive currents, of the capacitive currents of the low-side FET, and of the reverse recovery current of its body diode [20]. For simplicity, the contribution to losses caused by such current increase with respect to $I_{on} - \Delta I_{on}$ is usually calculated apart and the high-side channel current is assumed approximately constant. The operating point on the output characteristic of the FET drifts from point 2 to point 3 of Figure 5.17(a). The duration of this interval is given by Equation (5.18):

$$t_{3,on} \cong \frac{V_{off}}{V_{dr} - V_{sp,on}} R_{g,on} C_{rss@V_{off}} = R_{g,on} C_{rss@V_{off}} \Lambda_{3,on} \tag{5.18}$$

5.7.1.4 Interval t_{4ON}, $V_{sp,on} < v_{gs} < V_{dr}$

The gate-to-source voltage restarts rising exponentially from $V_{sp,on}$ up to gate drive voltage V_{dr}. The FET drain-to-source voltage falls below $v_{gs} - V_{th}$ and lands to $V_{on,min} = R_{ds,on}(I_{on} - \Delta I_{on}) = R_{ds,on} I_{on,min}$, and the drain-to-source current keeps almost constant at $I_{on} - \Delta I_{on}$. The gate current i_g finishes charging the input capacitance C_{iss}. The operating point on the output characteristic of the FET lands from point 3 to point 4 of Figure 5.17a. The duration of this interval is given by Equation (5.19):

$$t_{4,on} \cong R_{g,on} C_{iss@V_{on,\min}} \log\left[\frac{V_{dr} - V_{sp,on}}{0.01 V_{dr}}\right] = R_{g,on} C_{iss@V_{on,\min}} \Lambda_{4,on} \tag{5.19}$$

5.7.2 Turn OFF

5.7.2.1 Interval t_{1OFF}, $R_{ds,on}(I_{on} + \Delta I_{on}) < v_{ds} < v_{gs}\text{-}V_{th}$

The drain-to-source voltage rises from $V_{on} = V_{on,max} = R_{ds,on}(I_{on} + \Delta I_{on}) = R_{ds,on} I_{on,max}$ to $v_{gs}\text{-}V_{th}$, while the gate-to-source voltage drops exponentially from gate drive voltage V_{dr} to the Miller plateau value $V_{sp,off} = V_{th} + (I_{on} + \Delta I_{on})/g_{FS}$. The gate current i_g mostly discharges the input capacitance C_{iss}; the operating point on the output characteristic of the FET from point 4 takes off toward point 3 of Figure 5.17b. The duration of this interval is given by Equation (5.20):

$$t_{1,off} = R_{g,off} C_{iss@V_{on,max}} \log\left[\frac{V_{dr}}{V_{sp,off}}\right] = R_{g,off} C_{iss@V_{on,max}} \Lambda_{1,off} \tag{5.20}$$

where $R_{g,off}$ is the total gate resistance at turn OFF, including the pull-down driver resistance, the FET internal gate resistance, and eventual additional external resistance for gate current limiting.

5.7.2.2 Interval t_{2OFF}, $v_{gs}\text{-}V_{th} < v_{ds} < V_{off}$

The FET drain-to-source voltage rises from $v_{gs}\text{-}V_{th}$ to V_{off}. The S-FET body diode reverse voltage drops to zero. The gate current i_g mostly exits the gate-to-drain capacitance C_{gd}, which greatly decreases during this interval, thus causing the gate-to-source voltage v_{gs} to vary only slightly around the Miller plateau value $V_{sp,off}$. The operating point on the output characteristic of the FET drifts from point 3 to point 2 of Figure 5.17b. The duration of this interval is given by Equation (5.21):

$$t_{2,off} \cong \frac{V_{off}}{V_{sp,off}} R_{g,off} C_{rss@V_{off}} = R_{g,off} C_{rss@V_{off}} \Lambda_{2,off} \tag{5.21}$$

5.7.2.3 Interval t_{3OFF}, $V_{th} < v_{gs} < V_{sp,off}$

The gate-to-source voltage restarts dropping exponentially from the Miller plateau $V_{sp,off}$ to the threshold value V_{th}. The drain-to-source current falls from $(I_{on} + \Delta I_{on})$ to 0, while the diode current rises from 0 to $(I_{on} + \Delta I_{on})$. The drain-to-source voltage is clamped at $V_{off} + v_d$. The gate current i_g discharges the input capacitance C_{iss}. The operating point on the output characteristic of the FET lifts down from point 2 to point 1 of Figure 5.17b. The duration of this interval is given by Equation (5.22):

$$t_{3,off} = R_{g,off} C_{iss@V_{off}} \log\left[\frac{V_{sp,off}}{v_{th}}\right] = R_{g,off} C_{iss@V_{off}} \Lambda_{3,off} \tag{5.22}$$

5.7.2.4 Interval t_{4OFF}, $0 < v_{gs} < V_{th}$

The FET gate-to-source voltage falls exponentially from V_{th} to 0. The FET channel current i_{DS} is null. The current $(I_{on} + \Delta I_{on})$ flows through the diode. The voltage v_{ds} is clamped at $V_{off} + V_{F,S}$. The gate current i_g finishes discharging the input capacitance C_{iss}. The operating point on the output characteristic of the FET stands at point 1 of Figure 5.17b. The duration of this interval is given by Equation (5.23):

$$t_{4,off} = R_{g,off} C_{iss@V_{off}} \log\left[\frac{v_{th}}{0.01V_{dr}}\right] = R_{g,off} C_{iss@V_{off}} \Lambda_{4,off} \tag{5.23}$$

Using Equations (5.16) to (5.23) for switching times of Equation (5.7) and simplified current and voltage waveforms of Figure 5.18 provides the simplified formula (5.24) for switching loss calculation of C-FET:

$$\begin{aligned} P_{sw} &= 0.5(V_{off} + V_{L,S}) f_s \Big[I_{on,\min} R_{g,on} (C_{iss@V_{off}} \Lambda_{2,on} + C_{rss@V_{off}} \Lambda_{3,on}) \\ &\quad + I_{on,\max} R_{g,off} (C_{rss@V_{off}} \Lambda_{2,off} + C_{iss@V_{off}} \Lambda_{3,off}) \Big] \\ &\quad + f_s V_{dr}^2 C_{iss@V_{on,\min}} + f_s V_{dr}^2 C_{iss@V_{on,\max}} + 2 f_s V_{dr} V_{off} C_{rss@V_{off}} + f_s (V_{off} + V_{L,S}) C_{oss@V_{on}} \\ &= f_s K_s(\underline{tc}, \underline{p}, f_s, T_a) \end{aligned} \tag{5.24}$$

The value provided by Equation (5.24) must be interpreted as the global power loss determined by the two commutations of the C-FET over an entire switching period. It includes the losses caused by the C-FET in the part of the gate drive circuit external to the FET to charge and discharge the input capacitance C_{iss} and the loss determined by the charge and discharge of output capacitance C_{oss} in the external power circuit (typically these losses occur in the input or output filter capacitors of the power converter, depending on the topology). The loss formula (5.24) highlights the key role of the three FET capacitances. Regarding the values of these capacitances to be used in (5.24), accurate loss analysis requires to distinguish the value of C_{iss} to be used in Equations (5.16) and (5.17) (high $V_{ds} \rightarrow$ low C_{iss}) from the value to be used in Equation (5.19) (low $V_{ds} \rightarrow$ high C_{iss}), as well as to distinguish the value to be used in Equation (5.20) (high $V_{ds} \rightarrow$ low C_{iss}) from the value to be used in Equations (5.22) and (5.23) (low $V_{ds} \rightarrow$ high C_{iss}). In order to simplify the calculation of the loss contributions coming from charge and discharge of input and output capacitances C_{iss} and C_{oss}, the worst- case values of $C_{iss@Von,min}$, $C_{iss@Von,max}$, and $C_{oss@Von}$ have been used to obtain the simplified second, third, and fifth terms of (5.24), respectively. Unfortunately, the manufacturers either provide numerical values of the inter-electrode capacitances at a certain drain-to-source voltage rate (e.g., half the breakdown voltage of the device, where they are small) or do not provide numerical values at all. Thus the application of (5.24) requires the extraction of the values of inter-electrode capacitances from plots provided in the datasheets. It should be noted that, depending on the type of device, the values of inter-electrode capacitances at low voltage can be much higher than the values at high voltages (even five times bigger, for reverse and output capacitances). Some authors [10] suggest heuristic rules for the calculation of capacitances, starting from the numerical values provided in the numerical tables of the datasheets, based on the use of scaling factors. However, these formulas can lead to approximations that can still be not acceptable for loss estimation in very high-efficiency power converters dedicated to PV applications.

5.7.3 Thermal Analysis

An important element to take into account for correct calculation of conduction losses is the thermal dependence of $R_{ds,on}$ upon the junction temperature T_j. It must also be considered that the operating junction temperature of silicon devices directly influences the reliability of power converters [18]. In PV applications this is a main concern, especially for DMPPT solutions, which involve a much bigger number of parts, compared to FMPPT, where only a single centralized inverter is used. The value of the junction temperature T_j can be determined by means of an iterative resolution of Equations (5.25) to (5.27):

$$\underline{p} = \underline{p}(T_j) \tag{5.25}$$

$$P_d = P_d(\underline{tc}, \underline{p}, f_s, T_a) = K_c(\underline{tc}, \underline{p}, T_a) + f_s K_s(\underline{tc}, \underline{p}, T_a) \tag{5.26}$$

$$T_j = T_a + P_d R_{\vartheta ja} \tag{5.27}$$

where $R_{\vartheta ja}$ is the junction-to-ambient thermal resistance of the device, which depends on the type of package. Equations (5.25) to (5.27) highlight that conduction and switching losses depend on each other, as both influence the junction temperature, which in turn conditions the physical parameters. The two terms of Equation (5.26) are given by (5.12) and (5.24). Equation (5.25) provides the dependence of FET parameters upon the temperature. They are of the type shown in (5.28):

$$p_x = p_{x@25°C}(1 + \pi_x (T_j - T_a)) \tag{5.28}$$

where $p_{x@25°C}$ is the value of the parameter p_x at 25°C and π_x is the thermal coefficient of p_x. The main parameters of interest for the thermal model in Equations (5.25) to (5.27) are the channel resistance $R_{ds,on}$, the gate-to-source

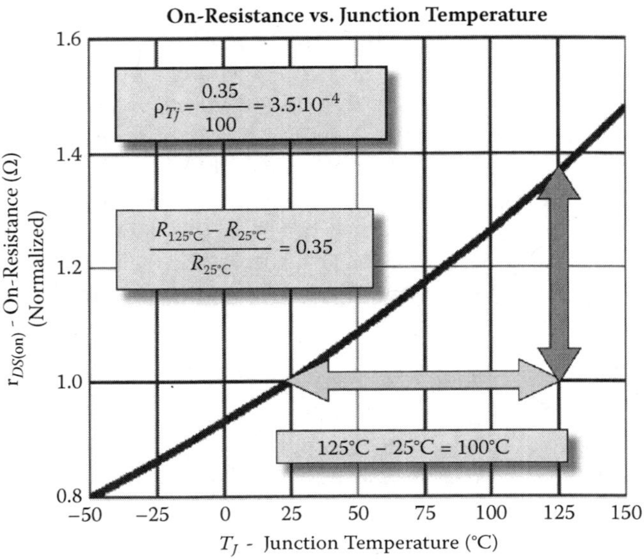

FIGURE 5.19
Normalized variation of channel resistance with respect to junction temperature.

threshold voltage V_{th}, and the transconductance g_{FS}. Both $p_{x@25°C}$ and π_x for these three parameters can be extracted from datasheets. Normally, $p_{x@25°C}$ is given in the main datasheet tables. The value of π_x, instead, has to be extracted from datasheet plots. Figure 5.19 shows a typical plot of the coefficient providing the channel resistance relative variation with respect to the junction temperature T_j, normalized to the 25°C value.

The thermal coefficient ρ_{Tj} appearing in the thermal equation for channel resistance (5.29)

$$R_{ds} = R_{ds@25°C}(1 + \rho_{T_j}(T_j - T_a)) \tag{5.29}$$

is then directly given by the ratio shown in the plot of Figure 5.19.

The calculation of the thermal coefficient ν_{Tj} appearing in the thermal equation for gate-to-source threshold voltage V_{th} (5.30),

$$V_{th} = V_{th@25°C}(1 + \nu_{T_j}(T_j - T_a)) \tag{5.30}$$

can be done using plots like the one shown in Figure 5.20. Such a plot normally provides the variance of V_{th} with respect to the temperature. So, the

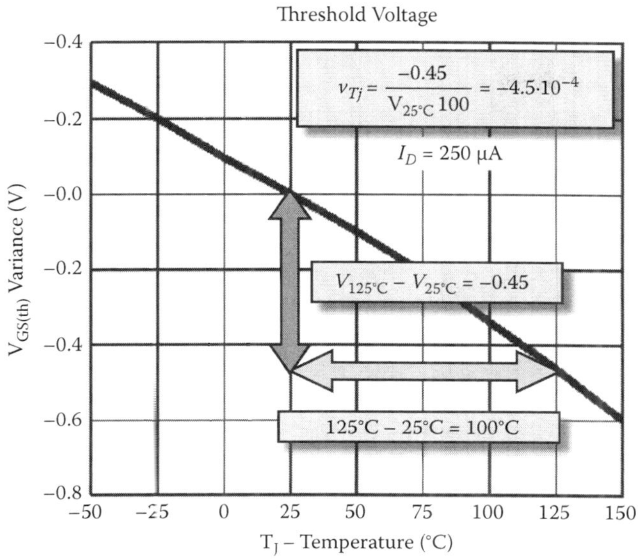

FIGURE 5.20
Variance of threshold voltage with respect to junction temperature.

variance over a 100°C temperature swing must be divided by the value of $V_{th@25°C}$ and 100 to get the thermal coefficient v_{Tj} of Equation (5.30).

Finally, the calculation of the thermal coefficient γ_{Tj} appearing in the thermal equation for transconductance g_{FS} (5.31),

$$g_{FS} = g_{FS@25°C}(1 + \gamma_{T_j}(T_j - T_a)) \tag{5.31}$$

can be done using plots like the one shown in Figure 5.21.

These plots normally provide the forward drain-to-source current I_{DS} as a function of gate-to-source voltage V_{gs} in the saturation region. Such a curve can be approximated by its tangent at the two temperatures of 25 and 125°C. The ratios between the current and voltage variations of each tangent provide the values of the transconductance for each temperature. From these values, the thermal coefficient γ_{Tj} can be easily calculated, as shown in Figure 5.21. If the curve of g_{FS} as a function of I_{DS} for different temperatures is available, the calculation of γ_{Tj} is simpler.

Equations (5.25) to (5.27) can be used to analyze the FET losses and converter energy efficiency over an entire daily operating cycle. In case the switching frequency and thermal resistance of devices are given, the system of Equations (5.25) to (5.27) must be iteratively solved, starting from a guess value of the temperature, for each irradiance level:

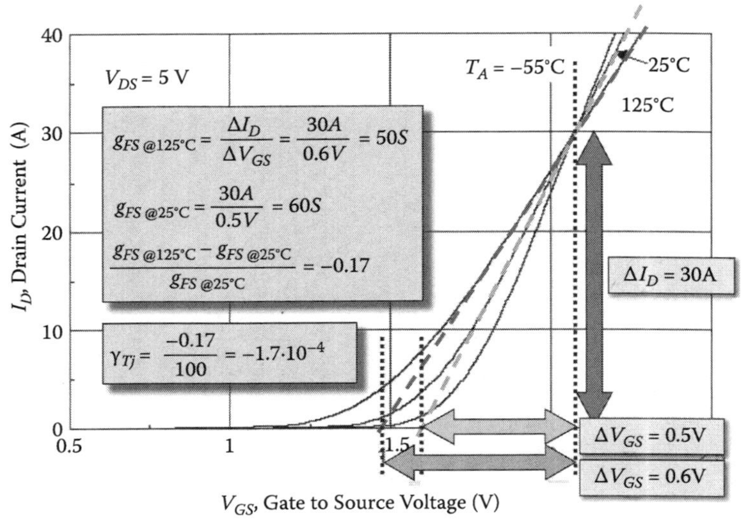

FIGURE 5.21
Drain current vs. gate-to-source voltage.

```
while |Tk-Tk-1|>TOL
      k = k+1
      calculate the parameters at Tj = Tk-1 [Equation (5.25)]
      calculate the total conduction and switching losses
      [Equation (5.26)]
      calculate the new value of temperature Tk [Equation
      (5.27)]
      end
```

where TOL is the required tolerance (1°C is mostly adequate). At the end of the loop the temperature and losses are obtained. If either the temperature is higher than the maximum allowed value T_{max} or the losses exceed the budget descending from the required efficiency, three possible actions are allowed. The first is to select different devices. The directions for such action are dictated by the specific loss analysis: If there is some evident heavy imbalance in one of the two FETs between conduction and switching losses and one of them is clearly responsible for loss excess, a device with either smaller channel resistance or smaller capacitances (i.e., smaller gate charge Q_g) must be selected, depending on which losses exceed the limit. It should be noted that equalizing the conduction and switching losses is often considered the best trade-off solution; however, losses balancing really produces minimization of total losses only in some cases. In most applications, depending on operating voltages, currents, and switching frequency and on the converter topology, conduction, and switching losses, ensuring minimization of their sum can be quite unbalanced. Unfortunately direct formulas to identify the correct loss distribution cannot be derived. So, although starting with equally distributed conduction and switching losses can help to perform a preselection of devices potentially suitable for the application, it is not convenient to rigidly polarize the search of devices subjected to the loss balancing constraints, as this can lead to solutions involving either total loss increase or the exclusion of devices with unbalanced but lower total losses. The best practical way to select the devices consists in selecting a family of feasible FETs, based on required voltage and current ratings and package, and then calculating the total losses and junction temperature for all of them, for both the C-FET and the S-FET function, based on datasheet parameters and on Equations (5.25) to (5.27), over a daily operation cycle. The direct comparison of FET combinations provides the best solution.

It may happen, and indeed it often happens in high-efficiency power supplies design, that no feasible solutions are found for a certain switching frequency or a certain package (thermal resistance). In this case, the model in Equations (5.25) to (5.27) can be used to calculate either the maximum switching frequency or the maximum thermal resistance allowing the temperature and loss limits compliance. In both cases the solution can be found in a straight way, by solving Equations (5.25) to (5.27) with respect to the required variable. If the problem is overtemperature, then given the desired maximum junction temperature T_{jmax} and the switching frequency f_s, Equation (5.25) provides the values of the parameters $\underline{p}(T_{jmax})$, Equation (5.26) provides the losses $P_d(T_{jmax}, f_s)$, and Equation (5.27) provides the maximum allowed thermal resistance $R_{\theta ja,max}$, given by (5.32):

$$R_{\vartheta ja,\max} = \frac{T_{j\max} - T_a}{P_d(T_{j\max}, f_s)} \tag{5.32}$$

A heat sink is required to lower the thermal resistance to (5.32). As an alternative, given the desired maximum junction temperature T_{jmax} and the thermal resistance of the device, Equation (5.25) provides the values of the parameters $\underline{p}(T_{jmax})$, Equation (5.27) provides the losses $P_d(T_{jmax}, R_{\theta ja})$, and Equation (5.26) provides the maximum allowed switching frequency f_{smax}, given by (5.33):

$$f_{s\max} = \frac{P_d(T_{j\max}, R_{\vartheta ja}) - K_c(\underline{tc}, \underline{p}, T_a)}{K_s(\underline{tc}, \underline{p}, T_a)} \tag{5.33}$$

Similar calculations can be done starting from the maximum allowed power dissipation of the device P_{dmax}, in case there are no feasible solutions because of excessive power dissipation.

When affording thermal analysis, and consequent device selection and operating conditions setting, it must be kept in mind that successful design of DC/DC power converters for DMPPT applications imposes a critical trade-off among cost, reliability, and energy efficiency.

5.7.4 Example

In the example of boost and buckboost converters for DC/DC DMPPT application with two 220 W panels illustrated in Section 5.3, we found that while buckboost topology imposes about 90 V maximum voltage stress to FETs, the boost topology imposes 80 V maximum voltage stress. The rms current stresses can be calculated starting from the average inductor currents I_{on} flowing through FETs when they are in the ON state using Equations (5.34) and (5.35):

$$I_{rms,C} = \sqrt{\frac{1}{T_s}\int_0^{T_s} i_C^2(t)\,dt} = I_{on}\sqrt{D} \tag{5.34}$$

$$I_{rms,S} = \sqrt{\frac{1}{T_s}\int_0^{T_s} i_S^2(t)\,dt} = I_{on}\sqrt{1-D} \tag{5.35}$$

where

$$I_{on} = \begin{cases} I_{MPP} & boost \\ \dfrac{I_{MPP}}{D} & buck/boost \end{cases} \tag{5.36}$$

Equations (5.34) and (5.35) allow the calculation of the nominal maximum current stress for C-FET and S-FET in boost and buckboost topologies. In STC the PV panel data are I_{MPP} = 7.63 A and V_{MPP} = 29.5 V. If both panels are uniformly and equally irradiated, then the output voltage of both converters in series is 40 V. This yields the values {D,Irms,C,Irms,S} at STC equal to {0.38, 4.7, 6.0} and {0.62, 9.7, 7.6}, respectively for boost and buckboost.

If 96% minimum efficiency is required at STC, the total allowed loss budget for the entire converter will be about 9 W. As a starting guess for preliminary FET selection, let us suppose that 50% of such 9 W loss budget is dissipated by FETs, and that such loss is equally distributed between C-FET and S-FET and between conduction and switching losses. Then a total 1.12 W conduction loss is allowed for each FET. Let us then

TABLE 5.3

MOSFETs Conduction Losses

	V_{DS} (V)	100V Part	R_{DS} @10V (mΩ)	I_D @ 100°C (A)	Q_g @ 10V (nC)	R_{thJC} (°C/W)	C-FET Boost Pc @ STC (W)	S-FET Boost Pc @ STC (W)	C-FET Buckboost Pc @ STC (W)	S-FET Buckboost Pc at STC (W)
1	100	IRLR3110Z	14	45	34.0	1.05	0.31	0.51	1.31	0.81
2	100	IRFR3710Z	18	39.00	69.0	1.05	0.40	0.65	1.69	1.04
3	100	IRFR540Z	29	25.00	39.0	1.64	0.63	1.03	2.68	1.64
4	100	IRFR3410	39	22	37.0	1.40	0.86	1.41	3.66	2.24
5	100	IRFR3411	44	23	48.0	1.20	0.97	1.59	4.13	2.53
6	100	IRLR3410	105	9.5	22.7	2.40	2.32	3.79	9.86	6.04
7	100	IRFR3910	115	9.5	29.3	2.40	2.54	4.15	10.80	6.62
8	100	IRLR120N	185	6.9	13.3	3.20	4.09	6.68	17.37	10.65
9	100	IRFR120Z	190	6.1	6.9	4.28	4.20	6.86	17.84	10.93
10	100	IRFR120N	210	5.8	16.7	3.20	4.65	7.58	19.72	12.09
11	150	IRFR4615	42	24.0	26	1.05	0.93	1.52	3.94	2.42
12	150	IRFR24N15D	95	17.0	30	1.10	2.10	3.43	8.92	5.47
13	150	IRFR18N15D	125	13.0	28	1.40	2.77	4.51	11.74	7.19
14	150	IRFR13N15D	180	9.8	19	1.75	3.98	6.50	16.90	10.36

assume that a D-PAK package is adopted to facilitate the heat transfer. Table 5.3 shows the main parameters of the commercial D-PAK *N*-channel 100 and 150 V FETs of Figures 5.9 and 5.10, whose current ratings comply with both application stresses, together with the conduction losses evaluated at STC.

The results show that only high-current-rating devices can be adopted to comply with expected conduction loss limitation. Some devices slightly exceed conduction loss limit. They cannot be excluded from the analysis, as their switching losses might be lower than the supposed budget. The global loss analysis, including thermal modeling, is required to assess feasible FET selections. Therefore FETs 1 to 5 and 11 pass the preliminary selection. Let us then analyze the switching losses of these FETs at STC. Let us assume that the gate driver resistances are $R_{gdron} = R_{gdroff} = 3\ \Omega$. The selection of gate driver is out of the goal of this example. Briefly, the gate driver should be selected, as all the other parts, to ensure the best trade-off among cost, efficiency, and reliability. High-current drivers have smaller source and sink resistances, which ensure shorter switching times and lower switching losses; however, they are more expensive. Let the driver voltage be $V_{dr} = 10$ V. Table 5.4 lists the values of parameters of the selected FETs extracted from manufacturer datasheets.

The value of V_{off} considered to extract the minimum capacitances of Table 5.4 is the highest one for which the curves of capacitances are plotted in the datasheets. As for boost and buckboost the operating V_{off} values at STC are different, namely, 40 V vs. 69.5 V; the minimum capacitances could be slightly refined by taking the values at the specific V_{off} rated voltage for each converter. The maximum capacitances are associated to zero voltage. Also in this case, a slight refinement could be achieved by considering that the V_{on} voltage is determined by the forward voltage of the device in the ON state, which depends on the I_{on} current and channel resistance. The real accuracy improvement achievable by both refinements is questionable: Indeed, the approximated model described in Section 5.5 allows just an evaluation, not an exact calculation, of switching losses, which is aimed at supporting a high-efficiency design feasibility investigation and a comparison among topologies in terms of stresses vs. losses.

Internal gate resistance is often not available for many FETs. Also for this parameter a dummy, yet realistic, value $R_{gint} = 3\ \Omega$ is adopted in this example. Based on (5.12) and (5.24), the total power dissipations for C-FET and S-FET are given by (5.37):

$$\begin{aligned} P_{C-FET} &= K_c^{C-FET} + K_s^{C-FET} f_s \\ P_{S-FET} &= K_c^{S-FET} \end{aligned} \tag{5.37}$$

Table 5.5 shows the values of coefficients $K_c^{C\text{-}FET}$, $K_s^{C\text{-}FET}$, and $K_c^{S\text{-}FET}$ for the preselected FETs, calculated according to Equations (5.12), (5.17), (5.18), (5.21), (5.22), (5.24), (5.34), (5.35), and (5.36).

TABLE 5.4

FET Parameters for Switching Loss Calculation

Part	C_{iss} @ 0 V (pF)	C_{iss} @ V_{off} (pF)	C_{oss} @ 0 V (pF)	C_{oss} @ V_{off} (pF)	C_{rss} @ 0 V (pF)	C_{rss} @ V_{off} (pF)	g_{FS} (S)	Vth_{min} (V)	Vth_{max} (V)
IRLR3110Z	4500	4000	1500	200	600	90	52	1	2.5
IRFR3710Z	3200	2800	1000	200	600	120	39	2	4
IRFR540Z	1900	1600	700	200	350	150	28	2	4
IRFR3410	2100	1700	1300	120	650	50	33	2	4
IRFR3411	2600	1900	1750	250	750	25	21	2	4
IRFR4615	2000	1800	1100	100	300	30	35	3	5

TABLE 5.5

Part	Boost			Buckboost		
	C-FET K_s	C-FET K_c	S-FET K_c	C-FET K_s	C-FET K_c	S-FET K_c
IRLR3110Z	8.540E-06	0.310	0.505	2.052E-05	1.315	0.806
IRFR3710Z	7.405E-06	0.398	0.650	1.798E-05	1.690	1.036
IRFR540Z	8.541E-06	0.630	1.029	2.129E-05	2.676	1.640
IRFR3410	3.324E-06	0.863	1.408	7.860E-06	3.662	2.244
IRFR3411	2.196E-06	0.973	1.588	4.842E-06	4.132	2.532
IRFR4615	2.000E-06	0.929	1.516	4.476E-06	3.944	2.417

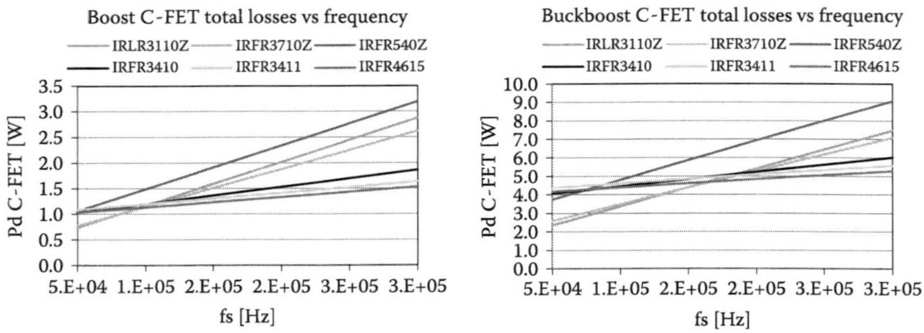

FIGURE 5.22 (See color insert)
C-FET losses vs. switching frequency.

The plots of Figure 5.22 show the total losses vs. the switching frequency for the six preselected FETs working as C-FET in boost and buckboost.

The plots show the much bigger losses involved by buckboost topology, requiring a more severe limitation of the switching frequency. The plots also show that high-current FETs are not suitable because of bigger capacitances (especially C_{rss}), causing much higher switching losses. Then a further FET selection can be done. The 150 V IRFR4615 FET should adopted for safe operation in buckboost topology, due to the higher voltage stress reaching about 90 V in mismatched conditions. Either the 150 V FET IRFR4615 or one of the two 100 V FETs IRLR3410 and IRFR3411 can be used for C-FET in the boost application, where maximum voltage stress in mismatched conditions is limited to 80 V. If 100 V FETs are used, a tight control of maximum voltage limiting is anyway required to guarantee a safe operation. Let us assume that the 150 V FET IRFR4615 is selected as C-FET ($P_c + P_{sw} = 4.4$ W at STC) and S-FET ($P_c = 2.4$ W at STC) for buckboost topology, whereas for boost topology IRFR3411 is selected as C-FET ($P_c + P_{sw} = 1.2$ W at STC) and IRLR3110Z as S-FET ($P_c = 0.5$ W at STC) to guarantee minimum losses. A 100 kHz switching frequency is adopted for both converters. The design can be completed now with the selection of the other power components. Regarding the inductors, the selection is particularly difficult in such an application, as high-current high-inductance parts are required. Indeed, if a 40% pk-pk maximum inductor ripple current is required at STC, based on duty cycle values of Table 5.3 and on (5.36), the inductances required for both converters are 37 μH, with about 9 A peak current for boost inductor and 15 A for buckboost inductor. Few commercial parts comply with saturation currents higher than these peak values and inductance close to 37 μH. A custom inductor can be the best solution to optimize size, cost, and losses. The design of power inductors is out of the topic of this book: Deep discussion of related issues can be found in several textbooks [16]. The fundamental issue is that it is difficult to keep the inductor size small, and then its cost, if power loss is too limited. The above FET selection leads to a residual loss budget at STC of 9 – 6.8 W = 2.2 W for buckboost and 9 – 2.9 W = 6.1 W for boost. Assuming that 1 W dissipation is reserved to all the parts other than the inductor, the residual loss budgets for the inductor will be 1.2 and 5.1 W, respectively, for buckboost and boost: As the size is roughly inversely proportional to the allowed power losses, such an imbalance between the two topologies would result in a buckboost inductor that is expected to be about four times bigger than the boost inductor. A possible alternative solution for the buckboost consists in using two FETs in parallel for C-FET and S-FET. Calculation of losses in this case leads to total losses of about $P_c + P_{sw} = 2.5$ W at STC for the two parallel C-FETs and $P_c = 1.2$ W at STC for the two parallel S-FETs. The residual loss budget for buckboost at STC then becomes 9 – 3.7 W = 5.3 W, 4.3 W of which can be reserved to the inductor, whose design leads to smaller size. The decision of doubling the number of buckboost FETs to reduce losses should be based on the evaluation of the real global benefits in terms of cost, size, and efficiency; once the use of four FETs is admitted, the four-switch bridge buckboost topology should also be considered as a further possible solution to facilitate loss management, thanks to its lower voltage stress. Such a solution is adopted in some commercial solutions for DMPPT applications. In spite of potential loss decrease, some singularity occurs in such a topology at 50% duty cycle, which may cause some decay of energy efficiency performances. In the following, the two buckboost solutions with single and double FETs will be compared with the boost one. The inductors for these three cases can be easily designed applying the method based on the geometric constant K_g illustrated in [16] and using ferrite cores to optimize the inductor design for 100 kHz operation. The inductors must be designed in all cases in such a way that they dissipate their nominal loss budget at STC and that their core

TABLE 5.6

Inductor Design Parameters

Loss Budget	Boost	Buckboost	
		Single FET	Parallel FETs
P_L at STC	5.1 W	1.2 W	4.3 W
I_{rmsL} at STC	7.7 A	12.4 A	12.4 A
R_W	86mΩ	8mΩ	28mΩ
Core	EE22	EE40	EE30
A_c (cm²)	0.410	1.27	1.09
W_a (cm²)	0.196	1.1	0.476
MLT (cm)	0.0804	1.40	0.519
L_m (cm)	3,99	8,5	6,6
Vol. (cm³)	1.62	9.78	6.29
Turns	33	17	20
AWG #	32	32	32
Nlitz	10	29	7

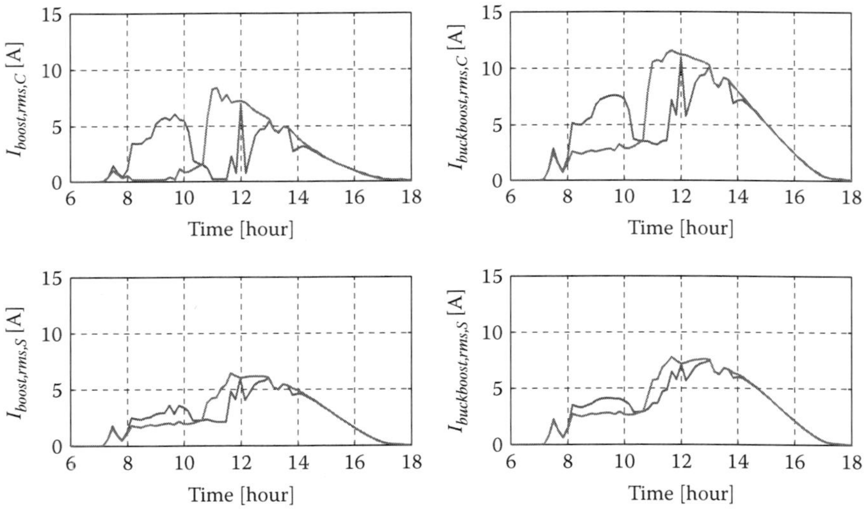

FIGURE 5.23 (See color insert)
Rms currents for C-FET and S-FET for boost and buckboost topologies: red line = east side PV panel, blue line = west side PV panel.

losses are small compared to conduction losses (about 1/10). The maximum allowed equivalent winding resistances are summarized in Table 5.6: These equivalent resistances include the core losses. They will be used to estimate the inductor losses over the daily operation. Using P-material magnetic cores the results shown in Table 5.6 are obtained for inductor design.

In these conditions let us now analyze the total FET energy losses over the shadowed day operation discussed in Section 5.3. The current I_{on} varies during the day, as shown in Figures 5.7c and 5.8c for boost and buckboost topologies. Based on (5.34) to (5.37), the rms current stresses for C-FET and S-FET over the entire day shown in Figure 5.23 are then obtained.

The plots of Figure 5.24 show the time diagram of losses for the three converters over the entire day. The plots clearly show that the double FET buckboost solution ensures a better limitation of global C-FET and S-FET losses. However, such FET loss reduction has been exploited to reduce the size of the inductor. The consequence is that in those days where the sun irradiation value is particularly high, like the one under analysis, the inductor dissipates more than what we save with FETs.

Figure 5.25 shows the inductor losses for west side and east side panels. Integrating the total loss of each converter provides the total FET energy loss over the day, as shown in Table 5.7. The results show that in the mismatch case under study, where there are only relatively short time intervals where the boost operation is disabled because of an unfeasible conversion ratio, the benefits of buckboost topology are not fully exploitable, because of higher losses. About 3.7% more energy is produced by the single FET buckboost

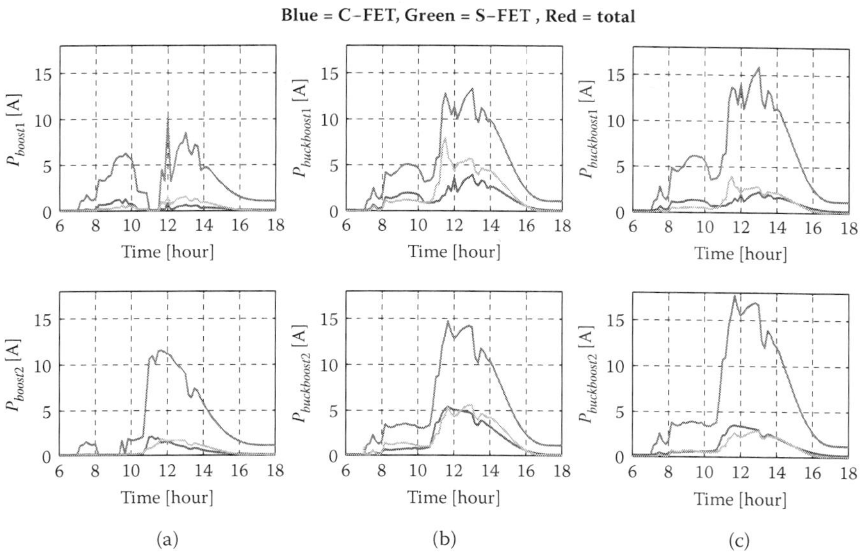

FIGURE 5.24 (See color insert)
Power dissipation for boost and buckboost topologies. (a) Boost. (b) Buckboost single FET. (c) Buckboost double FET.

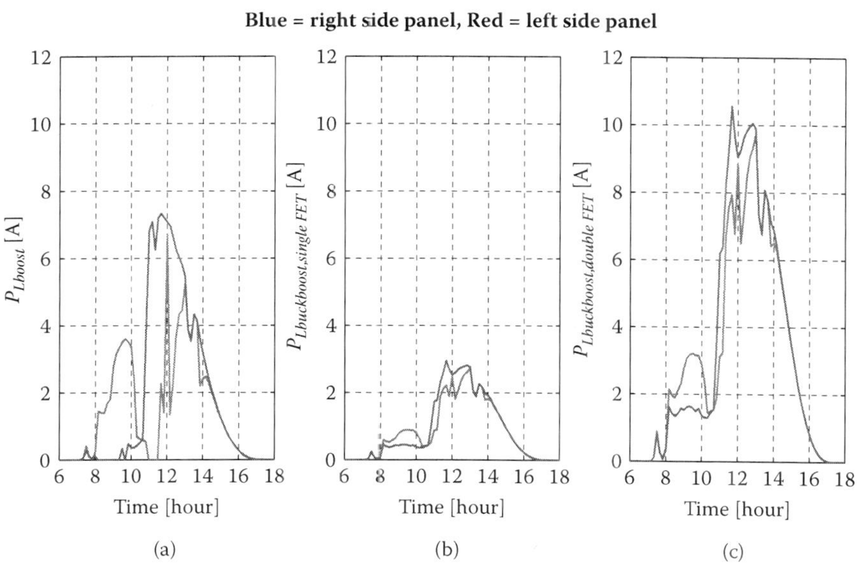

FIGURE 5.25 (See color insert)
Inductor power dissipation for boost and buckboost topologies: red line = east side PV panel, blue line = west side PV panel. (a) Boost. (b) Buckboost single FET. (c) Buckboost double FET.

TABLE 5.7

Thin Shadow Case, Conversion Efficiency, and Energy Productivity

PV Panel	Boost				Buckboost Single FET $U_{buckboost}/U_{boost} = 1.037$				Buckboost Double FET $U_{buckboost}/U_{boost} = 1.025$			
	U_{pan} (Wh)	U_d (Wh)	U_{out} (Wh)	U_{out}/U_{pan} (%)	U_{pan} (Wh)	U_d (Wh)	U_{out} (Wh)	U_{out}/U_{pan} (%)	U_{pan} (Wh)	U_d (Wh)	U_{out} (Wh)	U_{out}/U_{max} (%)
Left	803	37	767	95.4	846	62	785	92.8	846	71	775	91.6
Right	1038	42	995	95.9	1107	64	1043	94.2	1107	75	1032	93.2
Total	1841	79	1762	95.7	1953	126	1828	93.6	1953	146	1807	92.5

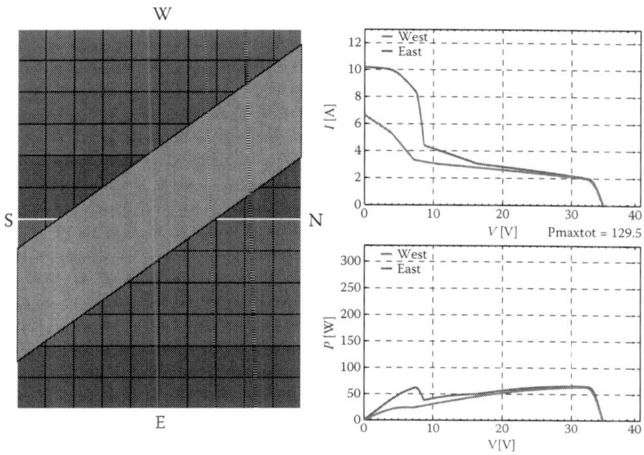

FIGURE 5.26 (See color insert)
PV panels subjected to a shadow twice larger than the one of Figure 5.5.

solution. The double FET buckboost solution, due to high inductor losses, shows only a 2.5% increase of energy productivity compared to boost topology. Let us analyze the same two panels in the presence of the bigger shadow, shown in Figure 5.26, sweeping over their surface during the same day.

Figures 5.27 to 5.32 show the relevant plots for this case. Integrating the total loss of each converter provides the total FET energy loss over the day, as shown in Table 5.8. The results show that when the mismatch conditions

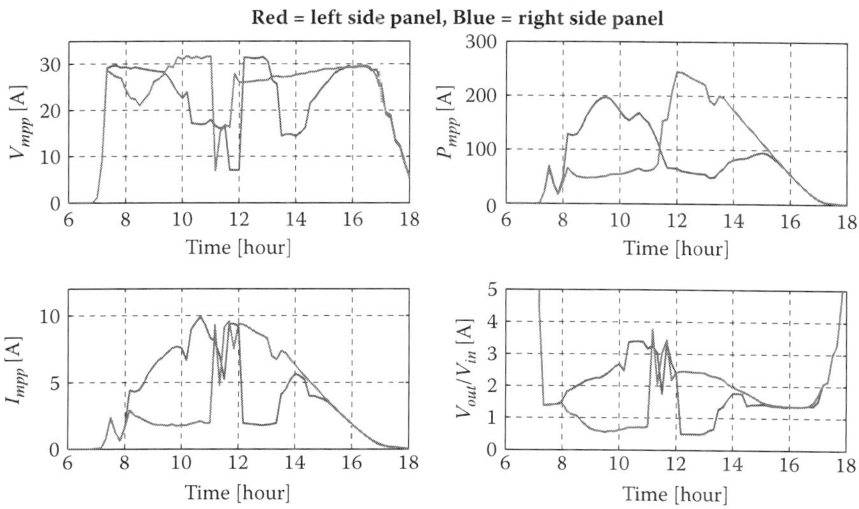

FIGURE 5.27
MPP voltage, current, and power and input-output voltage conversion ratio: red line = east side PV panel, blue line = west side PV panel.

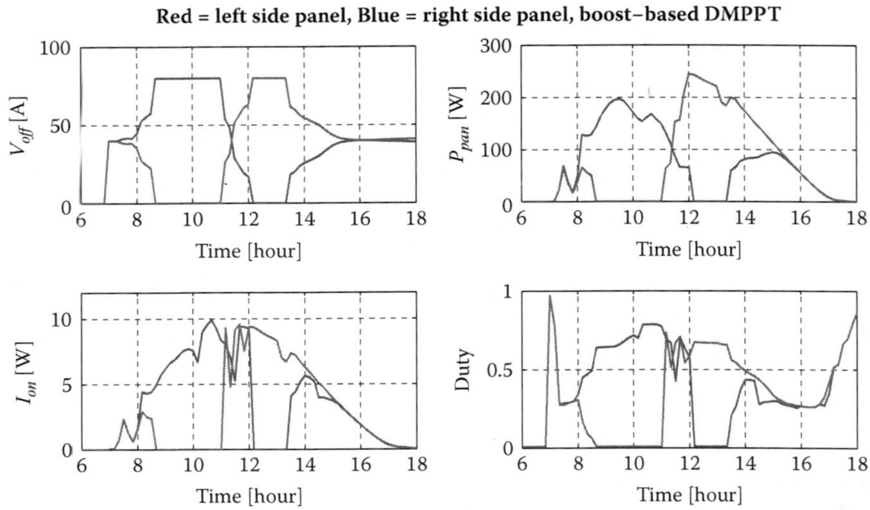

FIGURE 5.28 (See color insert)
Voltage and current stresses, power extracted from the panels, and converter duty cycle for boost topology: red line = east side PV panel, blue line = west side PV panel.

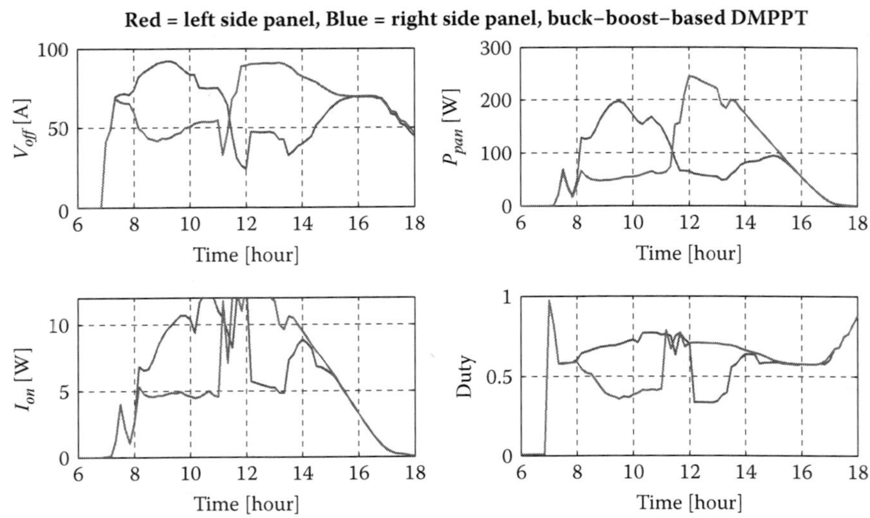

FIGURE 5.29 (See color insert)
Voltage and current stresses, power extracted from the panels, and converter duty cycle for buckboost topology: red line = east side PV panel, blue line = west side PV panel.

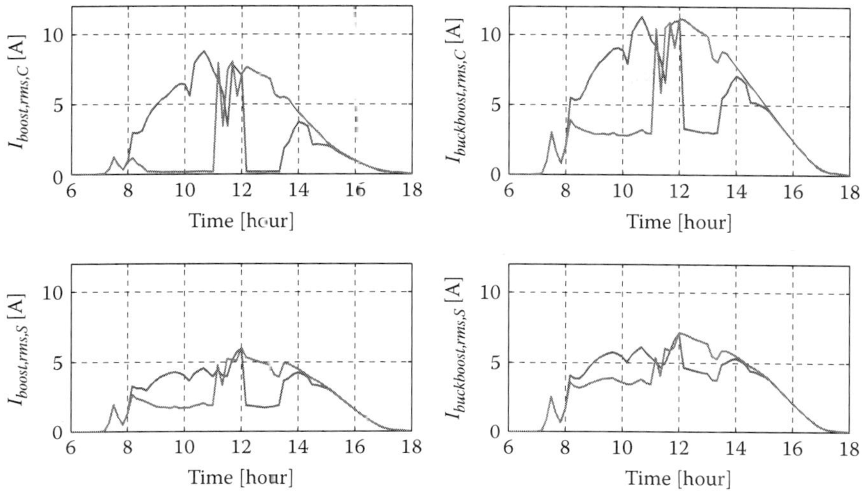

FIGURE 5.30 (See color insert)
Rms currents for C-FET and S-FET for boost and buckboost topologies: red line = east side PV panel, blue line = west side PV panel.

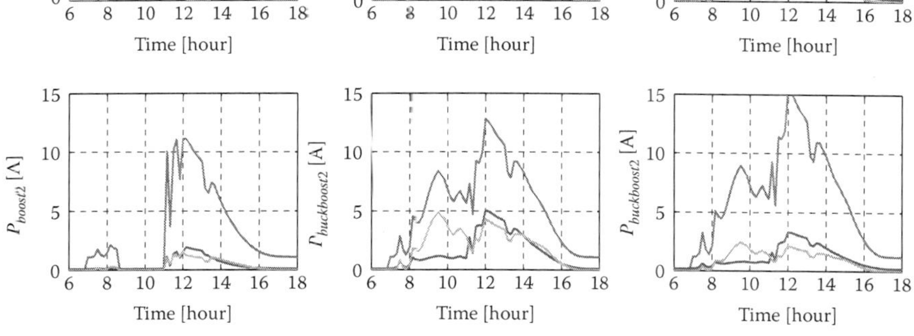

FIGURE 5.31 (See color insert)
Power dissipation for boost and buckboost topologies. (a) Boost, (b) Buckboost single FET, (c) Buckboost double FET.

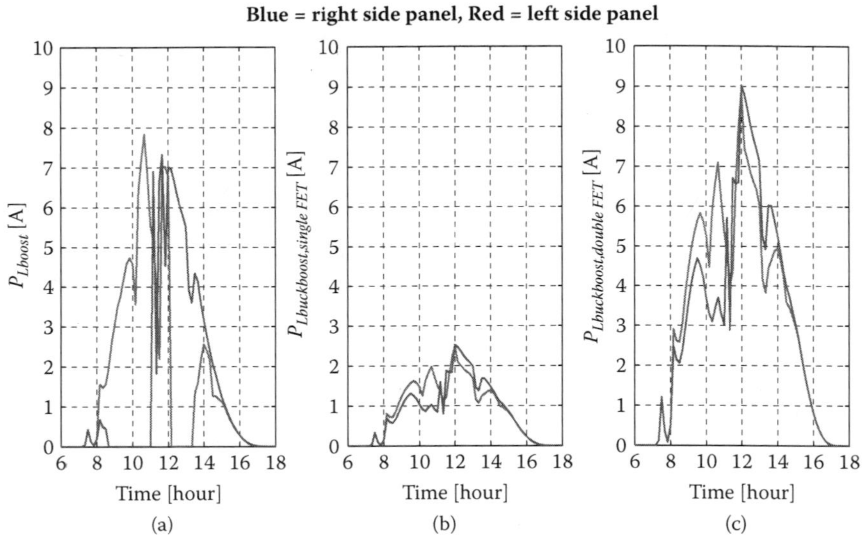

FIGURE 5.32 (See color insert)
Inductors' power dissipation for boost and buckboost topologies. (a) Boost, (b) Buckboost single FET, (c) Buckboost double FET.

are more severe, the benefits of buckboost topology are more exploitable, although there are higher losses. About 10% more energy is produced by the single FET buckboost solution compared to boost topology in this case. The double FET buckboost solution, due to high inductor losses, still shows a 3.3% increase of energy productivity compared to boost topology. In these cases, in spite of lower conversion efficiency of the buckboost power converter with respect to the boost power converter, the energy productivity of the buckboost can be much higher than the boost one. Based on the results of the previous examples, the double FET buckboost topology can be considered an alternative to a single FET solution only if the cost/size of the inductor is a real issue in the physical realization of the DMPPT buckboost converter.

As a final test, the conversion efficiency and energy productivity of the three design solutions are compared in the absence of mismatch. Figures 5.33 to 5.38 show the relevant plots for this case.

Integrating the total loss of each converter provides the total FET energy loss over the day, as shown in Table 5.9. In case of uniform irradiation, the energy efficiency decay caused by the presence of the DMPPT converters goes from –4% of the boost topology to –10% of the double FET buckboost topology.

In these conditions, the DMPPT converters should operate in pass-through mode. This is allowed to boost topology only, by turning permanently OFF the C-FET and permanently ON the S-FET. In this way only conduction losses of inductor and S-FET are involved. For buckboost topology, instead, both C-FET and S-FET should be turned permanently ON to connect permanently

TABLE 5.8

Thick Shadow Case, Conversion Efficiency, and Energy Productivity

	Boost				Buckboost Single FET $U_{buckboost}/U_{boost} = 1.100$				Buckboost Double FET $U_{buckboost}/U_{boost} = 1.033$			
PV Panel	U_{pan} (Wh)	U_d (Wh)	U_{out} (Wh)	U_{out}/U_{pan} (%)	U_{pan} (Wh)	U_d (Wh)	U_{out} (Wh)	U_{out}/U_{pan} (%)	U_{pain} (Wh)	U_d (Wh)	U_{out} (Wh)	U_{out}/U_{pan} (%)
Left	852	42	810	95.1	925	64	862	93.1	925	74	852	92.0
Right	875	37	838	95.8	1012	62	950	95.0	1012	72	940	92.9
Total	1727	79	1648	95.4	1937	126	1812	93.5	1937	146	1702	92.5

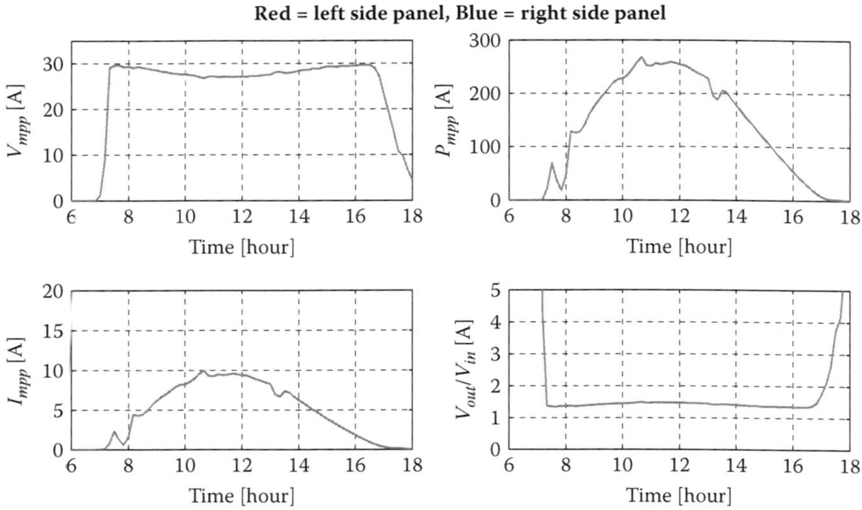

FIGURE 5.33
MPP voltage, current, and power and input-output voltage conversion ratio.

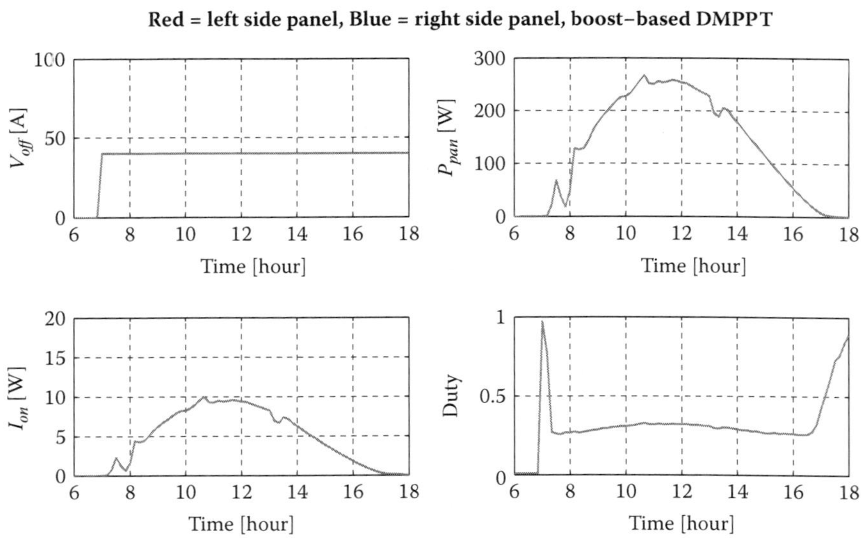

FIGURE 5.34
Voltage and current stress, power extracted from the panels, and converter duty cycle for boost topology.

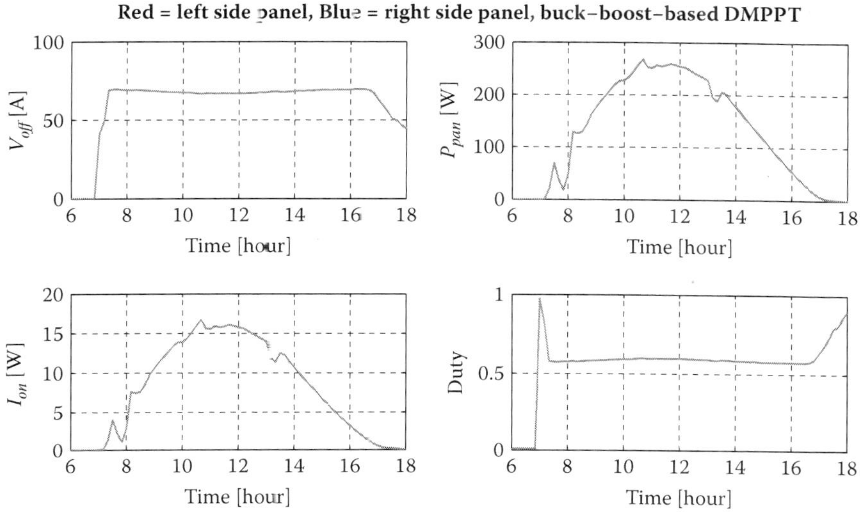

FIGURE 5.35
Voltage and current stress, power extracted from the panels, and converter duty cycle for buck-boost topology.

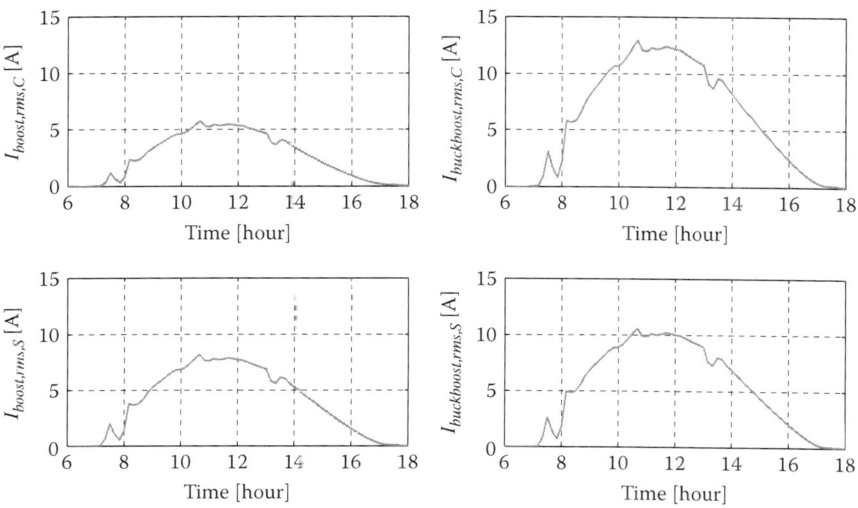

FIGURE 5.36
Rms currents for C-FET and S-FET for boost and buckboost topologies.

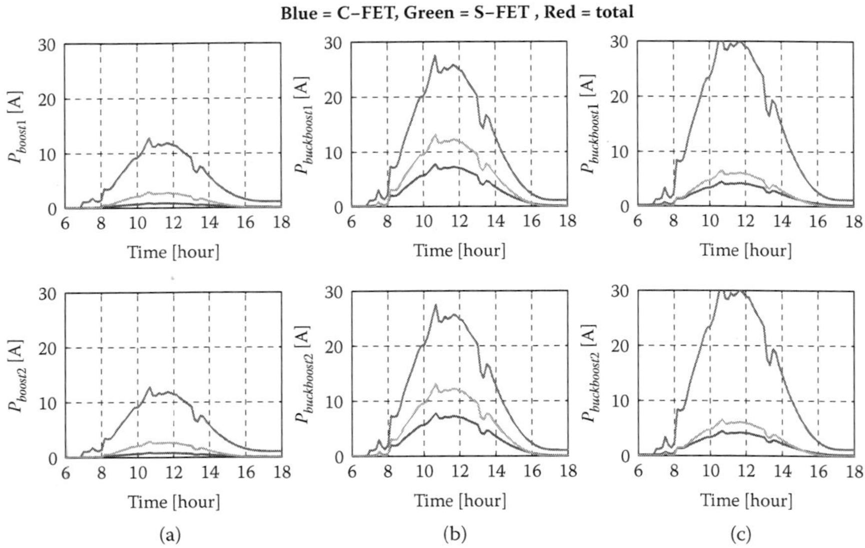

FIGURE 5.37 (See color insert)
Power dissipation for boost and buckboost topologies. (a) Boost, (b) Buckboost single FET, (c) Buckboost double FET.

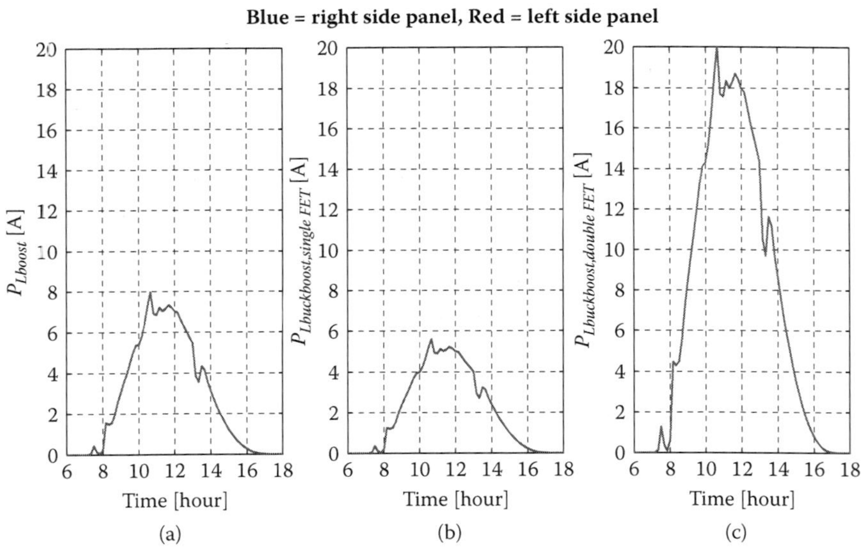

FIGURE 5.38
Inductor power dissipation for boost and buckboost topologies. (a) Boost, (b) Buckboost single FET, (c) Buckboost double FET.

TABLE 5.9

No Shadow Case, Conversion Efficiency, and Energy Productivity

PV	Boost				Buckboost Single FET $U_{buckboost}/U_{boost} = 0.954$				Buckboost Double FET $U_{buckboost}/U_{boost} = 0.941$			
Panel	U_{pan} (Wh)	U_d (Wh)	U_{out} (Wh)	U_{out}/U_{pan} (%)	U_{pan} (Wh)	U_d (Wh)	U_{out} (Wh)	U_{out}/U_{pan} (%)	U_{pan} (Wh)	U_d (Wh)	U_{out} (Wh)	U_{out}/U_{max} (%)
Left	1577	62	1515	96.1	1577	130	1446	91.7	1577	151	1425	90.4
Right	1577	67	1515	96.1	1577	130	1446	91.7	1577	151	1425	90.4
Total	3154	124	3030	96.1	3154	260	2892	91.7	3154	302	2850	90.4

the panel to the output; however, turning permanently ON the C-FET would involve shorting the PV panel through the inductor.

The examples discussed in this section, beyond the uncertainty limitations involved by the approximations adopted in the loss modeling, show that a real energy productivity of buckboost-based DMPPT converters is guaranteed in cases where the mismatch conditions are quite heavy. This happens, for example, in all those installations where shadows are generated by near buildings, chimneys, and other parts of the buildings or big trees. In all those applications where only thin shadows are involved, for example, due to piles and sparse dirtiness, boost topology certainly provides a cheaper and more reliable solution for DMPPT applications. More in general, a realistic costs vs. benefits comparative evaluation of PV plants equipped with either distributed MPPT or centralized MPPT power converters can only be done by means of a detailed energy productivity analysis taking into account the real installation irradiation and temperature conditions, the presence of systematic mismatch conditions, the MPPT limitations caused by the PV plant and power electronics architecture and by the conversion topology, and the power device losses.

References

1. M. Taherbaneh, A.H. Rezaie, H. Ghafoorifard, M. Mirsamadi, and M.B. Menhaj. A trade-off between the efficiency, ripple and volume of a DC-DC converter. *Journal of Power Electronics, KIPE*, 11(5):621–631, 2011.
2. M. Rodriguez, A. Rodriguez, P.F. Miaja, D.G. Lamar, and S. Zuniga. An insight into the switching process of power MOSFETs: An improved analytical losses model. *IEEE Transactions on Power Electronics*, 25(6):1626–1640, 2010.
3. F. Jizhen, Z. Zhiliang, L. Yan-Fei, and P.C. Sen. MOSFET switching loss model and optimal design of a current source driver considering the current diversion problem. *IEEE Transactions on Power Electronics*, 27(2):998–1012, 2012.
4. Z. Zhiliang, F. Jizhen, L. Yan-Fei, and P.C. Sen. Switching loss analysis considering parasitic loop inductance with current source drivers for buck converters. *IEEE Transactions on Power Electronics*, 26(7):1815–1819, 2011.
5. I.K. Budihardjo, P.O. Lauritzen, and H.A. Mantooth. Performance requirements for power MOSFET models. *IEEE Transactions on Power Electronics*, 12(1):36–45, 1997.
6. Y. Xiong, S. Sun, H. Jia, P. Shea, and Z.J. Shen. Physical insights on power MOSFET switching losses. *IEEE Transactions on Power Electronics*, 24(2):525–531, 2009.
7. G. Verneau, L. Aubard, J.C. Crebier, C. Schaeffer, and J.L. Schanen. Empirical power MOSFET modeling: Practical characterization and simulation implantation. In *Proceedings of the 37th IAS Annual Meeting Industry Applications Conference*, 2002, vol. 4, pp. 2425–2432.

8. L. Aubard, G. Verneau, J.C. Crebier, C. Schaeffer, and Y. Avenas. Power MOSFET switching waveforms: An empirical model based on a physical analysis of charge locations. In *Proceedings of the IEEE 33rd Annual Power Electronics Specialists Conference,* 2002, vol. 3, pp. 1305–1310.
9. HSPICE Reference Manual: MOSFET Models, Version D-2010.12, December 2010, Synopsis Inc., USA.
10. S. Maniktala. *Switching power supplies from A to Z*. Newness-Elsevier, Burlington, 2006.
11. J. Klein. *Synchronous buck MOSFET loss calculations*. AN6005. Fairchild Semiconductor, July 2003.
12. J. Brown. Modeling the switching performance of a MOSFET in the high side of a non-isolated buck converter. *IEEE Transactions on Power Electronics,* 21(1):3–10, 2006.
13. T. Eichhorn. Comparing the calculated vs measured efficiency loss in synchronous buck converters. *EE Times*, October 2006.
14. D. Graovac, M. Pürschel, and An. Kiep. *MOSFET power losses: Calculation using the data sheet parameters*. AN V 1.1. Infineon, July 2006.
15. Vishay. Si7806DN N-channel MOSFET datasheet.
16. R.W. Erickson and D. Maksimovic. *Fundamentals of power electronics*. Kluwer Academic/Plenum Publishers, New York, 2001.
17. G. Di Capua and N. Femia. The effects of shadows on power and reliability of PV panels. Presented at Proceedings of 2012 PCIM, Nurnberg, May 2012.
18. *Military handbook, reliability prediction of electronic equipment*. MIL-HDBK-217F. U.S. Department of Defense.
19. G. DiCapua, N. Femia, A Novel Approach to Transformers Design Based on Acceptability Boundary Curves of Magnetic Cores, *Proceedings of IEEE 12th Workshop on Control and Modeling for Power Electronics (COMPEL)*, pp.1–8, Boulder, Colorado, June 2010.
20. G. DiCapua and N. Femia. Modeling Switching Losses in MOSFETs Half-Bridges. *2012 International Conference on Synthesis, Modeling, Analysis and Simulation Methods and Applications to Circuit Design (SMACD)*, Seville, September 2012, pp.93–96.

Index

N

O

P

Q

R